Biotribology of Natural and Artificial Joints

Elsevier Series on Tribology and Surface Engineering

Biotribology of Natural and Artificial Joints

Reducing Wear Through Material Selection and Geometric Design with Actual Lubrication Mode

Teruo Murakami
Professor Emeritus, Kyushu University, Japan

ELSEVIER

Elsevier
Radarweg 29, PO Box 211, 1000 AE Amsterdam, Netherlands
The Boulevard, Langford Lane, Kidlington, Oxford OX5 1GB, United Kingdom
50 Hampshire Street, 5th Floor, Cambridge, MA 02139, United States

ISBN: 978-0-12-823669-7

For information on all Elsevier publications visit our website at https://www.elsevier.com/books-and-journals

Publisher: Matthew Deans
Acquisitions Editor: Dennis McGonagle
Editorial Project Manager: Sara Valentino
Production Project Manager: Maria Bernad
Cover Designer: Vicky Pearson Esser

Typeset by TNQ Technologies

Contents

Preface

Just a half-century ago, "Biotribology" as "those aspects of tribology concerned with biological systems" was introduced by Prof. Dowson and Prof. Wright at the Conference on the Rheology of Lubricants held in Nottingham in 1972 (Dowson & Wright, 1973). Biotribology includes various areas/topics on tribology in biological systems. Particularly, the biotribology of natural and artificial joints has been the most important issue since the introduction of biotribology. Here, tribology is defined as "the science and technology of interacting surfaces in relative motion and related subjects and practices" and its three pillars are friction, lubrication, and wear. In both natural and artificial joints, friction, lubrication, and wear play important roles in adequate load-carrying capacity and smooth movement in musculoskeletal systems.

As described in this book, the ingenious multimode lubrication mechanisms in human synovial joints are likely to bring extremely low friction property and long durability with minimal wear in various daily activities. However, under some severe conditions such as at mal-alignment and/or at edge loading, the occurrence of local direct contacts between rubbing cartilage surfaces may raise friction level and cause local wear and/or deterioration of surfaces. To sustain healthy functions of human synovial joints, the well-suited strategy from the viewpoint of biotribology becomes important.

In recent artificial joints such as total hip replacements and total knee replacements with ultrahigh molecular weight polyethylene in mixed lubrication regime, local direct contact between rubbing materials in insufficient fluid film lubrication may cause some wear problems. By applying improved highly cross-linked polyethylene and improved hard materials, which were optimized with geometric design, wear volumes were significantly reduced. However, wear debris-induced osteolysis problems have not been completely solved. Further improvements in wear reduction with biotribological viewpoints are required, although some successful cases have been reported.

As an effective method of solving general tribological problems, it is recommended to consider from the three viewpoints of lubrication (lubricants), design, and materials. For artificial joints in human body, it is predominantly applicable to adopt the improvements of material selections and geometric (macro-, micro-, and nanoscopic) designs, because there is the limitation to change periprosthetic lubricants optionally. It is considered that the improvement of lubrication mechanism may be applicable.

In strength design to prevent failures of materials, the criteria as [the maximum stress] $\leq$ [the permissible stress] are recommended, but in most designs for anatomical knee prostheses with UHMWPE, the plastic deformation can occur at concentrated contacts under high load conditions. In addition, in case of wear phenomena at direct contact areas, for example, $W \times L$ values (W: load, L: sliding distance) are applicable as wear criteria. Furthermore, the corrosive situation should be considered in biological environment. Biotribological survey with biomechanical viewpoint for clinical situation becomes important.

This book is composed of the following chapters:

Chapter 1 Importance of biotribology

Chapter 2 Biomechanics of human and artificial joints

Chapter 3 Biotribology of natural joints

Chapter 4 Biotribology of artificial joints

Chapter 5 Biotribology of artificial joints with artificial cartilage

Chapter 6 Biotribology of regenerated cartilage

Chapter 7 Summarization and future directions of biotribology of human and artificial joints

Each chapter contains fundamental issues and provides specific examples to give useful information on chapter topics. In Chapter 1, the outline of biotribology is described. Tribological behaviors of natural and artificial joints should be evaluated with general understanding of influence of biomechanical conditions, and the importance of biomechanics for both joints is indicated in Chapter 2. The important terms of biotribology for natural synovial joints and artificial joints are discussed in Chapter 3 and Chapter 4, respectively. The similarity of tribological properties to natural joints in artificial joints with artificial cartilage is indicated in Chapter 5. The tribological behaviors of regenerated cartilage are discussed in Chapter 6. In Chapter 7, all contents in this book are summarized and the future directions of biotribology of human and artificial joints are discussed.

It is expected for this book to contribute to sustaining of healthy situation in human joints, repairing of diseased/damaged joints, and improving of artificial joints in interdisciplinary biomedical collaborations with advanced technologies through mutual understanding of biotribology.

I would like to express my appreciation to all researchers/authors in referenced works including my colleagues and collaborators. I am grateful to all publishers for their kind permission to reuse previously published material. I would like to thank Elsevier for the publication of this book with professional supports.

Teruo Murakami
Professor Emeritus
Kyushu University, Japan
June 2022

Reference

Dowson, D., & Wright, V. (1973). Bio-tribology. In T. C. Davenport (Ed.), *The rheology of lubricants* (pp. 81–88). Applied Science Publishers Ltd, Institute of Petroleum.

Importance of biotribology

1

1.1 Introduction to biotribology

Prior to appearance of the term *biotribology*, *tribology* as "the science and technology of interacting surfaces in relative motion and related subjects and practices" was proposed in a report "Lubrication (tribology) – education and research" (The Jost Report) in 1966 (Department of Education and Science, 1966). This report quantified the impact of tribology on the economy and society in general. The term "tribology" derives from Greek "τριβω, tribos" to mean "rubbing," and tribology includes friction, lubrication, wear, bearing design, and so on. Therefore, tribology is an interdisciplinary science and technology encompassing physics, chemistry, material science, geology, biology, mathematics, and related engineering.

With worldwide awareness of tribology, the new term *biotribology* was introduced as "those aspects of tribology concerned with biological systems" by Dowson and Wright at the Conference on the Rheology of Lubricants (organized by the Institute of Petroleum, the Institution of Mechanical Engineers and the British Society of Rheology) held in Nottingham in 1972 (Dowson & Wright, 1973).

The main areas in this proposal are as follows:

1. The influence of microorganisms on lubricants employed in machinery.
2. The abrasive wear characteristics of human dental tissues.
3. The use of low friction protective coatings on razor blades.
4. Fluid transport in the body (e.g., (a) fluid mechanics of ureter, (b) the movement of sputum, and (c) the flow of blood).
5. The motion and lubrication by plasma of red blood cells in narrow capillaries.
6. Levitation by means of "aerostatic bearings" for the treatment of burns.
7. The action of saliva.
8. Tribological studies of performance of natural synovial joints
9. Tribological aspects of endoprostheses.

First, for areas 1 to 7 in the above list, short comments including current viewpoints are remarked.

1. The biodegradation of lubricants with an aqueous phase, such as cutting fluids by proliferation of microorganisms such as bacteria, yeasts, and fungi, is a most serious problem. Infection by microorganisms deteriorates the lubricants, and lowers the pH and depresses the rust-inhibiting and lubricating properties, accompanied with bad smells. Appropriate quality control and maintenance for lubricants is required.
2. Resistance to abrasive wear of materials is related to the indentation hardness. The enamel as the top layer of teeth is the hardest tissue in human body, and it has high wear resistance but can be worn by abrasive wear due to trapped hard particles in foods at occlusal load during

Biotribology of Natural and Artificial Joints. https://doi.org/10.1016/B978-0-12-823669-7.00007-7

mastication or due to suspended hard particles during toothbrushing (Zhou & Jin, 2015). Dental tribology is an important subject in oral tribology.

3. Coatings on razor blades (carbon steel or hardened stainless steel) should have low friction and protective properties to prevent damage and corrosion of the blades. For example, metallic hard coating is treated to improve the resistance to wear and corrosion of the acute cutting edge of blade. Furthermore, fluoroethylene resin is coated on metallic coating layer to enable the blade to glide smoothly over the skin surface. Various coating methods have been applied.
4. In the fields of biofluid transport and biofluid mechanics, various tribological phenomena are found. (a) In ureter the urine is transported by peristaltic action from the kidney to the bladder. Tribological viewpoints are required to establish appropriate transport by peristaltic action. (b) Sputum is matter coughed up and expectorated from the mouth, composed of saliva and discharges from the respiratory passages such as mucus, phlegm, or pus. The moving behavior of droplets containing sputum with virus from human mouth to surrounding atmosphere should be appropriately estimated to prevent the virus infection by considering the influence of oral and mask friction and air flow resistance. (c) The blood is mainly composed of blood cells (red blood cells, white blood cells, platelets) and plasma. The non-Newtonian viscous property of blood changes depending on concentration and state of aggregate of blood cells. A plasmatic lubricating layer along the capillary wall is effective in reduction of friction. The elucidation of mechanism of the formation of thrombi with consideration for influences of blood flow and mutual interaction of constituents is an important subject.
5. Red blood cell is a biconcave disc of about 8 μm outer diameter in a normal situation (Fig. 1.1a). When a red blood cell moves in narrow capillaries of 5−10 μm inner diameter, it as assumes a low frictional shape and goes through (Fig. 1.1b), where the elastohydrodynamic lubrication film thickness is estimated as 0.2 μm and constituents in blood plasma appear to play a boundary lubricating role. In addition, the important role of the endothelial surface layer lining the vessel walls while passing through smaller bore capillary is indicated (Secomb et al., 2006).
6. Levitation of burn-injured patient is applied by aerostatic bearings during the treatment of severe burns to control infection, to protect against contamination, and to provide suitable environment for the regeneration of tissues.
7. Saliva acts as starch digester and a solvent, and aids in cleaning action, moistening, and lubrication. It plays a protective and friction-reducing role of boundary lubricant between teeth and oral soft tissues and makes swallowing easier. In addition, saliva is a source of inorganic ions (calcium and phosphate ions) necessary for remineralization during repairing of erosive damage of teeth (Zhou & Jin, 2015).

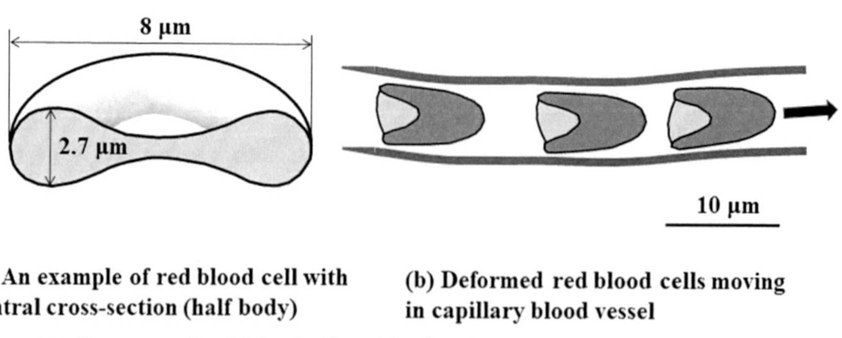

(a) An example of red blood cell with central cross-section (half body)

(b) Deformed red blood cells moving in capillary blood vessel

Figure 1.1 Geometry of red blood cell and its flow in narrow capillary blood vessel. Red blood cell as a biconcave disc can assume a less frictional form in a narrow capillary.

In the points 8 and 9, "tribology of natural and artificial joints" was the most active research topic at that time, as demonstrated in Proceedings of the Symposium on Lubrication and Wear in Living and Artificial Joints held in London in 1967. In natural and artificial joints, friction, lubrication, and wear play important roles to maintain and/or control their function and long durability.

Here, the mutual relationship between natural and artificial joints is shortly described for the typical cases of human hip joints (Fig. 1.2). In the healthy natural synovial joints composed of bone ends covered with compliant biphasic articular cartilage layers lubricated with synovial fluid, their superior lubricating ability and excellent load carrying capacity with appropriate range of motion bring the extremely low friction and minimal wear in various daily activities for a whole life (Fig. 1.2a). In some joints under severe loading/rubbing or diseased conditions, however, even articular cartilage can be worn and deteriorated. For example, in the case of osteoarthritis with worsening of wear and deterioration, pathognomonic changes such as osteophyte, bone hardening, and scar in joint capsule take place (Fig. 1.2b). The advanced osteoarthritis lowers the ability of walking and produces severe pain. For patients who are deficient in ability to move and suffer severe pain, it is recommended to apply total hip arthroplasty, in which damaged joint is replaced by artificial hip joint usually composed of ultrahigh molecular weight polyethylene (UHMWPE) acetabulum and metallic or ceramic femoral head with stem (Fig. 1.2c). Successful total joint arthroplasty brings significant benefits of recovery of activity and relief from pain to patients. In both natural and artificial joints, tribological properties play very important roles. The important matters of tribology in natural and artificial joints were described by Dumbleton in 1981 (Dumbleton, 1981). In this book, "biotribology of natural and artificial joints" is described as the main theme including the current viewpoints.

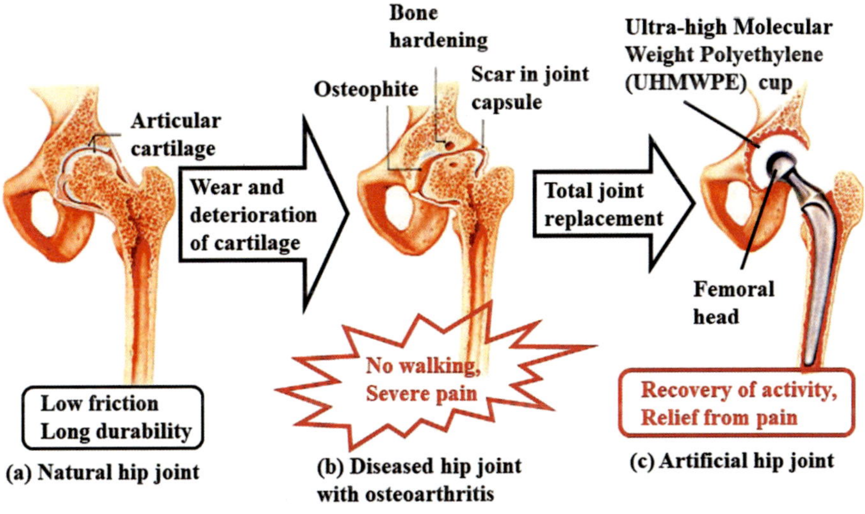

Figure 1.2 Natural and artificial hip joints. Total joint arthroplasty by artificial joint for diseased joint can bring the recovery of activity and relief from pain.
Modified with permission from Oka, M. (Ed.). (1990). Artificial Joints and Biomaterials. Medical View Co., Ltd.

After proposal of biotribology by Dowson et al., various topics on biotribology have been added. For example, in 1988, Sasada et al. (1988) gave additional topics of biotribology as shown below.

As fields of clinical medicine:

1. Friction of contact lens and eye.
2. Skin friction, particularly evaluation of gripping force by finger.
3. Friction and lubrication between tendon and tendon sheath.
4. Friction between genitals and condom.
5. Role of friction in mechanical balancing in living body.
6. Friction in artificial organs.
7. Surface deterioration of implant materials due to friction.
8. Friction in medical treatment such as friction between scalpel and biological tissue.

As fields of basic medicine and biology:

1. Role of frictional stimulus as physical and/or chemical forms on generation, growth, metabolism, pathological changes, and medical treatment in various organs.
2. Role of friction on vital functions.
3. Tribological approach on functional changes in biological evolution.
4. Frictional behaviors between cells.

As additional field:

1. Feedback of conception and knowledge on frictional phenomena in biological systems to science and engineering such as biomimetics and bio-inspired technology.

Furthermore, many topics in biotribology are familiar to all in our everyday lives, as discussed in "A tribological day" by Dowson (2009). Some examples in the morning activities are shown below.

1. Well-functioning synovial joints or morning stiffness on rising from our beds: effectiveness of lubrication mechanism of synovial joint.
2. Easiness to remove clothing prior to taking a morning shower and difficulty to pull on socks and undergarments after soaking the skin in water and then drying with a towel: effectiveness of skin hydration.
3. Slip at entering or leaving a bath: danger of low friction such as effect of unguent upon friction.
4. Care for combing, brushing, and washing hair-keratinous fibers: differential friction effect due to surface property of hair cuticle.
5. Comfortable shaving with low friction and protection: low friction PTFE coatings on the cutting flanks of the built-in blades in disposable razors.
6. Satisfactory transport of foods at breakfast which are processed by teeth, and then transported across tongue into the dietary tract: adequate cutting and grinding action of teeth facilitated by versatile mandibular joints and transport by tongue which are assisted in lubrication, moistening, and bringing out delicious taste by salivary fluids.
7. Maintenance of oral hygiene and avoiding serious erosion of the natural dental tissues or restorative dental materials: control of wear of teeth by appropriate selection of toothbrush and dentifrice.

If all topics are extended to full day activities, other various biotribological phenomena will be found. An example is known as well-functioning lungs, hearts, ocular surfaces, and gut in movement: low friction and effective protection for soft tissues by layers of surface-active phospholipid in pleura, pericardium, ocular surfaces, and guts in synergistic multimode lubrication of fluid-film lubrication and boundary lubrication.

In the extensive fields of biotribology as described previously, research and technologies of the joint, skin, and oral tribology have made further progress as current fundamental fields, but other biomedical and biological-related tribology has been developed. In addition, the importance of biotribology in further new fields including biomimetics (Dowson & Neville, 2006) and bio-inspired tribology (Shiddaiah & Menezes, 2016) has been explored. The development of further advanced artificial joints based on bio-inspired tribology is expected.

1.2 Friction

Friction is defined as "the resisting force tangential to the common boundary between two bodies when, under the action of an external force, one body moves or tends to move relative to the surface of the other" (Organisation for Economic Co-operation and Development (OECD), 1969). And coefficient of friction or friction coefficient is defined as the ratio obtained by dividing the friction force by normal force pressing these bodies together.

Friction for rubbing surfaces shows different behaviors depending on environment such as dry, boundary lubricated, or fluid film lubricated condition. First, friction under dry and unlubricated conditions is described.

1.2.1 Friction under dry and unlubricated conditions

It is well known that Amontons' laws (1699) or Amontons-Coulomb laws (1785) shown below are broadly applicable to sliding friction under dry and unlubricated conditions.

1. Friction force F is proportional to normal force (applied load) W.
2. Friction force F is independent of the size of the apparent contact area.

As shown in Fig. 1.3, when pulling force F_p is applied to a blockish body under normal load W, a block slides while its bottom surface is applied with friction force F in reverse direction. When pushing force is applied on the left, friction force acts in reverse direction to pushing force. For many cases friction force is proportional to normal force. The ratio of friction force to normal force is defined as friction coefficient or coefficient of friction: μ.

$$\mu = F/W \tag{1.1}$$

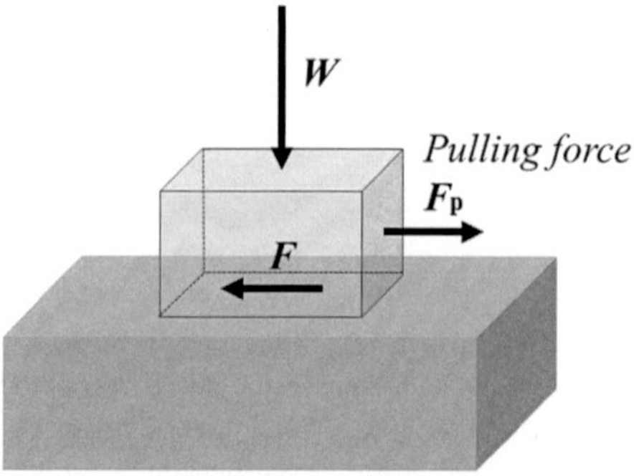

Figure 1.3 Friction force for a block matter on plate. Friction force F occurs when pulling force F_p is applied to block matter under normal load W.

Friction coefficient changes widely according to material combination and environment.

Here, in association with the first law, Coulomb's friction indicates that friction force is proportional to normal load.

Next, important comment is given in relation to the second law. Usual rubbing surfaces are not perfectly smooth but have various surface roughnesses with different irregularity with valleys and peaks. Therefore, when two bodies contact, the actual loading areas are not apparent contact areas but discretely distributed in real contact areas as shown in Fig. 1.4. At these contact points, the roughnesses or asperities touch the other, and thus the real area of contact is very narrow than the apparent contact area. If the load, the contact pressure, and the area of i-th contact point are assumed as W_i, p_i and A_i, respectively, the pressure at contact point is shown as

$$p_i = W_i/A_i \quad (1.2)$$

In usual cases for metal-metal contacts, the contact pressure p_i is large and equal to the yield pressure of softer metal p_m inducing the plastic deformation at the local contact points.

$$W_i/A_i = p_m \quad (1.3)$$

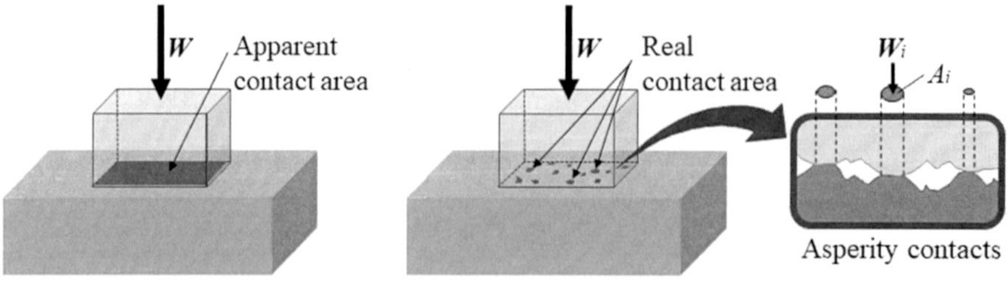

(a) Apparent contact area (b) Real contact area

Figure 1.4 Apparent contact area and real contact area. In contact between actual bodies with surface roughness, the actual loading areas are discretely distributed in real contact area.

Therefore, total real contact area A_R at load W is estimated as shown below for loading area with numerous contact points.

$$A_R = \sum_{i=1}^{n} A_i = \sum_{i=1}^{n} W_i/p_m = W/p_m \tag{1.4}$$

Thus, the real contact area is determined by the load and yield pressure, and independent of the size of the apparent contact area.

Although the proportional relationship between real contact area and load is derived for plastic contact at asperities, it is pointed out by Greenwood–Williamson that the real contact area is proportional to the load, even for contacting surfaces with surface roughness at elastic contact (Greenwood & Williamson, 1966). Therefore, for many contact states of two solid bodies, the real contact area is proportional to the load.

Coulomb added the third law:

3. kinetic friction force F is independent of sliding speed V.

The third law is not always applicable for very low or very high speed.

It should be noted that friction for sliding pair include some polymeric material changes with sliding speed depending on viscoelastic behavior (Myshkin & Kovalev, 2009). Furthermore, the friction force-sliding speed relationship depends on test temperature. In sliding test conducted near the glass-transition temperature the sliding speed has a remarkable effect on friction.

In general recognition, frictional resistance is caused by two main factors: adhesion and deformation (or plowing).

Adhesive force is defined as the attractive force between the adjacent surfaces in frictional contacts. When a solid body is moved along the interface, the force required to shear the adhesive junction formed between the rubbing surfaces (Fig. 1.5a) acts as the friction force. The friction force F_a as the molecular component due to the adhesive term is shown in the following formula, when the shearing stress at junction and total normal load is assumed as s and W, respectively.

$$F_a = A_R\, s = \left(\frac{W}{p_m}\right) s = \left(\frac{s}{p_m}\right) W \tag{1.5}$$

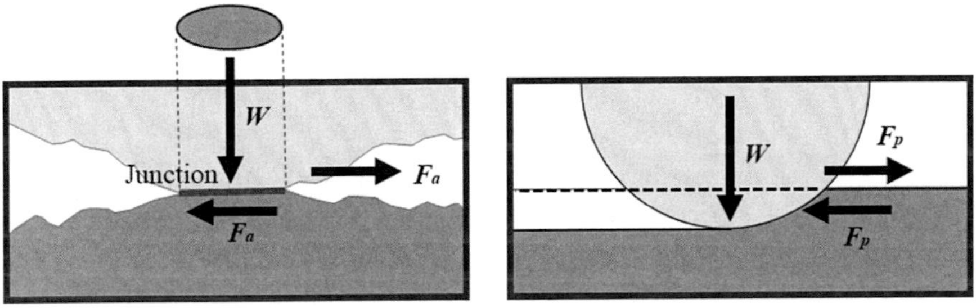

(a) Friction force by adhesion (b) Friction force by plowing

Figure 1.5 Friction forces due to adhesion and plowing. Adhesive friction force occurs due to shearing force at junction and plowing friction force occurs due to resistance to plowing.

As seen in Eq. (1.5), the friction force due to adhesive term is proportional to the load. For many sliding cases with infinitesimal friction due to deformation term, this formula corresponds to the first and second Amontons-Coulomb laws.

In the meanwhile, *plowing* is defined as the formation of grooves by plastic deformation of the softer of two surfaces in relative motion (Fig. 1.5b) where the upper cylindrical or spherical solid body is hard material (indenter) and lower solid is softer metal. The soft metal is plastically deformed at high contact pressure, and the resistant friction force of the softer material to the plowing by asperities of the harder one is shown in the following formula.

$$F_p = A'p' \tag{1.6}$$

here A' is the frontal projected contact area and p' is the yield pressure for softer material ($p' = p_m$).

In another example, when a polymer with viscoelastic property slides against a hard surface, the energy dissipation as deformation friction term is caused by the hysteresis loss for polymeric material.

For various rubbing pairs, the friction force F is composed of adhesive and deformation (or plowing) terms as shown below.

$$F = F_a + F_p \tag{1.7}$$

For most rubbing surfaces of machine elements and related parts with small roughness, the adhesion term is extremely larger than the deformation one. Therefore, friction force is approximated as below.

$$F \cong F_a \tag{1.8}$$

As described above, friction force for many rubbing surfaces is proportional to the normal load. Friction coefficient μ defined by Eq. (1.1) indicates the level of friction. Some examples in sliding pairs at dry condition in air are shown in Fig. 1.6 (Tribonet: Friction Coefficient Tables in Air and Vacuum, 2020).

1.2.2 Static friction and kinetic friction

In Fig. 1.3, if the pulling force is smaller than frictional resistance, the block stays stationary. But, when the pulling force exceeds the frictional resistant force, the block begins to move. In many cases, friction when a body starts moving is larger than friction when the body is sliding. To discriminate these friction types, the friction to act on a stationary body and friction to act on a moving body are called the *static friction* and the *kinetic friction*, respectively. Friction coefficient when a body starts moving is called the maximum static friction coefficient.

The static frictional behavior can be confirmed by observing the moving of a body on an inclined surface with changing the inclination. When a body of weight W puts on

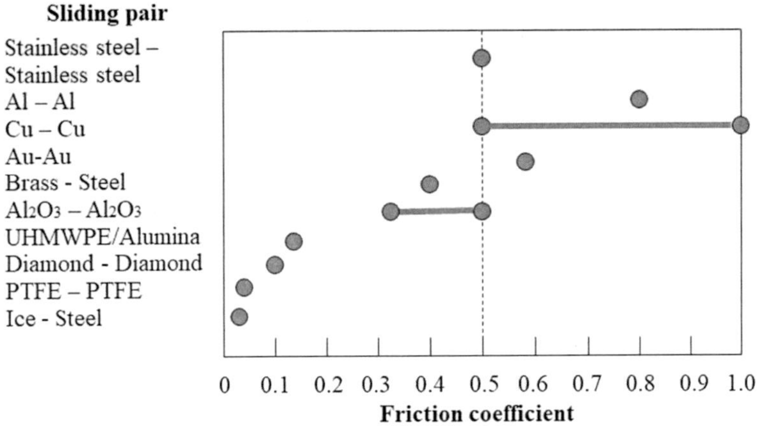

Figure 1.6 Examples of friction coefficient between various materials in air. Experimental data for friction coefficients are cited from Tribonet.
Data from Tribonet: Friction coefficient tables in air and vacuum. (2020). https://www.tribonet.org/wiki/friction-coefficients-in-atomosphere-and-vacuum/

an inclined surface with tilt angle θ, W is dissolved to the force normal to inclined surface F_n and the force parallel to the surface F_p, which tends to cause the sliding. Both forces are shown below.

$$F_n = W \cos\theta \tag{1.9}$$

$$F_p = W \sin\theta \tag{1.10}$$

A body stays at original location under low inclined condition when F_p is smaller than the static frictional force F_f (Fig. 1.7a), but at some angle θ, it will begin to slide when the force F_p is greater than the static frictional force F_f (Fig. 1.7b). At the onset of sliding, these forces are equal. The static friction coefficient is estimated as shown below.

$$\mu_s = F_f/F_n = W \sin\theta / W \cos\theta = \tan\theta \tag{1.11}$$

In the case of Fig. 1.7b ($\theta = 30°$), μ_s is about 0.577.

In many friction experiments, the frictional force can be directly measured by the load cell.

1.2.3 Friction under lubricated conditions

Friction under lubricating conditions is widely changed depending on the corresponding lubrication modes. In Stribeck curve, the relationship between friction coefficient and the lubrication parameter [viscosity × speed/load] as a similar parameter to Hersey number is shown for boundary, mixed, and fluid film lubrication regimes (Fig. 1.8).

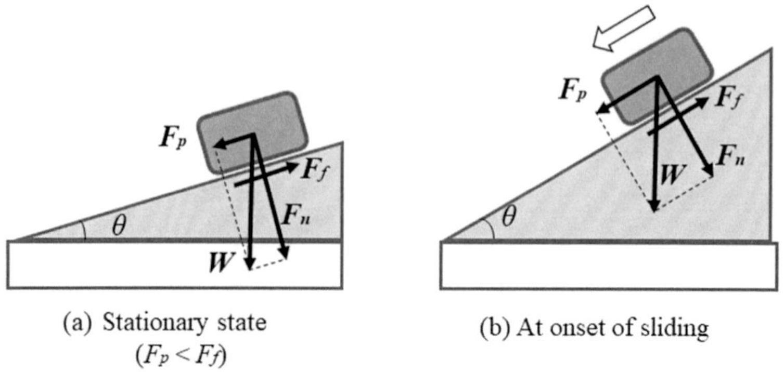

Figure 1.7 Sliding of a body on inclined surface. A block will begin to slide on inclined surface when the force F_p is greater than the static frictional force.

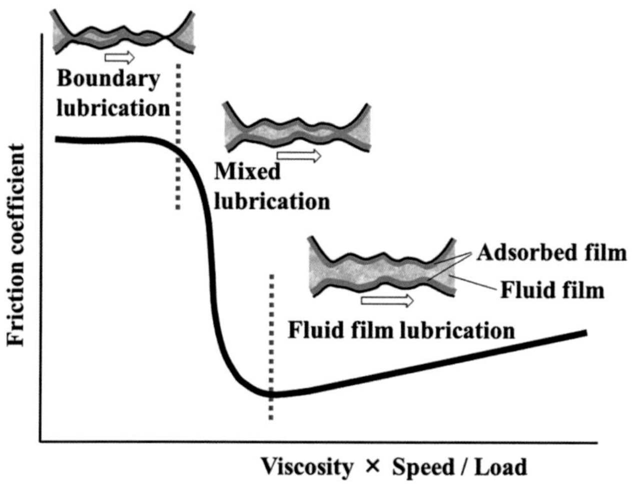

Figure 1.8 Stribeck curve. Fluid film thickness increases with an increase in lubrication parameter: Viscosity × Speed/Load. Depending on the relative ratio of fluid film thickness to surface roughness, the lubrication modes and corresponding friction levels change.

In general, the fluid film thickness of lubricant increases with an increase in the lubrication parameter. The detailed mechanisms for typical lubrication modes are described later in Section 1.3, Roles of Lubrication. In the region at low lubrication parameter, high friction is observed as a consequence of severe contact between solid surfaces covered with adsorbed films under thin film condition in boundary lubrication. In contrast, in the region at high lubrication parameter, low friction is maintained due to low shearing force of the thicker fluid film, which prevents the direct contact between rubbing surfaces, in fluid film lubrication. In this mode, friction gradually increases with an increase in the value of the lubrication parameter. In medium region at mixing of boundary and fluid film lubrication modes, friction changes from high to low levels with the lubrication parameter. This region is called *mixed lubrication*.

1.2.4 Rolling friction

As mentioned previously, the friction force gives resistance to relative motion in sliding contact and this mode is called *sliding friction*. In contrast, the rolling friction shows resistance to rolling motion as observed in rolling bearings. Some examples of rolling friction for regular polygons are shown in Fig. 1.9, where three cases in rolling motion of regular hexagonal cylinder, regular dodecagonal cylinder and circular cylinder are demonstrated. Here, G is center of gravity, h is its height, C_1 and C_2 are vertexes and edges of contact zone, O is center of $C_1 - C_2$ zone, x is the distance of OC_1, θ is angle for GC_1 and centroidal line, W is weight of cylinder, and F_R is driving force which is equilibrium to rolling friction force.

To roll the cylinders, the rolling moment by F_R at contact edge (vertex) C_1 is required to be in equilibrium to the moment produced by the weight. Therefore, the following equations are obtained.

$$F_R \, h = W \, x \tag{1.12}$$

$$\therefore \; F_R = W\frac{x}{h} = W \tan\theta \tag{1.13}$$

$$\mu_R = F_R/W = \tan \theta \tag{1.14}$$

Thus, rolling friction coefficients are estimated as 0.58 and 0.26 for hexagonal and dodecagonal cylinders, respectively. For polygonal cylinders, friction coefficient becomes lower with an increase in number of sides. For combination of rigid circular cylinder and rigid flat plane as ultimate polygon, it becomes zero. However, friction coefficient for usual circular cylinder becomes infinitesimal as 0.001 or so by very small rolling resistance, because usually narrow flattened contact area is formed for elastic bodies by elastic deformation due to normal load W.

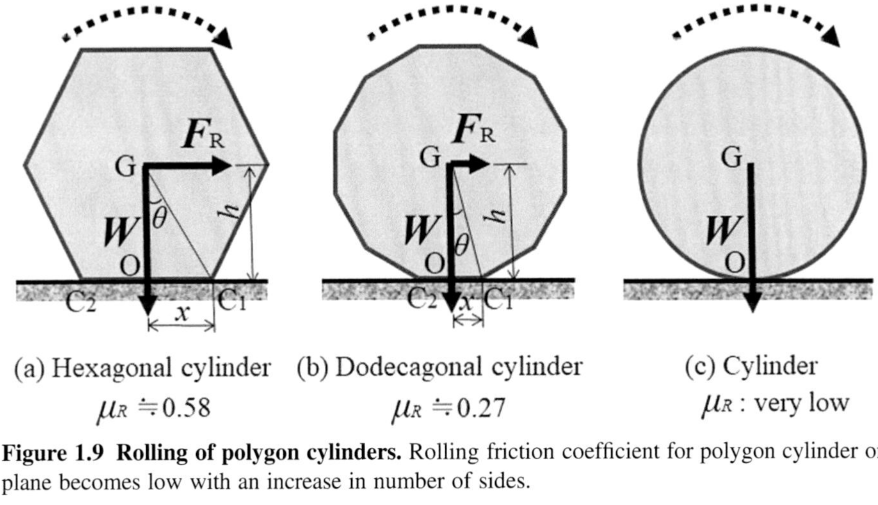

(a) Hexagonal cylinder $\mu_R \fallingdotseq 0.58$ (b) Dodecagonal cylinder $\mu_R \fallingdotseq 0.27$ (c) Cylinder μ_R : very low

Figure 1.9 Rolling of polygon cylinders. Rolling friction coefficient for polygon cylinder on plane becomes low with an increase in number of sides.

As described previously, the rolling friction for roller or ball is very low, and therefore the rolling bearings such as roller and ball bearings have been widely applied in various industrial uses.

In human history on technology, the use of round woods as rollers to move big and heavy stone statue in Assyria in about 700 BCE and the invention of ball bearing in Ca.50 CE in Italy is reported as old precedents (Dowson, 1979).

On the other hand, in animal history, the advantage of low friction in rolling motion has been taken to move bigger ball of dung by dung beetles (*Scarabaeus sacer*) (Fig. 1.10 (Wikipedia: Scrab (Artifact), 2020)). In human knee joints, the combination of sliding and rolling contact enables smooth and stable flexion-extension motion in wide range of motion.

1.2.5 Low friction and high friction in biosystems

The range of friction in biosystems is very large. In this section some examples are described.

As an example of low friction, the extremely low friction from 0.002 to 0.03 is observed for natural synovial joints even under severe operating conditions with high load, low viscosity, and slow velocity condition as seen in Stribeck curve (Fig. 1.8). This low friction is likely to be attained by superior lubrication mechanism with synergistic performance as rubbing a pair of compliant biphasic articular cartilages with lubricious surfaces lubricated with synovial fluid containing useful constituents (Fig. 1.11). Under operating conditions with different severities for natural joints in various daily activities, it is required for minimizing friction and wear that various

Figure 1.10 Dung beetles rolling a ball of dung. Dung beetles move the bigger ball of dung using rolling motion.
Modified from Wikipedia: Scrab (Artifact). (2020). https://en.wikipedia.org/wiki/Scrab_(artifact)

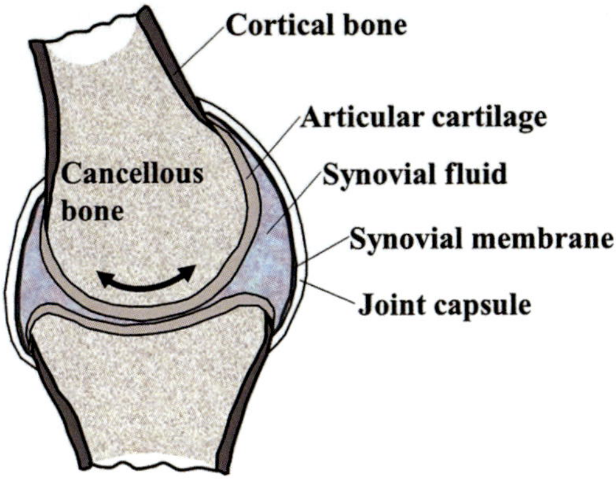

Figure 1.11 Simplified model of natural synovial joint. Mating articular cartilage surfaces are smoothly rubbed and lubricated with synovial fluid.

lubrication modes of not only fluid film lubrication but boundary lubrication and other modes can play effective roles (Dowson, 1966-67; Murakami, 1990; Murakami et al., 2017). The detailed mechanism of the multimode lubrication is discussed in the Chapter 3 of this book.

Next, an example of high friction is introduced on finger skin to easily and substantially hold various materials in daily life. It is expected that the wetting situation of finger skin surface plays an important role to control friction level because the openings of eccrine sweat glands are placed in the epidermal ridge of the finger.

In friction tester where the finger was moved toward body and the wetting contact situations were evaluated by observing the wetted areas on prism surface by total reflection as shown in Fig. 1.12 (Murakami et al., 1995), the maximum and minimum friction values of finger skin on glass prism were examined at each

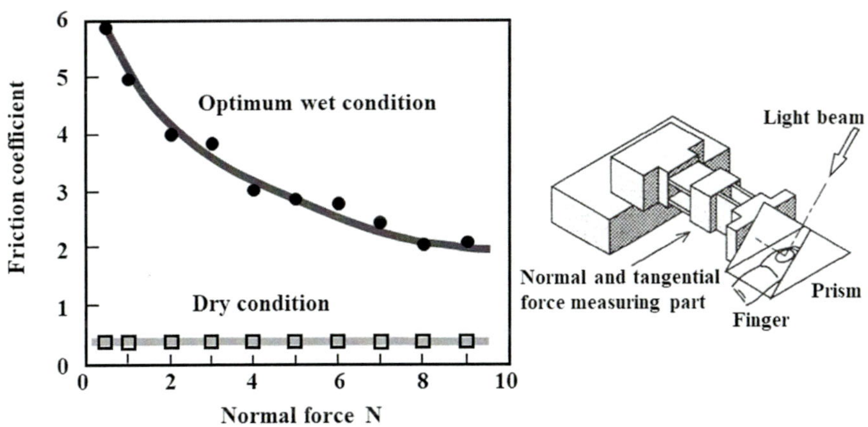

Figure 1.12 Frictional behavior of finger skin on glass plate. Maximum friction for finger is observed at optimum wet (sweated) condition and shows significant load-dependancy. In contrast, minimum friction with load-independancy is observed under dry condition.

load. It is noticed that very high friction coefficients as maximum values (≥ 2.0) with load-dependency were observed under the optimum wet condition, while low friction as minimum ones ($\fallingdotseq 0.4$) with load-independency were confirmed under dry condition which was monitored as little wetting area in total reflection images. The constant friction coefficient under dry condition is considered as solid or boundary lubrication mode. In contrast with this load-independency, friction coefficient under optimum wet condition is high at low load and tends to decrease with increasing load. This load-dependency is considered from two-term law of friction as follows:

$$F = A + \mu P \tag{1.15}$$

$$f = \frac{A}{P} + \mu \tag{1.16}$$

F: Friction force for skin on glass.
A: Load-independent adhesive term.
P: Normal load
μ: Coulomb's friction coefficient
f: Friction coefficient for skin on glass.

Adhesive term A is attributable to surface force such as surface tension, viscous force of intervening liquid (Bowden & Tabor, 1954), and others. At low load, A/P is high and f becomes high. Highest friction as optimum value at each load is attained by maximizing adhesive surface force, and excessive wetting reduces friction by fluid film effect.

Next, the influence of excessive wetting on friction was examined in similar tester under different wetting conditions at 6 N at 1 mm/s. At condition with sweating, the rubbing test was started after static loading accompanied with spontaneous sweating for 1 min. At condition of water supplied, the test was started after distilled water was supplied to bring on significant wetting. It is confirmed that higher friction is observed on finger wetted with sweating than the finger supplied with distilled water as shown in Fig. 1.13 (Murakami et al., 1995), where the images of total reflection at 6 s are given as dark area for wetted zone. For water supplied condition a certain level of fluid film effect appears to reduce friction but maintains higher friction except local drop than dry condition. This fact indicates that finger seems to have excellent high friction mechanism even under excessive wetted condition. The skin with dermal ridge including palm and sole has high friction property and called friction skin or sensing skin, where sweating appears to play an important role in controlling friction. By contrast, the other skins without ridges such as forehead and forearm show low friction and smooth feel compared with friction skins. Although skin friction varies with anatomical sites, for most of skins drier skin has a lowered friction and hydrated skin has high friction in cases without fluid film effect. The hydration increases friction due to an increase of adhesive force and an increment of contact area with skin softening (Sivamani & Maibach, 2006).

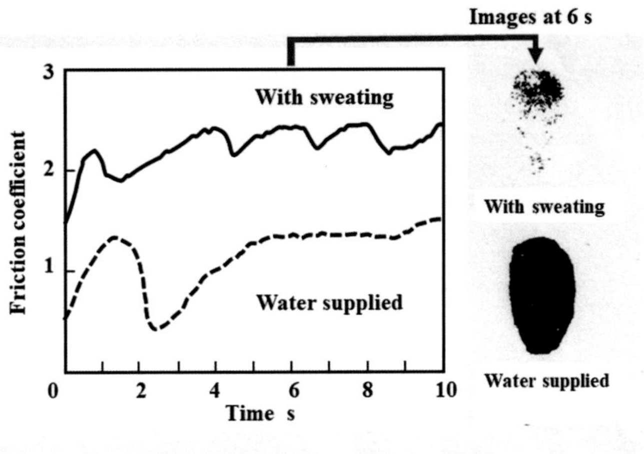

Figure 1.13 Influence of wetting on friction for finger on glass. At condition with sweating, high friction for finger skin is sustained. At condition with water supply, friction decreases but maintains higher friction than dry condition.

1.3 Roles of lubrication

1.3.1 Lubrication modes

Lubrication is defined as "the reduction of frictional resistance and wear, or other forms of surface deterioration, between two load-bearing surfaces by the application of a lubricant" (Organisation for Economic Co-operation and Development (OECD), 1969).

Main lubrication modes or regimes of lubricated rubbing surfaces are generally categorized as boundary, mixed, and fluid film lubrication as shown in Stribeck curve (Fig. 1.8). In addition, solid lubrication or solid film lubrication is another lubrication mode which is widely employed by applying powder or thin film solid lubricants as layer or lamellar structures such as graphite and molybdenum disulfide. Fundamental features for main lubrication modes are described as below.

1.3.2 Fluid film lubrication

In design of various bearings and machine elements with rubbing surfaces, it is recommended as the first choice for protecting surfaces and controlling friction and wear that the full fluid film should be formed and act as fluid film lubrication to prevent the direct contact between sliding and/or rolling mating surfaces. It is required for fluid film to support the operating loads with pressurized fluid. The pressurization of fluid is established by hydrostatic or hydrodynamic action. The former is called hydrostatic lubrication in which pressurized fluid is supplied by external pumping system. The latter is called hydrodynamic lubrication in which relative motion of mating surfaces brings in the pressurization of fluid. Fundamental features of hydrodynamic lubrication are mentioned below.

The fluid pressurization in sliding bearing with relative motion of bearing surface and shaft lubricated with oil in locomotive train was found as measurement of high pressure by Tower in 1883 (Tower, 1883). Reynolds applied Navier-Stokes equation for viscous fluid to the fluid flow in small clearance and proposed the mechanism of hydrodynamic pressurization as Reynolds equation (Reynolds, 1886). For simplified case of rubbing bodies with small clearance (fluid film thickness) h lubricated with fluid as shown in Fig. 1.14 where the velocity of lower and upper surfaces in x-direction are u_1 and u_2, respectively, and the velocity of upper surface in z-direction is w, the Reynolds equation are shown below.

$$\frac{\partial}{\partial x}\left(\frac{\rho h^3}{12\eta}\frac{\partial p}{\partial x}\right)+\frac{\partial}{\partial y}\left(\frac{\rho h^3}{12\eta}\frac{\partial p}{\partial y}\right)=\frac{u_1+u_2}{2}\frac{\partial(\rho h)}{\partial x}+\frac{\rho h}{2}\frac{\partial}{\partial x}(u_1+u_2)+\frac{\partial}{\partial t}(\rho h) \tag{1.17}$$

where p is pressure, ρ is fluid density, and t is time.

From this formula, main features such as pressure distribution, velocity distribution, fluid film thickness, load-carrying capacity, and friction coefficient are obtained under corresponding boundary conditions.

Next, the load-carrying mechanism with hydrodynamic lubrication based on fluid pressurization by relative motion between rubbing surfaces is described. To generate fluid pressurization, it is required to satisfy the following situation.

$$\frac{\partial^2 p}{\partial x^2}+\frac{\partial^2 p}{\partial y^2}<0 \tag{1.18}$$

Therefore, if the right hand value of Eq. (1.17) is negative, it is expected that the fluid pressurization is generated to carry the load.

$$\frac{u_1+u_2}{2}\frac{\partial(\rho h)}{\partial x}+\frac{\rho h}{2}\frac{\partial}{\partial x}(u_1+u_2)+\frac{\partial}{\partial t}(\rho h)<0 \tag{1.19}$$

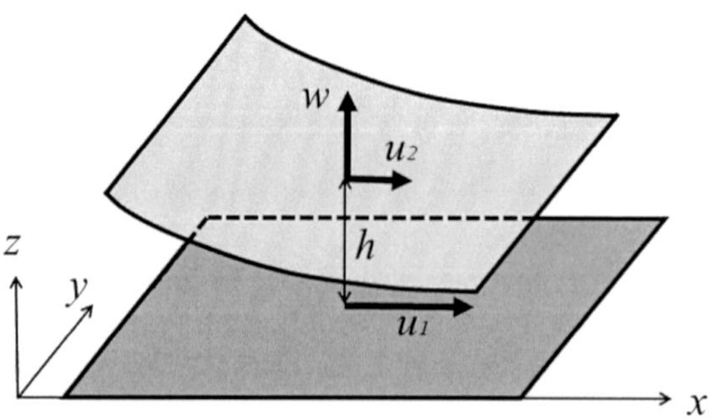

Figure 1.14 Simplified model of mating surfaces with velocity boundary condition. In subject on fluid film lubrication, the fluid film thickness h is very thin.

The fluid pressurization mechanisms derived from these three terms are identified as the wedge action, the stretch action, and the squeeze action, respectively. The stretch action caused by negative second term is observed in only rare eventuality such as the hydroplaning phenomenon for decelerating tire on wet road; therefore, the detailed description is focused on the wedge and squeeze actions.

1.3.2.1 The wedge action

The fluid pressurization is produced by the wedge action for $\partial(\rho h)/\partial x < 0$. For incompressible fluid, the hydrodynamic pressure is generated when there is a converging wedge. The corresponding simplified example at $u_1 = u$ and $u_2 = 0$ is shown in Fig. 1.15a, where the hydrodynamic pressure and fluid flow distributions are depicted. If the rubbing surfaces are parallel, the fluid pressurization for incompressible fluid by wedge action is not substantialized. In a diverging wedge, the negative pressure is generated and it is difficult to carry load by fluid film.

1.3.2.2 The squeeze action

As another mechanism, the fluid pressurization is produced by the squeeze action as $\partial(\rho h)/\partial t < 0$ even for parallel mating planes. An example for parallel planes which close distance with approaching velocity of $\partial h/\partial t$ for incompressible fluid is shown in Fig. 1.15b. In the loading process or in cases which the fluctuating load acts to the rubbing surfaces, the squeeze action plays a predominant role to generate the hydrodynamic pressure.

In appropriate design of hydrodynamic sliding bearings for industrial uses, the specifications are arranged so that the full fluid film can be formed by the wedge and/or squeeze actions and the direct contact between rubbing surfaces can be prevented. In these cases at fluid film lubrication mode, friction is low as shown in Stribeck curve (Fig. 1.8), because the frictional resistance is caused by low shearing resistance of lubricating fluid. Furthermore, the wear is infinitesimal due to no direct interaction between rubbing surfaces and their longer durability is expected. To establish the fluid film lubrication, it is required that the minimum fluid film thickness is larger than the surface roughness of rubbing surfaces.

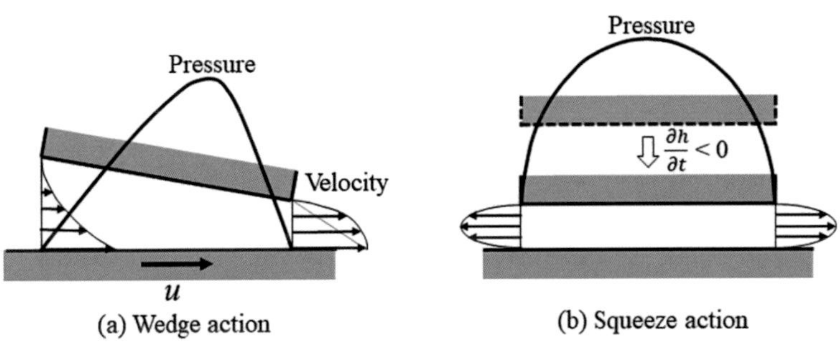

Figure 1.15 Wedge and squeeze actions for hydrodynamic pressurization. For incompressible fluid, the hydrodynamic pressure as wedge action is generated for converging wedge. The hydrodynamic pressure as squeeze action is generated for approaching process.

Reynolds indicated that the self-acting hydrodynamic lubrication plays an important part in animal joints (Reynolds, 1886).

1.3.3 Elastohydrodynamic lubrication

In previously mentioned mechanism, the rubbing surfaces are assumed as nondeformed ones of rigid bodies as observed in sliding bearing such as metallic thrust bearing. In contrast, the influence of elastic deformation of rubbing surfaces on fluid film formation should be taken into consideration for the rubbing surface situations of ball/roller rolling bearing and gear at high contact pressure in line or point contact, and compliant rubbing surfaces such as natural synovial joints and elastomeric seals. This lubrication mode is called the elastohydrodynamic lubrication (EHL), where thicker fluid film can be formed by the effects of elastic deformation of elastic bodies and/or lubricant viscosity increase due to high pressure.

As shown in Fig. 1.16, the rubbing pairs of elastic body and rigid one can generate thicker fluid film as EHL modes than the rigid-rigid rubbing pair. Here, EHL modes are divided broadly as hard-EHL for piezo-viscous and elastic bodies (for rolling bearing, gear, etc.) and soft-EHL for iso-viscous and elastic bodies (for elastomeric seal, natural synovial joints, etc.). Detailed explanations on soft-EHL for natural and artificial joints are described in Chapters 3–5.

1.3.4 Boundary lubrication

In rubbing conditions lubricated with low viscosity lubricant under high load and slow speed, the hydrodynamic fluid film is likely to become thinner than the surface roughness of rubbing surfaces. At this situation, the penetration of the lubricant film will occur, and direct contact will take place between the surface asperities. As a result of direct contact of rubbing surfaces, friction increases and surface failures such as wear and scuffing tend to take place.

In order to sustain low friction and protect the surfaces under thin film condition with considerable direct contacts between rubbing surfaces, it is required that some protective molecular films with lubricity on solid surfaces can serve as a barrier against severe rubbing actions. This lubricating boundary film is formed by boundary

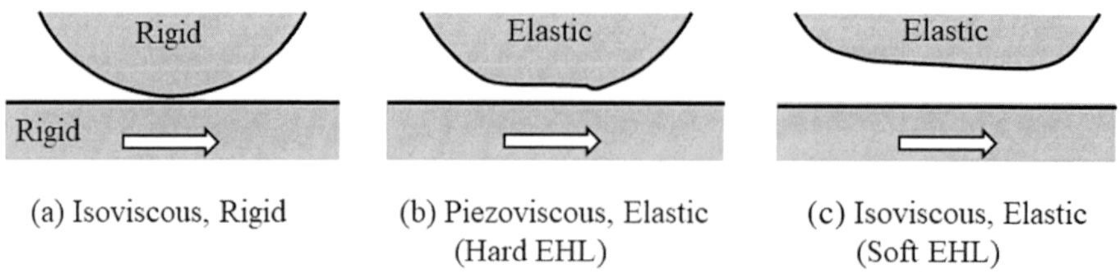

Figure 1.16 Simplified models of lubricating fluid film profiles for sliding pairs of rigid or elastic cylindrical surface on rigid plane. Lubricant viscous property is iso-viscous or Piezo-viscous. Elastic deformation of rubbing surface and/or lubricant viscosity can increase due to high pressure increase in fluid film thickness in elastohydrodynamic lubrication (EHL).

lubricant. The main function of a boundary lubricant is to interpose between the rubbing surfaces a protective film that is able to reduce the amount of solid-solid interaction and that is itself easily sheared. The best protection is provided by a boundary film as a solid film consisting of long chain molecules possessing the following properties: (a) strong attraction between the chains to resist penetration by surface asperities, (b) low shear strength to give a low friction, (c) high melting point so that it provides solid film protection up to high temperature (Bowden & Tabor, 1974).

For example, when we use mineral oil as lubricant for metallic rubbing surfaces in machine element, a small quantity (usually less than 1%) of certain active organic compounds such as alcohols (R−OH), amines ($R-NH_2$), and fatty acids (R−COOH) are added as additives. Here, R means the hydrocarbon group as long chain. These surface active molecules usually form surface films as adsorbed films of mono-molecular layer which are strongly attached to the metal surfaces by electrostatic (dipole) force as physisorption. If the metal is reactive, fatty acids have the additional merit to react with the metal to form the metal soaps as chemisorption, which have desirable shear properties and higher melting points than that of the original fatty acids. Therefore, stearic acid on platinum as inert metal shows friction increases at melting point of stearic acid, but stearic acid on steel surface maintains low friction until melting (softening) point of metal soap. Examples of adsorbed films of hexadecane as physisorption and stearic acid on iron oxide layer as chemisorption are shown in Fig. 1.17 (Murakawa, 1969).

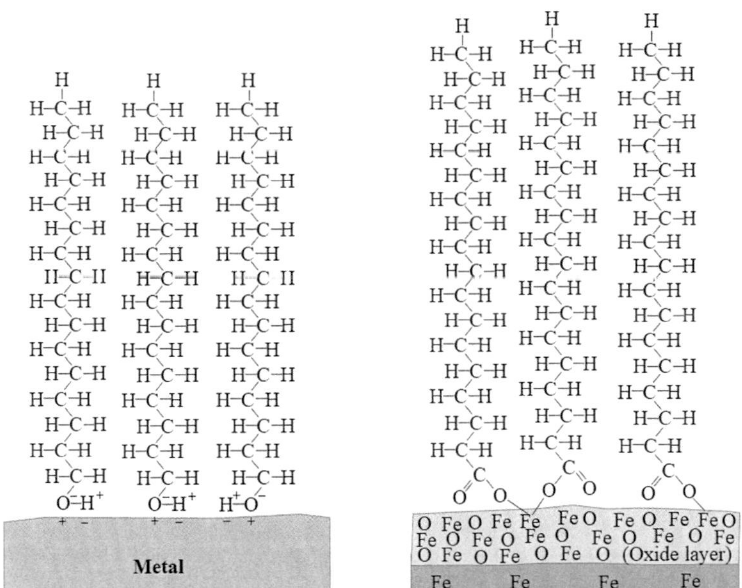

(a) Physically adsorbed film (b) Chemically adsorbed film

Figure 1.17 Models of adsorbed films. (a) *n*-Hexadecanol on metal as physically adsorbed film and (b) stearic acid on iron oxide film as chemically adsorbed film.
Modified from Murakawa, T. (1969). Metal surface chemistry and lubrication (in Japanese). *Journal of Japan Society of Lubrication Engineers (Junkatsu)*, *14*(4), 195−202.

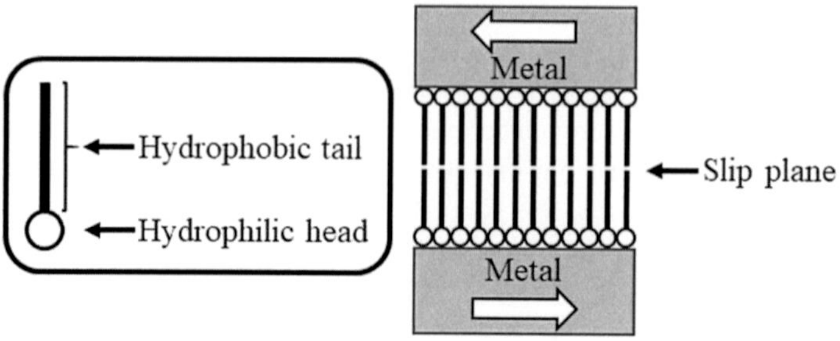

Figure 1.18 Sliding model for adsorbed film covered metal surfaces in hydrocarbon oil. In sliding between adsorbed film covered surfaces, the shearing occurs in slip plane.

The lubrication mechanism provided by these boundary layers is known as boundary lubrication (Hardy & Doubleday, 1922a,b).

A model for adsorbed films of polar molecules on metallic surfaces lubricated with hydrocarbon oil is shown in Fig. 1.18. For example, the fatty acid is composed of hydrophilic head and hydrophobic CH_2 tail. In sliding between adsorbed film covered surfaces, the shearing occurs in slip plane between top surfaces of adsorbed films. Friction force in boundary lubrication is caused by this shearing force. The friction level depends on the property of adsorbed film.

Fig. 1.19 shows a sliding model for adsorbed film covered rubbing surfaces in aqueous media. The examples of adsorbed films of phospholipid in bilayer and bilayer on monolayer are illustrated on hydrophilic and hydrophobic rubbing surfaces, respectively. In this case, the sliding occurs in upper or lower slip plane.

1.3.5 Mixed lubrication

In actual lubricated rubbing surfaces, there are large variations in surface roughness from mirror-finish to highly roughened surfaces. If fluid film thickness is thinner than the surface roughness, some direct contact occurs between rubbing surfaces. In

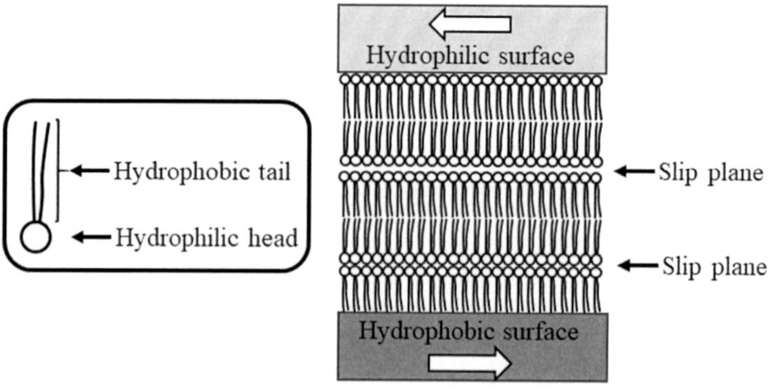

Figure 1.19 Sliding model for adsorbed film covered rubbing surfaces in aqueous media. In this case, the sliding occurs in upper or lower slip plane.

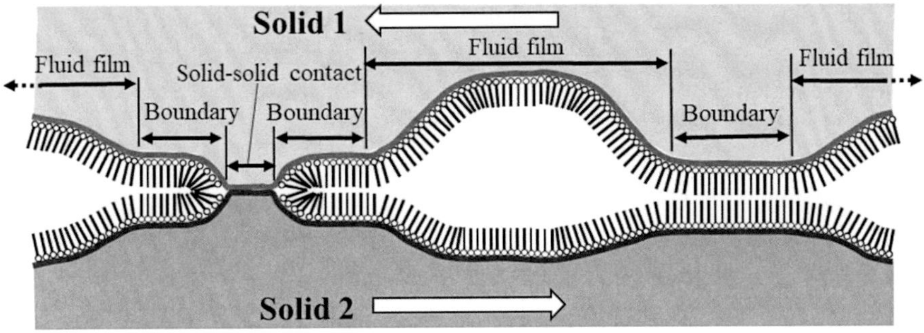

Figure 1.20 Rubbing surface model in mixed lubrication. The actual rubbing surfaces in mixed lubrication are composed of various zones with different lubrication modes such as fluid film lubrication, boundary lubrication, and solid-solid contact.

severer contact zone, the adsorbed films are removed and the solid-solid contact occurs. Therefore, the actual rubbing surfaces are composed of various zones with different lubrication modes such as fluid film lubrication, boundary lubrication, and solid-solid contact as shown in Fig. 1.20.

In such situation as mixed lubrication, friction shows intermediate value between boundary and fluid film lubrication regimes depending on severity of operating conditions as shown in Fig. 1.8. Friction force in mixed lubrication F_{mix} is composed of three components which are dependent on their friction coefficient μ_{fluid}, $\mu_{boundary}$, and μ_{solid} and load-carrying values W_{fluid}, $W_{boundary}$, and W_{solid}.

$$F_{mix} = \mu_{fluid} W_{fluid} + \mu_{boundary} W_{boundary} + \mu_{solid} W_{solid} \tag{1.20}$$

Therefore, friction coefficient in mixed lubrication μ_{mix} under total load of W is shown in the following formula.

$$\begin{aligned} \mu_{mix} &= \frac{F_{mix}}{W} \\ &= \frac{\mu_{fluid} W_{fluid}}{W} + \frac{\mu_{boundary} W_{boundary}}{W} + \frac{\mu_{solid} W_{solid}}{W} \\ &= \mu_{fluid}\beta_{fluid} + \mu_{boundary}\beta_{boundary} + \mu_{solid}\beta_{solid} \end{aligned} \tag{1.21}$$

Here, β_{fluid}, $\beta_{boundary}$, and β_{soild} are load-carrying rates for fluid film, boundary lubrication, and solid-solid contact, respectively.

Examples for $\mu_{fluid} = 0.01$, $\mu_{boundary} = 0.2$, and $\mu_{solid} = 0.5$ are shown;

For load-carrying rates with $\beta_{fluid} = 0.3$, $\beta_{boundary} = 0.6$ and $\beta_{solid} = 0.1$, $\mu_{mix} = 0.173$.

For load-carrying rates with $\beta_{fluid} = 0.86$, $\beta_{boundary} = 0.12$ and$\beta_{solid} = 0.02$, $\mu_{mix} = 0.0426$.

As shown above, friction in mixed lubrication changes widely depending on the severity of rubbing surfaces estimated as load-carrying rates.

1.3.6 Solid lubrication

In many cases where usual fluid lubricants become ineffective under extreme environment conditions such as in vacuum, at high/low temperature and in specific environment, solid lubricants are applied as a powder or thin film on a surface to protect the rubbing surfaces and to reduce friction and wear. Many solid lubricants have layered or lamellar structures such as graphite (Fig. 1.21) and molybdenum disulfides. These layers themselves are strong but the bonding between the layers is weak. Consequently, the layered structure is strong in compression and provides low resistance to relative motion. As other usage, the solid lubricants are used as additives in lubricating oils, water, and greases. Some materials with solid lubrication ability are used as self-lubricating surfaces. In biological systems, the oligolamellar phospholipids are likely to behave like solid lubricant or adsorbed film on articular cartilage surfaces (Hills, 1989). This lubrication mode is described as solid lubrication or solid-film lubrication.

1.4 Wear

Wear is defined as "the progressive loss of substance from the operating surface of a body occurring as a result of relative motion at the surface" (Organisation for Economic Co-operation and Development (OECD), 1969). Wear is usually detrimental, but may be beneficial in mild form such as during running-in. In many wear processes, small-scale viscoelastic or plastic deformation may occur. In evaluation of wear, the appropriate discrimination of wear and deformation is required.

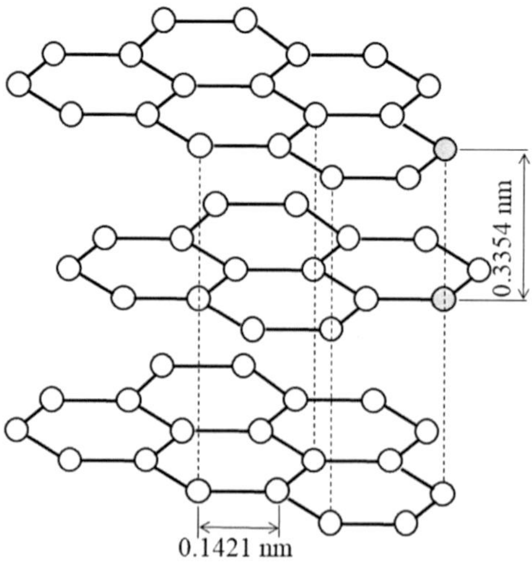

Figure 1.21 Crystal structure of graphite. The layered structure is strong in compression and provides low resistance to relative motion between layers.

Wear mechanisms are classified as the following types. In this section, each wear type is defined on the OECD definition (Organisation for Economic Co-operation and Development (OECD), 1969). In cases where several mechanisms combine to occur, complicated wear failure may be observed.

1.4.1 Wear mechanisms

1.4.1.1 Adhesive wear

Adhesive wear is defined as "wear by transference of material from one surface to another during relative motion, due to a process of solid-phase welding." As shown in Fig. 1.22a, the asperities of rubbing surfaces adhere strongly in real contact area at high contact pressure and form the junction. If the junction is weaker than the material on either side of it, the shearing occurs in the interface itself, which means no wear. If the junction is stronger than one of materials, the shearing occurs not in interface itself but within the softer or weaker material (lower material in case for Fig. 1.22). Therefore, worn area is found on softer material and the removed part of softer material is transferred to mating material surface (Fig. 1.22b). However, in some cases where the surface asperity of harder material is rubbed under severer conditions, due to geometrical or structural severity, a part of harder material of rubbing material can be removed and transferred. In repeated rubbing process, mixing of small transfer pieces and forward-and-backward transfer occur and transferred pieces pile up to larger transfer film. If the transferred film integrated on surface cannot withstand to the shearing force during rubbing, it is released as wear debris. The elastic recovery of transferred part appears to contribute to the release of debris.

The incidence of adhesion changes depending on material combinations, geometrical properties of surface asperity, and existence or nonexistence of adsorbed films and oxide films. When similar materials without protective surface films slide together, adhesion tends to occur. The avoiding of combination of similar materials is recommended in sliding systems.

In rubbing process accompanied with adhesion, all the junctions contribute to friction but not all of them contribute to wear. Wear volume V is estimated by considering

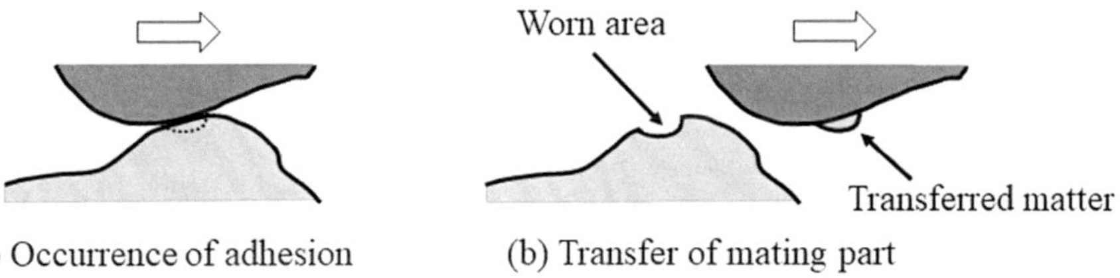

Figure 1.22 Model for adhesive wear. (a) The asperities of rubbing surfaces adhere strongly in real contact area at high contact pressure and form the junction. (b) If the junction is stronger than one of the rubbing materials, the shearing occurs within weaker material. The removed part of weaker material is transferred to mating material surface. The wear debris is produced in repeated rubbing process accompanied with piling up of transferred pieces.

the factor k as the fraction of junctions producing wear (debris) for simplified junction model assuming hemispherical wear particles of the same radius as contact areas proposed by Archard (Archard, 1953; Cameron, 1981).

$$V = (k/3)WL/p_{\mathrm{m}} \tag{1.22}$$

where, W: load, L: sliding distance, p_{m}: flow pressure of softer material.

In actual wear process, more complicated behaviors accompanied with transfer phenomena are likely to produce wear debris, but the phenomena as $V \propto WL$ have been widely observed in various experimental works. Therefore, the following formula with wear factor of K is widely used to evaluate wear volume.

$$V = KWL \tag{1.23}$$

1.4.1.2 Abrasive wear

Abrasive wear occurs when a hard rough surface or hard particles run against soft surface as shown in Fig. 1.23. In many cases, abrasive wear produces larger wear than adhesive wear; therefore, the attention to suppress the abrasive wear is required in adequate selection of rubbing surfaces and in prevention of contamination by foreign hard materials.

In the wear mode as two-body abrasive wear, hard protuberances penetrate into soft material and plow the soft surface (Fig. 1.23a), or act as microcutting tool. In three-body abrasive wear (Fig. 1.23b), hard particles indent, groove, and cut material out of the soft rubbing surfaces.

1.4.1.3 Fatigue wear

Fatigue wear is defined as "removal of particles detached by fatigue arising from cyclic stress variation." On the rubbing surfaces under cyclic and/or fluctuating load, we should pay careful attention to the possibility of fatigue wear. The fatigue wear mechanism for polyethylene delamination in artificial knee joint is described in Chapter 4.

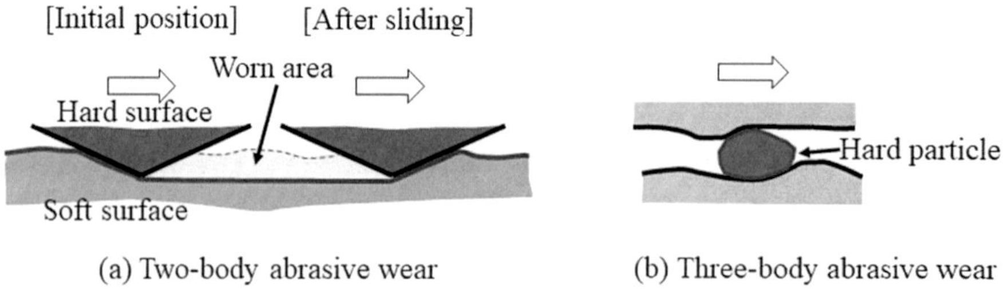

Figure 1.23 Models for abrasive wear. (a) Two-body abrasive wear occurs when a hard rough surface runs against soft material. (b) Three-body abrasive wear occurs when hard particles run against soft material surface.

1.4.1.4 Erosive wear (erosion)

Erosive wear is defined as "loss of material from a solid surface due to relative motion in contact with a fluid which contains solid particles." Erosive wear occurs likewise due to the action of streaming fluid alone. There is different usage for this terminology in dental tribology, where erosion is defined as "surface loss of dental hard tissues due to a chemical process without the involvement of microorganisms" (Zhou & Jin, 2015). Acid-eroded teeth enamel is more susceptible to abrasion and attrition than pristine enamel.

As special case, the cavitation erosion means the erosion caused by collapse of cavities as cavitation phenomena in fluid machinery or fluid flow system with negative pressure.

1.4.1.5 Corrosive wear (corrosion)

Corrosive wear is defined as "a wear process in which chemical or electrochemical reaction with the environment predominates." In many cases, the mechanochemical mechanism, a combination of chemical reaction and mechanical action, enhances the corrosive wear. The inside of the human body is a corrosive environment; hence it is required for medical implants such as artificial joints to have adequate corrosion resistance.

1.4.2 Importance of wear

In actual rubbing surfaces, various types of wear occur as single or combined type. Excessive wear and/or wear above the critical level bring about deterioration in functioning of rubbing materials or devices. To solve wear problems, it is important to elucidate the real wear mechanism and explore the appropriate methods to reduce wear. To extend the durability of rubbing elements or devices, the minimization of wear is required. But, in some cases, the surface modification during running-in process contributes to reducing the rubbing severity. Furthermore, it is important to recognize the influence of wear debris at rubbing area and surroundings. In artificial joints, wear debris-induced osteolysis as biological reaction (Ingham & Fisher, 2005; Willert & Semlitsch, 1977) brought severe clinical problems such as joint loosening. Improvement of rubbing materials with wear resistance and design in recent artificial joints significantly reduced wear and wear-debris-induced complications. However, the occurrence of local direct contact between rubbing materials in insufficient fluid film lubrication can cause wear problems for long usage. Therefore the role of lubrication mechanism becomes important to reduce wear. For reduction of wear, the consideration to sustain the function of rubbing elements or devices from the tribological viewpoints of design, material, and lubrication is required. In addition, in medical devices/implants and biomaterials, the influence of biological environment on wear phenomena should be taken into consideration.

References

Archard, J. F. (1953). Contact and rubbing of flat surfaces. *Journal of Applied Physics, 24*(8), 981–988. https://doi.org/10.1063/1.1721448

Bowden, F. P., & Tabor, D. (1954). *The friction and lubrication of solid I.* Oxford University Press.

Bowden, F. P., & Tabor, D. (1974). *Friction an introduction to tribology*. Heinemann Educational Books Ltd.

Cameron, A. (1981). *Basic lubrication theory* (3rd ed.). Ellis Horwood Limited.

Department of Education and Science. (1966). *Lubrication (tribology) – Education and research. A report on the present Position and Industry's Needs*. HMSO.

Dowson, D. (1966–67). Modes of lubrication in human joints. Proceedings of Institute of Mechanical Engineers, 181(Pt3J), 45–54.

Dowson, D. (1979). *History of tribology*. Longman.

Dowson, D. (2009). A tribological day. *Proceedings of the Institution of Mechanical Engineers, Part J: Journal of Engineering Tribology, 223*(3), 261–273. https://doi.org/10.1243/13506501JET557

Dowson, D., & Neville, A. (2006). Bio-tribology and bio-mimetics in the operating environment. *Proceedings of Institute of Mechanical Engineers, Part J: Journal of Engineering Tribology, 220*(3), 109–123. https://doi.org/10.1243/13506501JET156

Dowson, D., & Wright, V. (1973). Bio-tribology. In T. C. Davenport (Ed.), *The Rheology of lubricants* (pp. 81–88). Applied Science Publishers Ltd.

Dumbleton, J. H. (1981). *Tribology of natural and artificial joints*. Elsevier Scientific Publishing Company.

Greenwood, J. A., & Williamson, J. B. P. (1966). Contact of nominally flat surfaces. *Proceedings of the Royal Society of London. Series A, 295*, 300–319.

Hardy, W. B., & Doubleday, I. (1922a). Boundary lubrication - the paraffin series. *Proceedings of the Royal Society of London. Series A, 100*, 550–574.

Hardy, W. B., & Doubleday, I. (1922b). Boundary lubrication - the temperature coefficient. *Proceedings of the Royal Society of London. Series A, 101*, 487–492.

Hills, B. A. (1989). Oligolamellar lubrication of joints by surface active phospholipid. *Journal of Rheumatology, 16*(1), 82–91.

Ingham, E., & Fisher, J. (2005). The role of macrophages in osteolysis of total joint replacement. *Biomaterials, 26*(11), 1271–1286. https://doi.org/10.1016/j.biomaterials.2004.04.035

Murakami, T. (1990). The lubrication in natural synovial joints and joint prostheses. *JSME International Journal, Series III, 33*(4), 465–474. https://doi.org/10.1299/jsmec1988.33.465

Murakami, T., Ohtsuki, N., & Higaki, H. (1995). The adaptive multimode lubrication in biotribological systems. *Proceedings of International Tribology Conference Yokohama, 1995*, 1981–1986.

Murakami, T., Yarimitsu, S., Sakai, N., Nakashima, K., Yamaguchi, T., & Sawae, Y. (2017). Importance of adaptive multimode lubrication mechanism in natural synovial joints. *Tribology International, 113*, 306–315. https://doi.org/10.1016/j.triboint.2016.12.052

Murakawa, T. (1969). Metal surface chemistry and lubrication (in Japanese). *Journal of Japan Society of Lubrication Engineers (Junkatsu), 14*(4), 195–202.

Myshkin, N. K., & Kovalev, A. V. (2009). Adhesion and friction of polymers. In *Polymer tribology* (pp. 3–37). Imperial College Press. https://doi.org/10.1142/9781848162044_0001

Organisation for Economic Co-operation and Development (OECD). (1969). *Friction, Wear and Lubrication = Tribology =, Glossary of terms and definitions in the field of friction, wear and lubrication = tribology* -. OECD.

Reynolds, O. (1886). On the theory of lubrication and its applications to Mr. Beauchamp Tower's experiments including an experimental determination of the viscosity of olive oil. *Philosophical Transactions of the Royal Society of London, 177*, 157–234.

Sasada, T., Tsukamoto, Y., & Mabuchi, K. (1988). *Biotribology - Friction and lubrication of synovial joints (In Japanese)*. Sangyo Tosyo.

Secomb, T. W., Hsu, R., & Pries, A. R. (2006). Tribology of capillary blood flow. *Proceedings of the Institution of Mechanical Engineers, Part J: Journal of Engineering Tribology, 220*(8), 767–774. https://doi.org/10.1243/13506501JET121

Shiddaiah, A., & Menezes, P. L. (2016). Advances in bio-inspired tribology for engineering applications. *Journal of Bio- and Tribo-Corrosion, 2*(4), 1–19. https://doi.org/10.1007/s40735-016-0053-0. Article 23.

Sivamani, R. K., & Maibach, H. I. (2006). Tribology of skin. *Proceedings of the Institution of Mechanical Engineers, Part J: Journal of Engineering Tribology, 220*(8), 729–737. https://doi.org/10.1243/13506501JET85

Tower, B. (1883). First report on friction experiments. *Proceedings of Institute of Mechanical Engineers, 34*(a), 632–659.

Tribonet. (2020). *Friction coefficient tables in air and vacuum*. https://www.tribonet.org/wiki/friction-coefficients-in-atomosphere-and-vacuum/.

Wikipedia. (2020). *Scrab (Artifact)*. https://en.wikipedia.org/wiki/Scrab_(artifact).

Willert, H. G., & Semlitsch, M. (1977). Reaction of the articular capsule to wear products of artificial joint prostheses. *Journal of Biomedical Materials Research, 11*(2), 157–164. https://doi.org/10.1002/jbm.820110202

Zhou, Z. R., & Jin, Z. M. (2015). Biotribology: Recent progresses and future perspectives. *Biosurface and Biotribology, 1*, 3–24. https://doi.org/10.1016/j.bsbt.2015.03.001

Biomechanics of human and artificial joints

2

2.1 Importance of biomechanics in human and artificial joints

Skeletal system of human body as living organism is composed of more than 200 bones (for adult) which are connected by various joints. Human body has over 300 joints which are classified as fibrous joints, cartilaginous joints, and synovial joints. Synovial joints such as hip, knee, and finger joints as diarthrosis have a cavity containing synovial fluid between the bones and are smoothly moved in three dimensions by muscular systems. In this book, the synovial joints are focused as the main theme. The representative synovial joints are indicated with human skeletal system composed of about 200 bones in Fig. 2.1 (Yamamoto et al., 2005). To facilitate various human movements, synovial joints have different types of structure, geometry, size, and degree of freedom. For example, hip and shoulder joints operate as ball and socket (spheroidal) joints. When the socket is well rounded it is called cotyloid or cup-shaped. When less well rounded it is called glenoid, as in the shoulder (Wright, 1981). In contrast, knee joint operates as condylar joint. In this type, one bone articulates with the other by two distinct surfaces whose movements are not dissociable. These surfaces are termed condyles. These different synovial joints are operated under various loading and movement conditions at daily activities.

From the viewpoint of mechanics, human body and its constituent units such as skeletal, muscular, circulatory, and visceral systems function in harmony with gravitational field and various external and internal forces. Particularly, musculoskeletal system including synovial joints is often subjected to higher loads or stresses in various daily activities. Healthy synovial joints can sustain the excellent load-bearing performance with smooth movement at low friction for whole human life. In considerable synovial joints operated under harsh conditions, however, degenerative changes of articular cartilage occur and their function as synovial joints is deteriorated. In severely diseased joints, the patients have restrictions in movement and load-support abilities with severe pains.

As effective medical treatment for patients with severely diseased joints, these joints were replaced by the artificial joints, which mimic natural joints but have usually simplified shapes and are composed of artificial materials with different properties. Artificial joints are subjected to similar or severe loading and similar or limited movement compared with natural joints.

The functions of both natural tissues/organs and artificial materials/structures are profoundly affected by mechanical conditions. Therefore, in order to sustain the function of natural and artificial joints, it is important to incorporate the viewpoint of biomechanics to be defined as "Mechanics applied to biology"

Biotribology of Natural and Artificial Joints. https://doi.org/10.1016/B978-0-12-823669-7.00008-9

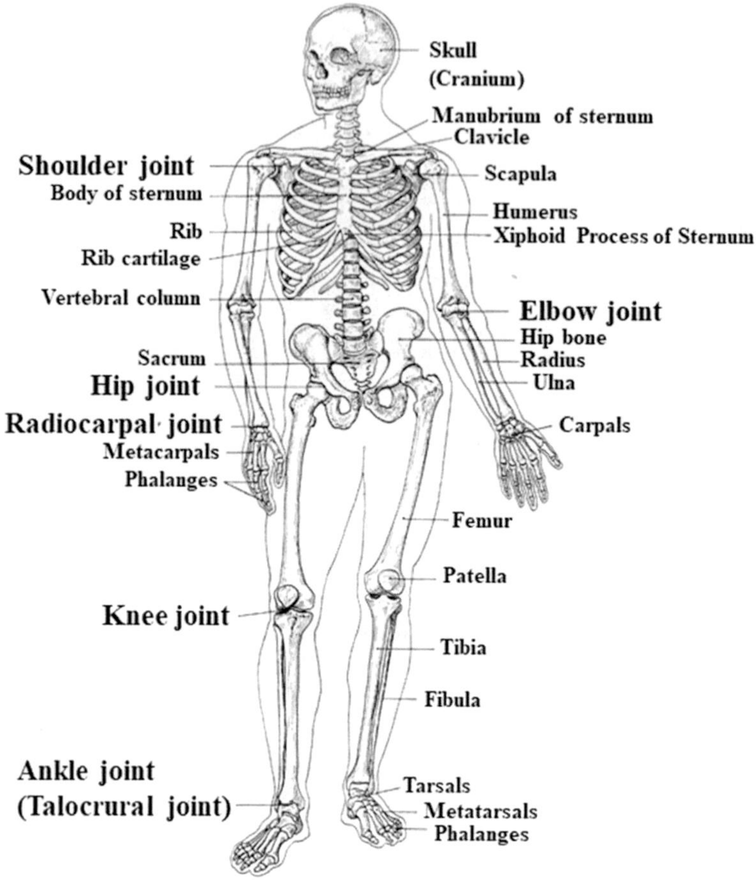

Figure 2.1 Skeletal structure of human body. Representative bones and synovial joints are indicated.
Reprinted and translated with permission from Yamamoto, T., Suzuki, T., & Tasaki, K. (2005). *Modern anatomy and physiology. 11th edition* (pp. 62). Nankodo Co., Ltd. Tokyo

(Fung, 1968). Thus, biomechanics of natural and artificial joints are discussed in this chapter.

In biomechanics, the structure and function of living organisms are examined and analyzed from the viewpoint of mechanics such as statics, dynamics, and kinematics. The accumulated biomechanical knowledge and information can be reflected in the elucidation of the roles of structure and property in living organism, the finding of the cause of diseases, and the development of biomedical technologies for diagnosis and medical treatments at clinical fields. The clinical application of artificial joints as an implant arthroplasty is the most successful biomedical technology in orthopedic surgery, in which biomechanics with biotribology play important roles in the optimization of implant design and evaluation of the performance of artificial joints.

The target levels of biomechanics are ranged from human body, the organs, tissues, and cells to molecules. For example, the natural synovial joint itself corresponds to one of organs, but in order to estimate the force and deformation of cartilage/bone construct in natural joint, it is required to consider not only the geometry, structure, and material properties of cartilage/bone construct but the mechanical situations of muscular forces, ligament forces, and inertia forces in related musculoskeletal system and/or whole body. In another example, to evaluate the influence of mechanical stimulus on metabolic activities of chondrocyte such as cell viability, proliferation, and production of cellular proteins, it is required to examine not only the response of cells and their components but the cell-extracellular matrix interaction accompanied with fluid flow containing nourishment through cartilage tissue. In biomechanical modeling of living organism with hierarchical structure, the single-scale or multi-scale analysis should be applied by considering the importance of mutual relation between various levels of hierarchy.

In addition, the changes with growth or with aging in living organism should be considered in biomechanical modeling. It takes usually more than 15–18 years for human musculoskeletal system to grow as an adult body. In analogous way, it takes more than 15 years for human joints to grow to matured structure as shown by X-ray images of hand including finger joints and wrist joint in Fig. 2.2 (Sugiura, 1985). After birth, babies and infants have very compliant synovial joints with thicker articular cartilage (Fig. 2.2a). At next step with growing of skeletal system, the ossification occurs at growth cartilage area and articular cartilage becomes thinner (Fig. 2.2b–c). At fifteen-and-a-half years old, the synovial joint attains the structure and morphology as matured joint (Fig. 2.2d). Therefore, we should pay attention to the fact that typical human joint model shown in Fig. 1.10 corresponds to adult joint.

As indicated by this example, the synovial joint appears to acquire ideal structure, morphology, size, and material properties by adaptive changes under various daily

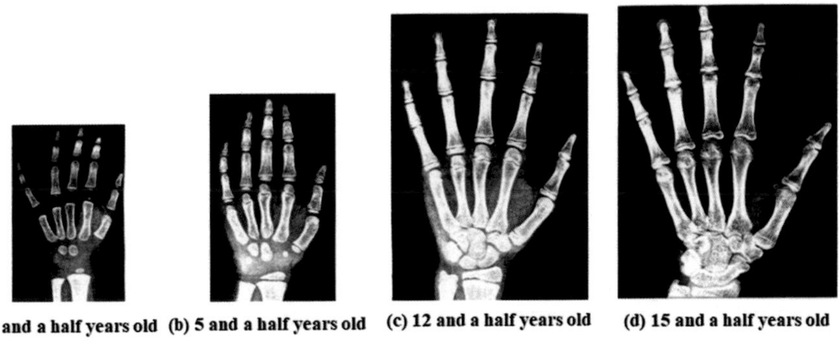

(a) 1 and a half years old (b) 5 and a half years old (c) 12 and a half years old (d) 15 and a half years old

Figure 2.2 Changes of bones and synovial joints in hand of the same person with growing. The structure, morphology, and size of bones and synovial joints are drastically changed with growing for over one-and-a-half decades.
Modified with permission from Sugiura, Y. (1985). *Bone age of Japanese*. Chugai Igakusya Press.

activities in the process of growing for over one-and-a- half decades. As the results of the optimization of structures, shapes, and material properties, natural synovial joints sustain the suitable load-carrying capacity and superior tribological properties with smooth movement for a whole life.

As mentioned previously, human synovial joints acquire excellent tribological properties with extremely low friction and minimal wear with long durability even in weight bearing joints such as hip, knee, and ankle joints under severe operating conditions at high load and slow speed. In certain cases, however, the synovial joints are damaged under various severe conditions. For example, in diseased joints with osteoarthritis, the protective cartilage surfaces are worn and deteriorated, and then some functions of entire joints including underlying bone become ineffective in the load-support and moving abilities. The patients suffer seriously from inability of daily activities and severe pain. The higher load with obesity, the repetitive high stresses with various activities on the joint and joint injuries are considered as the main risk factors of osteoarthritis, and the biomechanical evaluation with force/stress analyses provides basic important thinking behind the guideline to suppress the development of osteoarthritis. As the dominant medical treatment for severely damaged natural joints with osteoarthritis or other diseases, the rubbing surfaces and the damaged parts of diseased joints have been replaced by artificial joints composed of artificial materials such as polymers, metals, and/or ceramics to recover the daily activities of patients and provide relief from pain. The loading situation and stress distribution in artificial joints of various designs are usually different from those of healthy natural joints. Abnormal loading or excessive high stress in artificial joints may lead to surface failure, wear, loosening, and related problems. Therefore, appropriate biomechanical analysis and measurement are required to attain the successful recovery of joint function with long durability.

As described previously, tribological behaviors of natural and artificial joints should be evaluated with general understanding of influence of biomechanical conditions.

2.2 Anatomy and kinematics of human joints

2.2.1 Anatomy of human joints

Human body has over 300 joints and each joint has different structure in skeletal system (Fig. 2.1). Joints to connect two or more bones are divided into fibrous, cartilaginous, and synovial joints. Fibrous joints have no joint cavity and are connected by fibrous connective tissue. These are unmovable or slightly movable as suture of skull bone, distal tibiofibular joint, and dental alveolus. Cartilaginous joints such as connection through cartilaginous end-plate between vertebral disc and intervertebral disc and pubic joint have the limited motion. On the contrary, the synovial joints have a joint cavity containing synovial fluid between bones and are freely movable in various daily activities. The simplified model of natural synovial joint is shown in Fig. 1.9. In this book, the detailed explanation for joint is focused on the synovial joints.

The synovial joints are classified anatomically and physiologically in structural and motional types as shown in Table 2.1 and Fig. 2.3. For examples, hip and shoulder joints are classified as ball and socket joints in which the round end of one bone fits into the cavity of another bone and rotates around three axes as spherical joints (Fig. 2.3a). The degree of freedom for ellipsoidal joint is restricted in biaxial motion shown in Fig. 2.3b. The condylar joints such as knee and temporomandibular joints are special types of ellipsoidal joints, where one bone articulates with the other by two distinct surfaces (condyles) whose movements are not dissociable.

Fig. 2.3 exhibits the simplified synovial joint models in which the mating bone bodies only are extracted from whole joints to indicate the morphological features with relative movement. The actual synovial joints have a joint cavity containing synovial fluid between bones covered with articular cartilage. Next, the detailed models for hip and knee joints are shown as main examples in Figs. 2.4 and 2.5.

Fig. 2.4 shows a simplified model of human hip joint in frontal plane. Femoral head and acetabulum form a ball-and-socket or spherical joint structure to connect femur and pelvis. The rubbing surfaces of femoral head and acetabulum are covered with compliant articular cartilage. Hip joint has wide range of movement with three axes rotation. The hip joint is enclosed by the joint capsule, in which the inner surface as the synovial membrane provides the synovial fluid in joint cavity. The synovial fluid containing the filtrated serum constituents, hyaluronic acid, and others play roles of lubricating fluid and source of nutritious substances for articular cartilage. In daily activities such as running and jumping, heavy loads such as 5–6 times of body weight are applied to the hip joint. Hip joint is a conforming joint as ball-and-socket, which provides stability allowing a moderate range of motion. In this conforming joint, the average contact stress level becomes 2–3 MPa in normal walking which is low compared with nonconforming knee joint. Healthy hip joint has high load support capability and long durability with superior lubrication ability of low friction and minimal wear for a whole life. However, for example, if the excessive load is applied or the articular surfaces are deteriorated

Table 2.1 Types of synovial joints.
Typical examples and degree of freedom are shown.

Type	Examples	Degree of freedom
(a) Ball-and-socket (spheroidal)	Hip and shoulder joints	3 axes
(b) Ellipsoidal	Metacarpophalangeal and radiocarpal joints	2 axes
(b') Condylar	Knee and temporomandibular joints	2 axes
(c) Saddle	First carpometacarpal joint	2 axes
(d) Hinge	Humeroulnar joint	1 axis
(e) Pivot	Superior radioulnar and atlantoodontoid joints	1 axis
(f) Plane	Intermetacarpal joints	2 directions in a plane

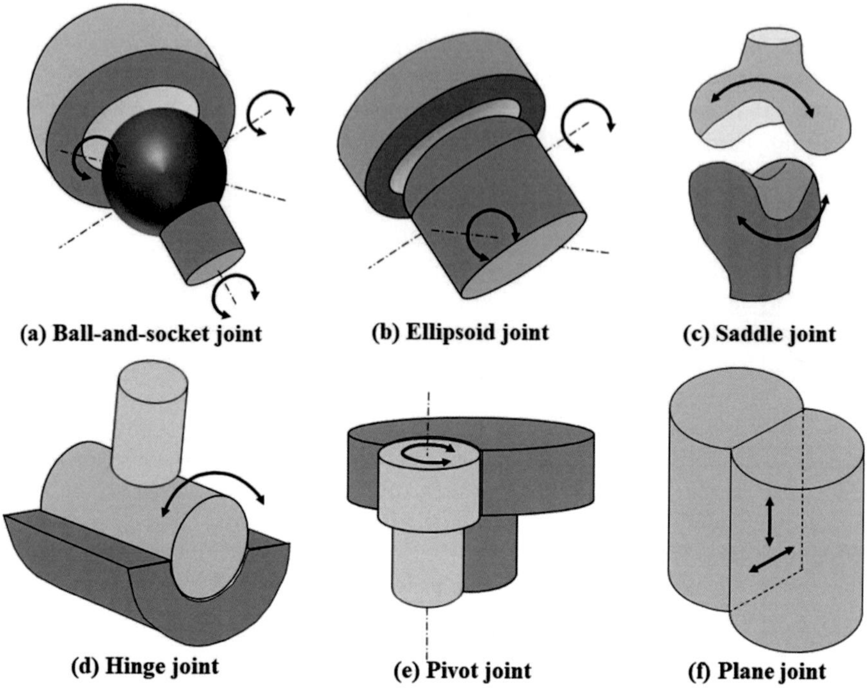

Figure 2.3 Types of synovial joints. In six types of simplified synovial joint models, the morphological features with relative movement are shown.

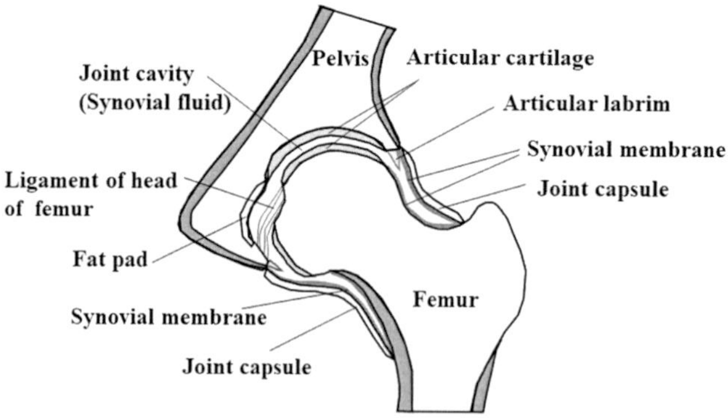

Figure 2.4 Simplified model of human hip joint (frontal plane). Hip joint is a conforming joint as ball-and-socket type which is composed of femoral head and acetabulum. The joint cavity is filled with synovial fluid.

or damaged in osteoarthritis, the hip joint declines in function of synovial joints. For diseased hip joints, various medical treatments such as osteotomy, partial or total joint replacement with artificial joints are applied.

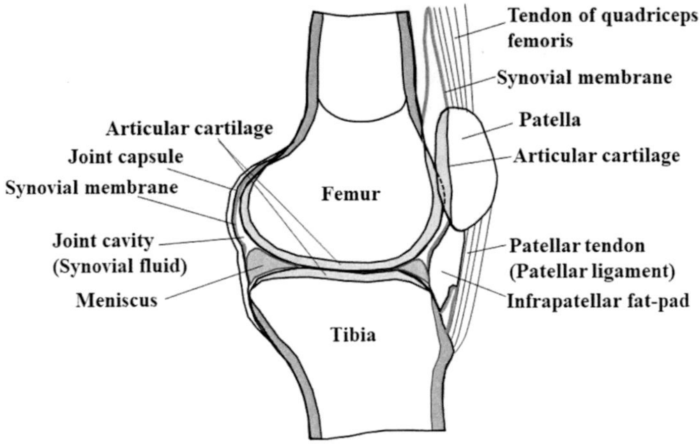

Figure 2.5 Simplified model of human knee joint (sagittal plane). Knee joint has two kinds of articular areas, i.e., tibiofemoral and patellofemoral joints. The loads on tibiofemoral joint are supported by articular cartilage surfaces and meniscus surfaces. The joint cavity is filled with synovial fluid.

Fig. 2.5 shows a simplified model for human knee joint in sagittal plane. These rubbing surfaces are covered with compliant articular cartilage and lubricated with synovial fluid similarly to hip joint. Fig. 2.6 shows frontal view. Knee joint is not a conforming joint with large range of motion and has two kinds of articular areas,

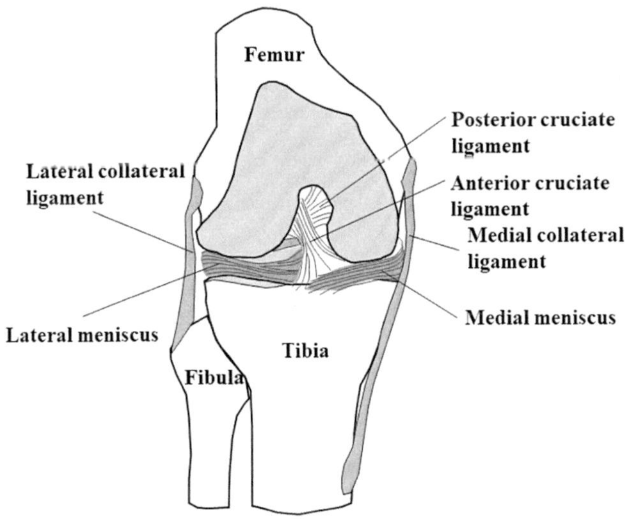

Figure 2.6 Simplified model of human knee joint (frontal view of right knee at flexed position). The tibiofemoral joint is constructed by medial and lateral compartments. The array of ligaments and menisci provide the stability of knee joint.

i.e., tibiofemoral and patellofemoral joints. For natural knee, a large range of flexion is required to actualize the various daily activities. The range of flexion is from the position of full extension to 150–160 degrees flexion. To facilitate large range of motion, the tibiofemoral joint is nonconforming. The tibiofemoral joint is constructed by medial and lateral compartments. Medial and lateral femoral condyles have convex geometry and varying radii that increase posteroanteriorly. The medial tibial plateau is slightly concave, but the lateral tibial plateau is convex in the anteroposterior direction and almost flat in the mediolateral direction. In tibiofemoral joint, the medial and lateral menisci attached on each tibial plateau relieve the contact pressure on the articular cartilage surfaces and provide the stability to compensate the lack of congruity. The stabilization is provided by an array of ligaments. Main ligaments, anterior cruciate ligament (ACL), posterior cruciate ligament (PCL), medial collateral ligament (MCL), and lateral collateral ligament (LCL) are shown in Fig. 2.6.

2.2.2 Motion of natural joints

To consider commonly about the motions of natural joints and artificial joints, it is required to define the planes and directions with a coordinate system (Fig. 2.7).

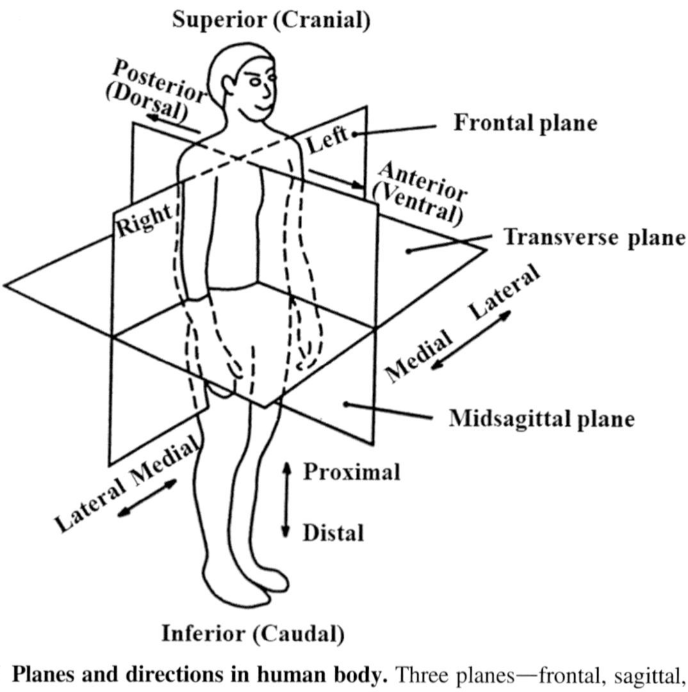

Figure 2.7 Planes and directions in human body. Three planes—frontal, sagittal, and transverse planes—are mutually perpendicular.
Modified from Murakami, T. (1997). *Motion and walking, in*: *Biomechanical Engineering: A First Course* (pp. 122–130). JSME.

For the human body in three-dimensional space, three kinds of anatomical planes—the frontal, sagittal, and transverse planes are defined as follows:

- Frontal plane (or coronal plane): Any vertical plane which divides the body into ventral and dorsal sections.
- Sagittal plane: Any plane that divides the body into right and left parts. Midsagittal plane divides the body exactly in half.
- Transverse plane (or horizontal plane): Any plane that divides the body into superior and inferior parts.

These three planes are mutually perpendicular.

The principal directions are described as "right versus left", "superior (cranial) versus inferior (caudal)" and "anterior (ventral) versus posterior (dorsal)."

Human motions are categorized as (a) angular motions to change angles of bones, (b) rotational motions to rotate along a longitudinal axis, and (c) gliding motions between mating surfaces.

Synovial joints allow the various paired movements as the following motions in human body.

1. Flexion and extension:

Flexion is the angular motion of flexing a limb at joint or to decrease the angle between two bones.

Extension is the angular motion of moving a limb from a flexed position to a straight one or to increase the angle between two bones. Angular motion over a normal straight position is called hyperextension.

Ranges of flexion-extension motions are different for different joints. Flexion-extension motions for hip joint and knee joint are shown in Fig. 2.8a. It is noted that the flexion of hip is movement of leg forward but the flexion of knee is movement of lower leg backward.

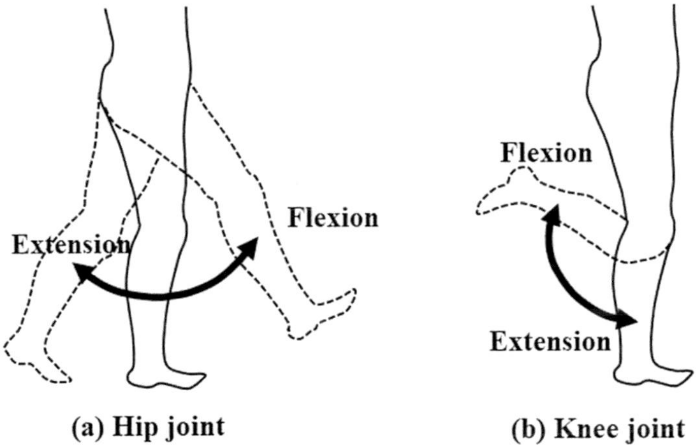

Figure 2.8 Flexion and extension motions for hip and knee joints. (a) Flexion and extension motions for hip joint, (b) Flexion and extension motions for knee joint.

In shoulder joint, flexion-extension motions are defined on sagittal plane, where the motion of shoulder joint as the angular movement of upper arm from the lowest position (at 0°) to top position (at 180°) is defined as flexion or forward up, and the angular movement of upper arm from the lowest position to backward (at 50°) is defined as extension or rear up. On transverse (horizontal) plane, transverse flexion is the angular motion of flexing an upper arm at shoulder joint anteriorly from the lateral position. Transverse extension is straightening the shoulder joint by moving the upper arm posteriorly (to the back).

Actual flexion-extension motion for various synovial joints should be specified as appropriate definition for corresponding joint motions.

2. Internal rotation and external rotation:

Internal rotation (or medial rotation) is the inward rotation of a part around the axis.

External rotation (or lateral rotation) is the outward rotation of a part around the axis.

Internal rotation and external rotation for lower limb are shown in Fig. 2.9a.

3. Abduction and Adduction:

Abduction is the motion of the structure away from the midline.

Adduction is the motion of the structure toward the center of body as midsagittal plane.

Abduction and adduction for lower limb are shown in Fig. 2.9b.

4. Dorsiflexion, palmer flexion, and planter flexion:

Dorsiflexion of hand is angular movement of hand in direction of the back at the wrist, toward the dorsal side of forearm.

Dorsiflexion of foot is angular movement of foot in direction of the back at the ankle.

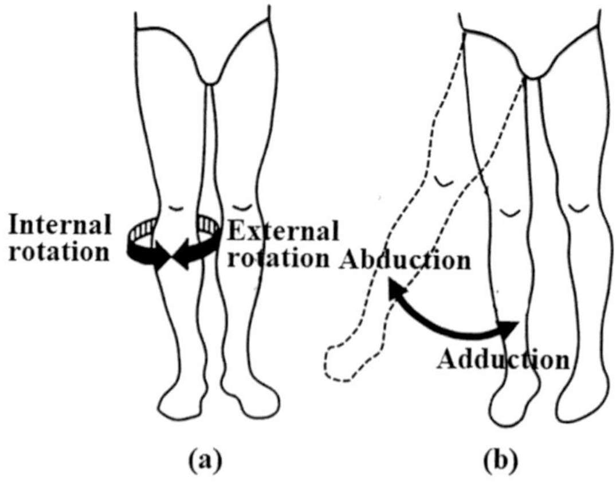

Figure 2.9 Internal/external rotations, abduction, and adduction for lower limb. (a) Internal rotation and external rotation for lower limb, (b) Abduction and adduction for lower limb. Modified from Murakami, T. (1997). *Motion and walking. In Biomechanical Engineering: A First Course* (pp. 122–130). JSME.

Palmar flexion of hand is angular movement of hand toward the palm at the wrist.

Plantar flexion of foot is angular movement of foot in direction of the sole of foot at the ankle.

Detailed movement behaviors and range of motion of upper and lower limbs at various daily activities are demonstrated by Kapandji (2019a,b).

2.3 Loads in human joints

In consideration of load-carrying capacity, tribological performance and durability of human synovial joints, the load is considered to be the most important parameter. Excessive loading may induce mechanical failure and/or excessive deformation of articular cartilage, meniscus, ligaments, tendons, bones, and related tissues in synovial joints. Although the contact pressure changes depending on the geometrical congruity of joints, the load determines the mechanical severity in joints with similar congruity. As indicated in Stribeck curve for "Friction coefficient versus Viscosity×Speed/Load" (Fig. 1.8), high load usually suppresses the lubricating fluid film formation and it follows that an increase in direct contacts enhances friction. Wear volume or wear rate appears to be controlled by the product value of load and rubbing speed. As mentioned above, loads/forces are important factors to influence the performance of human joints in daily activities.

To sustain the healthy functions of synovial joints, it is required to appropriately understand the actual joint forces on different types of joints in various daily activities. Kinetics involves both static and dynamic analysis. Statics is the study of forces acting on a body in equilibrium, and dynamics is the study of forces acting on a body which do not sum to zero. On the basis of the kinetic analysis with related data, it is possible to determine the magnitude of the forces on joints produced by muscles, body weight, connective tissues, and externally applied loads, either static or dynamic (Frankel & Nordin, 1980).

The most sophisticated calculation method would be a dynamic, three-dimensional analysis involving all the muscle groups and constraint forces by ligaments and soft tissues. In order to make the analysis complete the following are taken into account (Dumbleton, 1981):

1. mass and acceleration of body segments (inertia effects).
2. muscle group activity as a function of time so that at any one instant the muscles which are active can be taken into account.
3. location of muscle groups, line of action of muscles, effect of ligament and joint capsule constraints.

Prior to dynamic analysis, it is useful to understand the fundamental mechanism of joint loads by simplified analyses such as static analysis for two-dimensional models with selected muscle and constraint forces. In next subsections, some examples are shown as static and dynamic analyses.

2.3.1 Static analyses

Static analyses can be applied for estimation of static forces that act on the human bodies at static conditions. In simplified cases at motions with very slow movement, the estimated static forces may be used as approximate values.

In general, force has an influence tending to change the motion of a body or produce motion or stress in a stationary body. Here, we consider an example where the external forces act on the stationary rigid body (Fig. 2.10a). Forces are vectors which have magnitudes with specified directions, and are depicted as $\boldsymbol{F}_1$, $\boldsymbol{F}_2$ and so on for corresponding arrows. A force acts at the point of application on the rigid body in the direction of arrow on the line of action.

To sustain a rigid body applied with forces in equilibrium in three-dimensional space, the equilibrium in both forces and moments should be satisfied as follows:

To prevent the translation of a rigid body by forces $\boldsymbol{F}_i$, i.e., to attain the translational equilibrium;

$$\sum_{i=1}^{n} \boldsymbol{F}_i = 0 \tag{2.1}$$

According to Newton's second law, if the sum of all force vectors is zero, the stationary rigid body keeps the original state.

To prevent the rotation of a rigid body by moments $\boldsymbol{M}_i$ with moment arms around the center of rotation, i.e., to attain the rotational equilibrium;

$$\sum_{i=1}^{n} \boldsymbol{M}_i = 0 \tag{2.2}$$

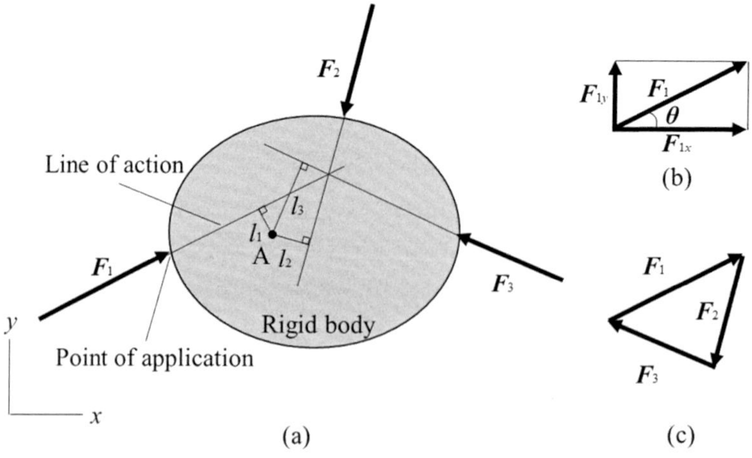

Figure 2.10 A model of stationary rigid body with applied forces at equilibrium. (a) All lines of action meet at the same point and the rigid body does not rotate. (b) Horizontal and vertical components separation of force vector. (c) Vector diagram for forces at equilibrium.

where the moment arm l_i is the orthogonal distance from the center of rotation to the line of action.

For the two-dimensional case in Fig. 2.10, all forces are divided as horizontal and vertical components, for example, $\boldsymbol{F}_1$ is divided as $\boldsymbol{F}_{1x}$ and $\boldsymbol{F}_{1y}$ (Fig. 2.10b).

$$F_{1x} = F_1 \cos\theta \tag{2.3}$$

$$F_{1y} = F_1 \sin\theta \tag{2.4}$$

$$F_1 = \sqrt{F_{1x}^2 + F_{1y}^2} \tag{2.5}$$

Therefore, the equilibrium conditions are shown in the horizontal and vertical components.

For the horizontal forces;

$$F_{1x} - F_{2x} - F_{3x} = 0 \tag{2.6}$$

For the vertical forces;

$$F_{1y} - F_{2y} + F_{3y} = 0 \tag{2.7}$$

At the force equilibrium conditions, all force vectors show the closed state (Fig. 2.10c).

The equilibrium condition for moment is shown below.

$$M_z = F_1 \times l_1 + F_2 \times l_2 - F_3 \times l_3 = 0 \tag{2.8}$$

Usually, the forces such as joint force, muscle force, ligament force, body force, the ground reaction force and other external force are applied to the human body. For the human body or part of mass m on the earth, the body weight $\boldsymbol{W}$ acts on each center of gravity,

$$\boldsymbol{W} = m\boldsymbol{g} \tag{2.9}$$

where $\boldsymbol{g}$ is acceleration of gravity ($g \cong 9.81\ \text{m/s}^2$).

As an example to estimate the force on femoral head in hip joint, the static analysis is performed in the case of single-leg standing as shown in Fig. 2.11.

To estimate the force acting on the femoral head in hip joint at single-leg standing, the equilibrium conditions for forces and moments for free body of rigid matters combined of femur, tibia, fibula, and foot in right lower limb is considered in two-dimensional model in the frontal plane (Fig. 2.11a).

For the single-leg standing, if the abductive muscles do not work, the upper body with left lower limb tends to rotate clockwise around hip joint with extremely low friction. Therefore, to stabilize the upper body without falling, the effective function of the

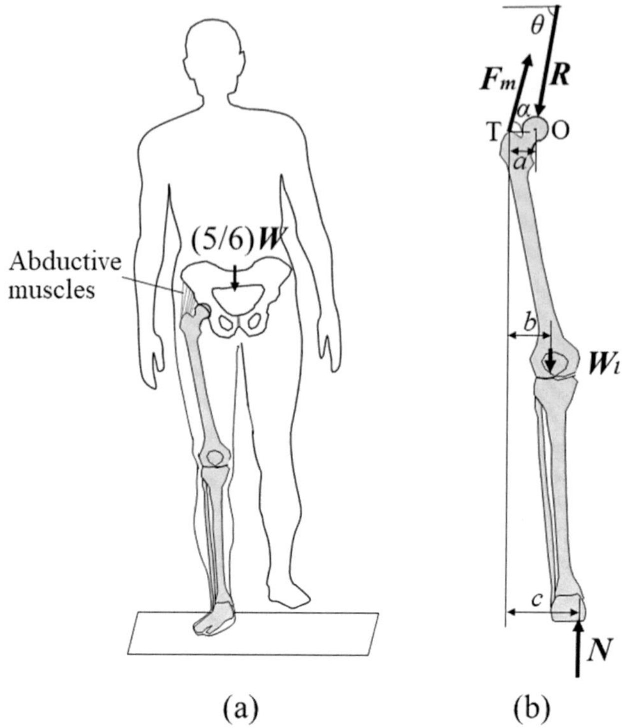

Figure 2.11 Estimation of force applied on femoral head in hip joint. (a) Single-leg standing condition, (b) Forces applied to free body for single lower limb as a unified rigid model.

abductive muscles is required. Thus, there are main four kinds of forces for right lower limb (Fig. 2.11b) as follows:

1. the normal load $\boldsymbol{R}$ on the femoral head in hip joint,
2. the abductive muscle force $\boldsymbol{F}_m$ in Gluteus medius muscle, Gluteus minius muscle, and tensor fasciae latae muscle (tensor fasciae latae muscle is omitted in Fig. 2.11),
3. the reaction force $\boldsymbol{N}$ from the floor to support the weight of body $\boldsymbol{W}$, and
4. the single leg's own weight $\boldsymbol{W}_l$ as combined weight of the leg, foot, and thigh which is about (1/6) W.

As the equilibrium conditions, the following equations should be satisfied.
For equilibrium of horizontal forces;

$$F_{mx} - R_x = 0$$

$$\therefore F_m \cos\alpha - R\cos\theta = 0 \tag{2.10}$$

For equilibrium of vertical forces;

$$F_{my} + N - R_y - W_l = 0$$

$$\therefore F_m \sin \alpha + W - R \sin \theta - (1/6)W = 0 \tag{2.11}$$

For equilibrium of moment around point T;

$$\text{Rsin } \theta \times a + W_l \times b - N \times c = 0$$

$$\therefore \text{Rsin } \theta \times a + (1/6)W \times b - W \times c = 0 \tag{2.12}$$

Or otherwise, for equilibrium of moments around the center of femoral head O;

$$F_m a \sin \alpha + W_l(b-a) - N(c-a) = 0$$

$$\therefore F_m a \sin \alpha + (1/6)W(b-a) - W(c-a) = 0 \qquad (2.12)$$

For case with specified direction of abductive muscle vector as $\alpha = 75°$ and anatomical data as $a = 5$ cm, $b = 8$ cm, and $c = 14$ cm, the solutions for F_m, R, and θ were obtained by simultaneous linear equations with three unknowns as (2.10), (2.11), and (2.12) or (2.12′), as follows;

$$F_m = 1.76W \tag{2.13}$$

$$R = 2.57W \tag{2.14}$$

$$\theta \cong 80° \tag{2.15}$$

As shown above, it is noticed that the larger forces than body weight are applied for the hip joint and the abductive muscles at single-leg standing depicted in Fig. 2.11.

As another approach, these forces can be estimated by considering the equilibrium for upper body side. As shown in Fig. 2.12, the force $\boldsymbol{R}$ applied to the acetabulum in hip joint is decided by the equivalence among $\boldsymbol{R}$, the weight for upper body and single lower limb: (5/6) $\boldsymbol{W}$ and the abductive muscle force: $\boldsymbol{F_m}$. For the same posture to Fig. 2.11, the same solutions as $R = 2.57\ W$ and $F_m = 1.76\ W$ are obtained.

By considering the moment balance around the center of femoral head O in Fig. 2.12, the abductive muscle force can be estimated as shown below.

$$F_m l_1 = (5/6)W l_2$$

$$\therefore F_m = (5/6)(l_2/l_1)W \tag{2.16}$$

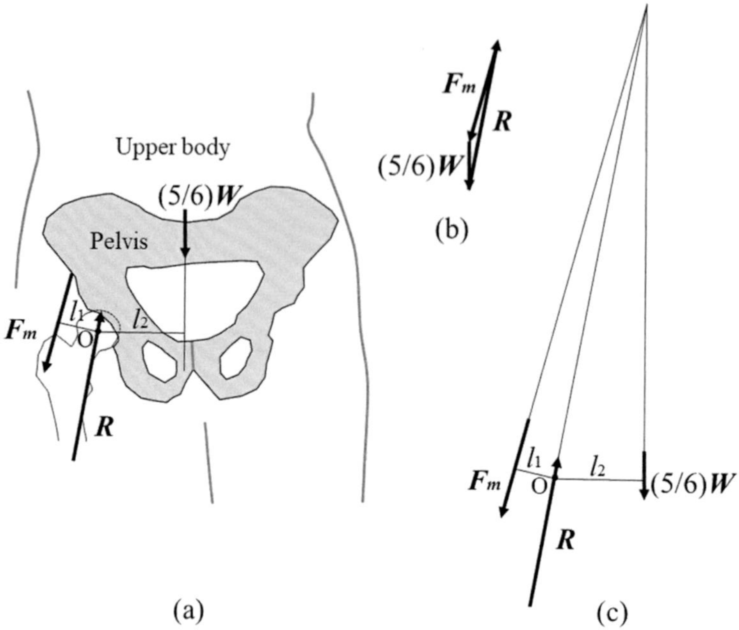

Figure 2.12 Equilibrium around femoral head for upper body and single lower limb. (a) The moment balance between the force by upper body weight including another lower limb (5/6) ***W*** and the abductive muscle force $\boldsymbol{F}_m$ indicates the changes in hip joint force depending on different postures. (b) The sum of force vectors for equilibrium of forces and the meeting of all lines of action for equilibrium of moment are shown.

Therefore, depending on the ratio of moment arm (l_2/l_1), the abductive muscle force can change. Thus, the forces for hip joint can remarkably change depending on the shift in posture or the center of gravity of the upper body with single lower limb. If the center of gravity shifts further away from the femoral head center *O*, the abductive muscle force becomes larger and then the hip joint force also becomes larger. On the contrary, if the center of gravity approaches to the femoral head center, the abductive muscle force and hip joint force can be significantly reduced.

To restrain the hip joint force as appropriate level after hip joint replacement and hip joint osteotomy, the alignment of hip joint should be adjusted to keep the moment arm length l_1 above adequate distance to suppress the abductive muscle force.

2.3.2 Dynamic analyses

In fundamental analysis of forces of human body and parts of body in various activities, the following Newton's laws of motion are applied for translation motion.

1. First law: (Law for Inertia) An object at rest remains at rest, and an object in motion remains in motion at constant speed and in a straight line unless it is acted on by an applied force.

2. Second law: (Law for Force) The rate of change of momentum of an object is directly proportional to the force applied. The acceleration of an object depends on the mass of the object and the amount of force applied.

The second law is written in the following formula;

$$\boldsymbol{F} = \frac{d(m\boldsymbol{v})}{dt} \tag{2.17}$$

or

$$\boldsymbol{F} = m\boldsymbol{a} \tag{2.18}$$

where $\boldsymbol{F}$ is the net force applied, m, $\boldsymbol{v}$, $m\,\boldsymbol{v}$, and $\boldsymbol{a}$ are mass, velocity, momentum, and acceleration of the body, respectively.

3. Third law: (Law of action and reaction) When one object exerts a force on a second object, the second object exerts a force that is equal in magnitude and opposite in direction to the first object.

Most daily movements of human body consist of linear and angular motion. Newton's law of angular motion is described as below (Davidovits, 2013).

The laws governing angular motion are analogous to the laws of translational motion. Torque is analogous to force, and the moment of inertia is analogous to mass.

1. First law: A body in rotation will continue its rotation with a constant angular velocity unless acted upon by an external torque.
2. Second law: The mathematical expression of the second law in angular motion is analogous to Eq. (2.18). It states the torque T is equal to the product of the moment of inertia I and angular acceleration $\ddot{\theta}$;

$$T = I\ddot{\theta} \tag{2.19}$$

where torquc T is given by the product of the perpendicular force to body axis and the distance (moment arm or lever arm) l from the point of application to the axis of rotation. The moment of inertia I of an element of mass m located a distance r from the center of rotation is

$$I = mr^2 \tag{2.20}$$

In dynamic analysis for walking motion of human body, the motion equations described above are applied for multisegment model. First, two-dimensional simplified model composed of foot, lower limb, thigh, and upper body as rigid bodies connected with low friction joints of ankle, knee, and hip joints (Fig. 2.13) (Makikawa & Yoshida, 2008). The center of gravity, mass and inertia moment for each segment are gotten from measurement and related data base. In this example, x-axis and y-axis indicate

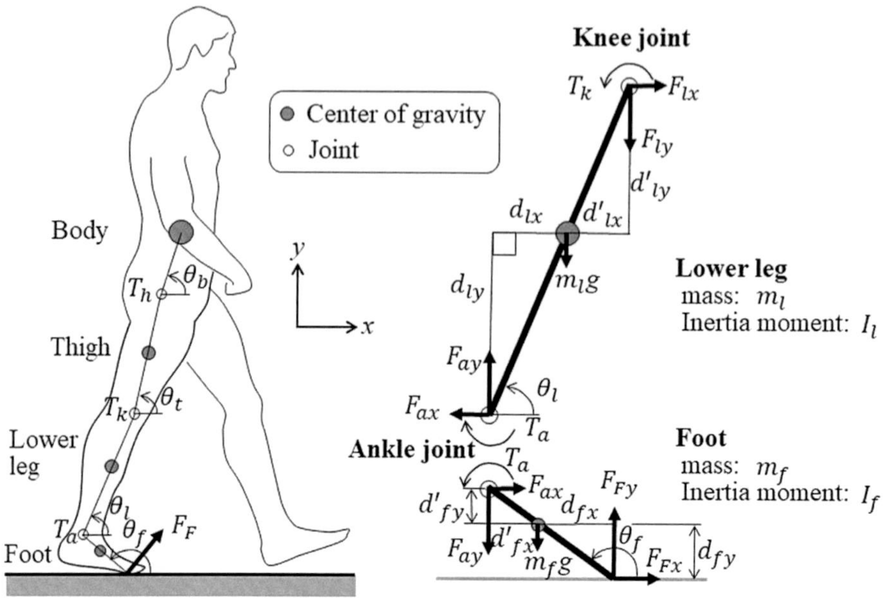

Figure 2.13 Simplified modeling of human walking. Detailed illustration for segments of thigh and upper body are omitted.
Modified from Makikawa, M. & Yoshida, M. (2008). *Biomechanics of human motion—Its hardware and software*. Corona Publishing Co., Ltd.

forward and upper directions as positive, respectively. On the angular motion, the counterclock direction is defined as positive. The variables are defined as shown below:

m: mass.
I: inertia of moment.
F: force.
T: joint moment
α: acceleration for translation.
$\ddot{\theta}$: angular acceleration
d: distance.

The following subscripts are used to discriminate common variables.

f: foot, l: lower leg, t: thigh, b: body, F: floor, a: ankle, k: knee, h: hip, x: x-direction, y: y direction

1. Motion equations for foot segment

(For translational motion)

$$m_f \alpha_{fx} = F_{Fx} + F_{ax} \tag{2.21}$$

$$m_f \alpha_{fy} = F_{Fy} - F_{ay} - m_f g \tag{2.22}$$

(For angular motion)

$$I_f\ddot{\theta}_f = F_{Fx}d_{fy} + F_{Fy}d_{fx} - F_{ax}d'_{fy} + F_{ay}d'_{fx} + T_a \tag{2.23}$$

Unknown variables: F_{ax}, F_{ay}, T_a

2. Motion equations for lower leg segment

(For translational motion)

$$m_l\alpha_{lx} = -F_{ax} + F_{tx} \tag{2.24}$$

$$m_l\alpha_{ly} = F_{ay} - F_{ty} - m_lg \tag{2.25}$$

(For angular motion)

$$I_l\ddot{\theta}_l = -F_{ax}d_{ly} - F_{ay}d_{lx} - F_{tx}d'_{ly} - F_{ty}d'_{lx} - T_a + T_k \tag{2.26}$$

Unknown variables: F_{tx}, F_{ty}, T_k

3. Motion equations for thigh segment

(For translational motion)

$$m_t\alpha_{tx} = -F_{tx} + F_{bx} \tag{2.27}$$

$$m_t\alpha_{ty} = F_{ty} - F_{by} - m_tg \tag{2.28}$$

(For angular motion)

$$I_t\ddot{\theta}_t = -F_{tx}d_{ty} - F_{ty}d_{tx} - F_{bx}d'_{ty} - F_{by}d'_{tx} - T_k + T_h \tag{2.29}$$

Unknown variables: F_{bx}, F_{by}, T_h

4. Motion equations for upper body segment

(For translational motion)

$$m_b\alpha_{bx} = -F_{bx} \tag{2.30}$$

$$m_b\alpha_{by} = F_{by} - m_bg \tag{2.31}$$

(For angular motion)

$$I_b\ddot{\theta}_b = -F_{bx}d_{by} - F_{by}d_{bx} - T_h \tag{2.32}$$

Unknown variables: F_{bx}, F_{by}

In usual gait analysis for human body with appropriate marker, the measurement data for translation, velocity, acceleration, and reaction forces on floor are obtained. The center of gravity, mass, and inertia moment for each segment are gotten from measurement and related data base.

Therefore, unknown variables for motion of foot segment are F_{ax}, F_{ay}, and T_a, which can be solved from three equations from (2.21) to (2.23). If the variables such as F_{ax}, F_{ay}, and T_a have been fixed, unknown variables for motion of foot segment are F_{lx}, F_{ly}, and T_k, which can be solved from three equations from (2.24) to (2.26).

In similar calculation process, the joint forces and joint moments can be specified for ankle, knee, and hip joints. Data for joint forces and joint moments are applicable for the estimation of acting muscular forces. Here, it should be taken into consideration that the estimated joint force is minimum value and the actual joint force can be larger than the estimated value. If cocontraction of antagonistic muscles occur, the joint force could be enhanced.

Detailed gate analysis should be performed for three-dimensional model. In normal walking, the example of reaction forces on floor is shown in Fig. 2.14 (Miyazaki,

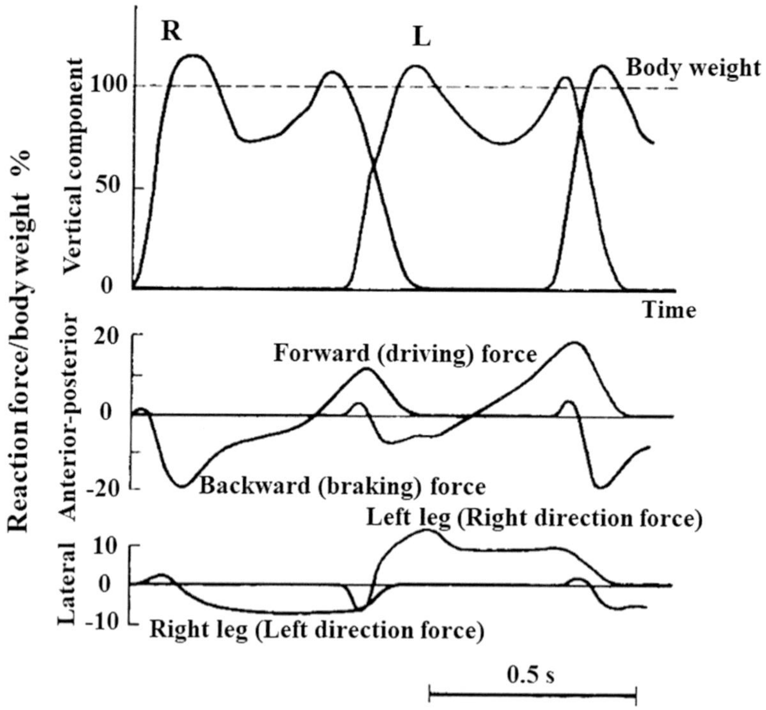

Figure 2.14 Reaction forces on floor during walking motion. Vertical, anterior-posterior, and lateral reaction forces for right foot and left foot are shown.
Modified from Miyazaki, S. (2002). Walking motion. In *Biomechanics of human movement* (pp. 102–125). Corona Publishing Co., Ltd.

2002). It is noticed in vertical component of reaction force at floor that the vertical force changes up and down over body weight level according to the downward and upward movement of human body, respectively. Two peak loads are shown after heel strike and before toe off. In anterior-posterior direction, the backward force acts as braking force in early phase of stance phase and the forward force acts as driving force. In lateral direction, the reaction force occurs according to lateral movement of human body.

Changes in forces transmitted by joints in human body during walking were estimated in three-dimensional model by Paul (1966–67). In this study the values of the force components transmitted at the hip or knee were calculated by the use of equilibrium equations considering the reaction forces and moments on the floor and active muscular forces. The following situations were assumed. The friction resistance of the joint is a negligible value in comparison with the other forces acting. The antagonistic activity of muscle groups can be ignored.

Typical results for joint forces are shown in Fig. 2.15.

In hip joints, two major peaks are noticed at the time of transfer of support from one foot to the other. At these times the trunk is passing through the lowest point in the vertical direction, and there will be a downward inertia force present. After heel strike, body weight is taken increasing on a foot in a front of hip and a moment resisted by the extensor muscles causes the first peak joint load. Near toe off the second peak load occurs with an increase in a moment by the flexed muscles. Throughout the stance phase the abductor muscle group must be active, since the hip joint is offset from the line of gravity between the trunk and foot, similar to the situation in single leg standing. This muscle tension corresponds to an increase in the y- and z-components of joint force throughout stance phase. In this study, the average cycle time was 1.13 s.

The lower graph of Fig. 2.15 shows the variation with time of the resultant force at the knee joint for one subject. It is noticed that three peak joint forces occur and the greatest peak is either the first or the last of the stance phase. The average of the greatest peak values of the ratio of joint force to body weight was 3.39 in 15 tests.

Morrison (1970) mentioned that the average maximum joint force in knee joint during walking was 3.03 times body weight in 12 subjects, and the main three peaks correspond to the peaks of muscle forces for hamstrings, quadriceps femoris and gastrocnemius. During stance phase, greater part of the joint force was transmitted by the medial condyle. In addition, he indicated the correspondence of forces of anterior and posterior cruciate ligaments, and medial and lateral collateral ligaments during walking motion.

To validate the contact forces on human joints of three-dimensional mathematical models, in vivo measurements of joint forces using instrumented joint prostheses with sensors have been performed. Stansfield et al. (2003) conducted direct comparison of calculated hip joint contact forces in three-dimensional musculoskeletal model of the lower limb with those measured using instrumented hip joint prostheses with strain gauges in two subjects with total hip arthroplasties at 3 years ago. Three-dimensional motion analysis with simultaneous ground reaction force recording was performed. A three-dimensional musculoskeletal model of the hip, knee, and tibiotalar

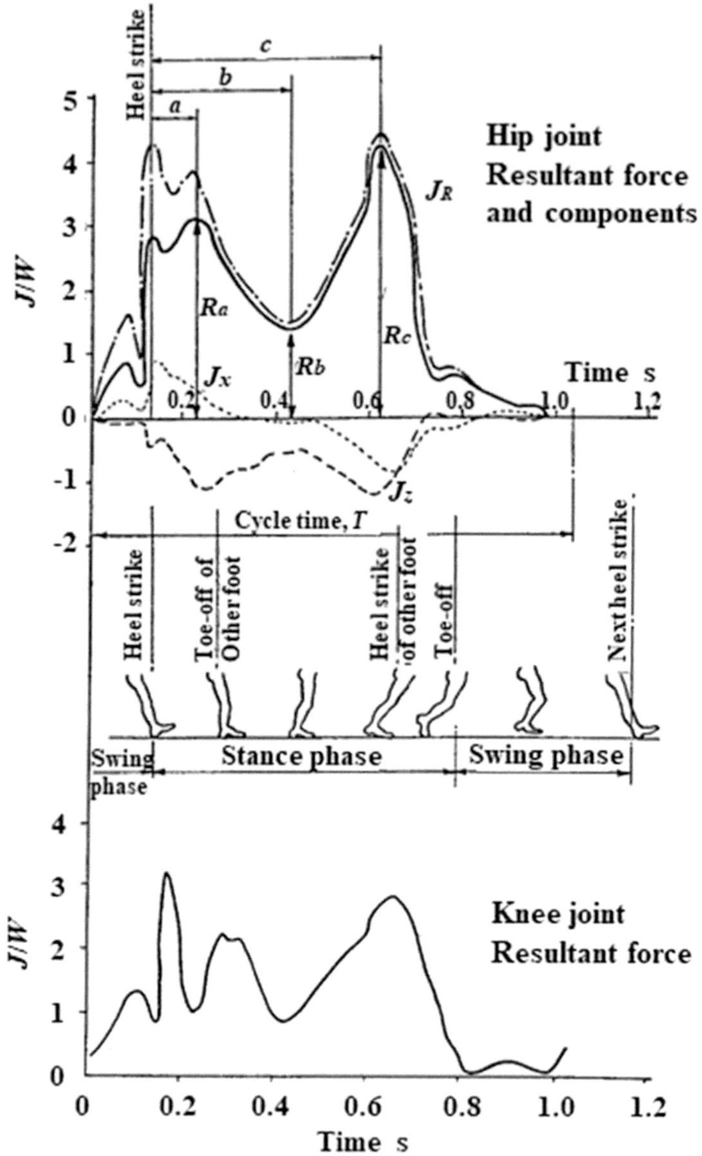

Figure 2.15 Typical patterns for hip and knee joints of variation with time of ratio of joint force/body weight during normal walking. In hip joint, the two curves labeled J_R correspond to the possible extreme values of resultant joint force. The components of J_X and J_Z correspond to the lower of two resultant forces. The component J_Y is not plotted since it lies very close to throughout the cycle. *W*: Body weight.

Modified from Paul, J. P. (1966-67). Forces transmitted by joints in the human body. *Proc. Instn. Mech. Engrs*, *181*, 8–15.

(ankle) joints was created with 48 muscle elements. The hip joint was defined as a ball-and-socket joint. Knee joint contact movement was defined ensuring physiological representation of the contact center throughout flexion. Knee joint has discrete medial and lateral compartmental loads and is applied at action of quadriceps muscles via patellar mechanism. And anterior and posterior cruciate ligaments and medial and lateral collateral ligaments were incorporated at the knee. The tibiotalar joint was defined as a single axis joint. Three-dimensional force and moment equilibrium were stipulated at all joints. The forces were distributed in the load bearing structures of the lower limb using a two-stage optimization routine.

Calculated (thick line) and measured (thin line) hip joint contact force components (×Body weight) during walking were shown in Fig. 2.16 for the two subjects at normal and fast walking conditions. It is noticed that peak loads for fast walking are higher than those for normal walking. There was general agreement between the calculated and measured hip joint contact forces in both pattern and magnitude, although some discrepancies are noticed.

On loading of the knee joint, Kutzner et al. (2010) examined the tibiofemoral contact forces and moments (three forces and three moments) acting on the instrumented

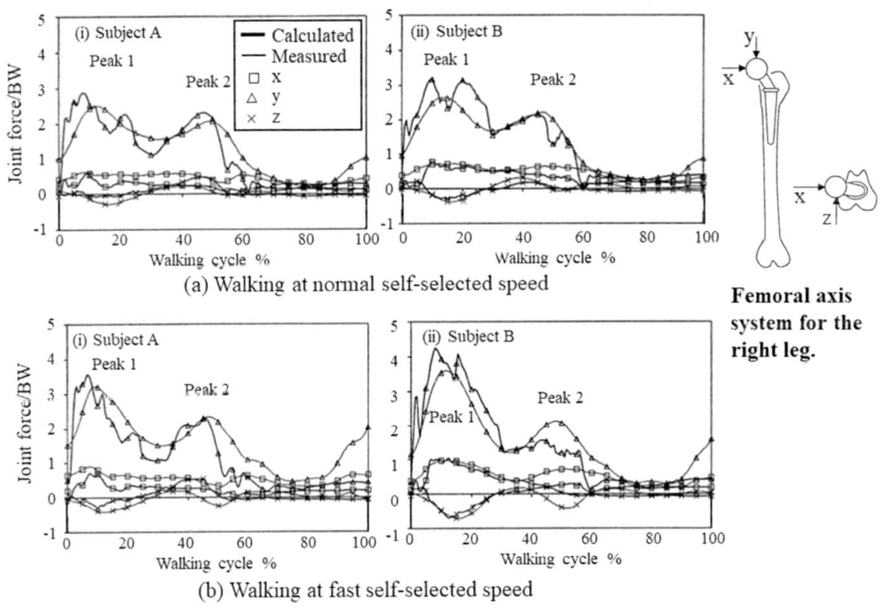

Figure 2.16 Calculated and measured hip joint contact force components for the two subjects. Femoral axis system is shown in right side.
Modified from Stansfield, B. W., Nicol, A. C., Paul, J. P., Kelly, I. G., Graichen, F., & Bergmann, G. (2003). Direct comparison of calculated hip joint contact forces with those measured using instrumented implants. An evaluation of a three-dimensional mathematical model of the lower limb. *Journal of Biomechanics*, *36*(7), 929–936. https://doi.org/10.1016/S0021-9290(03)00072-1.

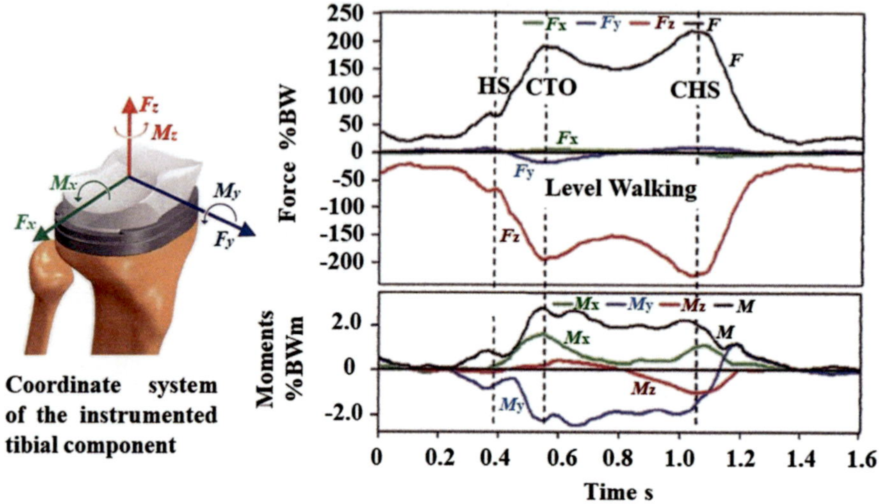

Figure 2.17 Load pattern in knee joint during walking. Peak forces are observed during CTO and prior to CHS.

Modified from Kutzner, I., Heinlein, B., Graichen, F., Bender, A., Rohlmann, A., Halder, A., Beier, A., & Bergmann, G. (2010). Loading of the knee joint during activities of daily living measured in vivo in five subjects. *Journal of Biomechanics*, *43*(11), 2164–2173. https://doi.org/10.1016/j.jbiomech.2010.03.046.

tibial trays with six semi-conductor strain gages in the knee joints of five subjects during daily life in vivo. The total knee system has an ultra-congruent tibial insert and a standard femoral component. These forces and moments are caused by external forces, active muscles, soft tissue deformations and contact forces, and moments acting directly between condyles and the tibial plateau. The coordinate system of instrumented tibial component and load patterns during level walking are shown in Fig. 2.17. It is noticed that peak forces can act either during the instants of contralateral toe off (CTO) or prior to contralateral heel strike (CHS). At CTO small shear force $-F_y$ was acting in posterior direction and at CHS $+ F_y$ was acting in anterior direction.

Flexion moment $+ M_x$ reached their peak values around CTO and CHS. An adduction moment $-M_y$ acted in the frontal plane during stance phase, indicating a medial load shift. The axial torque changed from initial external rotation moment $+ M_z$ at CTO to pronounced internal rotation moments $-M_z$ at CHS.

2.4 Importance of kinematics in human joints

Human synovial joints have different types of structure, geometry, size, and degree of freedom to facilitate various human movements, as mentioned in Sections 2.1 and 2.2.

In load-bearing joints of lower limb, hip joint is classified as ball-and-socket joints in which the round femoral head fits into the cavity of acetabulum and rotates around three axes as spherical joint with a wide range of movement and a greater degree of stability even under high loading. In the usual articulation for various daily activities, sliding motion sustains at conformal contact conditions in hip joint.

In contrast, knee joint operates as condylar joint as a special type of ellipsoidal joint, which is composed of tibiofemoral joint and patellofemoral joint. At flexion-extension motion for tibiofemoral joint, not only sliding but rolling motions exist at nonconformal contact conditions. In this section, the kinematic features of knee joint are focused.

In relation to tibiofemoral movement, Martelli and Pinskerova (2002) investigated the shapes of the tibial and femoral articular surfaces of cadaver knees in sagittal, frontal, and coronal planes with the aid of computer interpolation of sections of the articular surfaces acquired by a three-dimensional (3D) digitizer and the magnet resonance imaging (MRI). Fig. 2.18 shows 3D digitization at 0° and MRI images at 0° and 120° flexion for sagittal section through the medial compartment. Fig. 2.19 shows 3D digitization at 0° and MRI images at 0° and 120° flexion for sagittal section through the center of lateral compartment. In MRI images, the meniscus is visualized. They divided articular surface in sagittal section to anterior horn facet (AHF), extension facet (EF), flexion facet (FF), and posterior horn facet (PHF).

The observed shapes of the surfaces and their dimensions were very similar between 3D digitization and MRI image for sagittal, frontal, and coronal sections, except the arcs for medial femoral anterior horn facet (AHF) contacting with meniscus in sagittal section. The range of movement was limited to 120° which represents the approximate limit of active flexion. Flexion beyond 120° occurs passively but by 160° neither femoral condyles is in contact with the tibia.

It is noticed in contact situation that in medial compartment the femoral condyle has convex curved surface against concave tibial surface, but in lateral compartment the femoral condyle has convex curved surface against convex tibial surface which induces instability and high contact stress. In medial compartment, the radial difference is large compared with that in hip joint. To stabilize the movement of

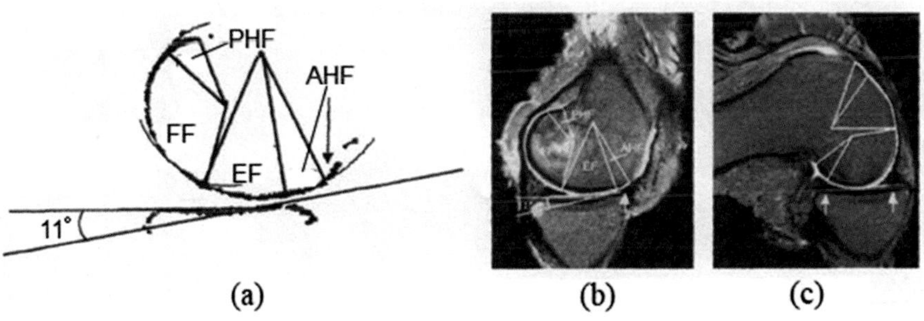

Figure 2.18 Computer interpolation with digitiser and MRI images for a sagittal section of knee joint through the medial compartment. (a) 3D digitization at 0°, (b) MRI image at 0°, (c) MRI image at 120 degrees flexion. The arrows point to the meniscal horns.
Modified with permission from Martelli, S. & Pinskerova, V. (2002). The shapes of the tibial and femoral articular surfaces in relation to tibiofemoral movement. *Journal of Bone and Joint Surgery - Series B*, *84*(4), 607–613. https://doi.org/10.1302/0301-620X.84B4.12149.

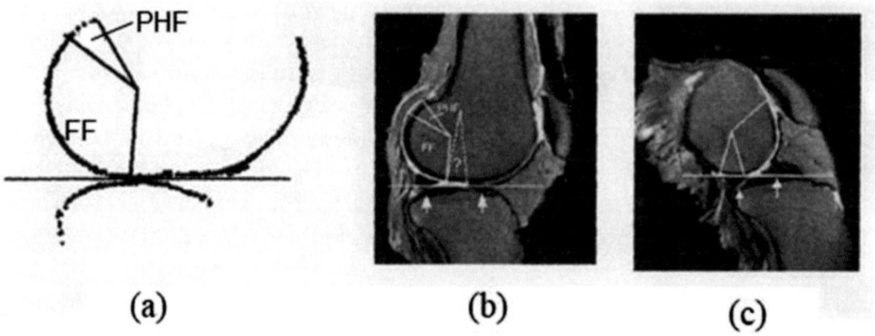

Figure 2.19 Computer interpolation with digitizer and MRI images for a sagittal section of knee joint through center of the lateral compartment. (a) 3D digitization at 0°, (b) MRI image at 0°, (c) MRI image at 120° flexion. The arrows point to the meniscal horns. Modified with permission from Martelli, S. & Pinskerova, V. (2002). The shapes of the tibial and femoral articular surfaces in relation to tibiofemoral movement. *Journal of Bone and Joint Surgery - Series B*, *84*(4), 607−613. https://doi.org/10.1302/0301-620X.84B4.12149

tibiofemoral joint and suppress the contact stress, it is expected that medial and lateral menisci perform the important function to stabilize the tibiofemoral joint and support the significant part of joint loading. In addition, anterior and posterior cruciate ligaments and collateral ligaments contribute the smooth motion in tibiofemoral joint.

The movements of synovial joint are driven by corresponding muscular forces but restricted by the binding effect of the ligaments and the geometry of articular surfaces which prevent the instability of joint motion. The main ligaments for knee joint, i.e., anterior and posterior cruciate ligaments and medial and lateral collateral ligaments, are shown in Fig. 2.6 and Fig. 2.20a. Anterior and posterior cruciate ligaments combined with femur and tibia construct the 4-bar linkage to control the motion of the knee joint as shown in Fig. 2.20b. For example, backward gliding of the tibia relative to the femur (or forward gliding of femur relative to the tibia) is controlled by the posterior cruciate ligament (PCL), and forward gliding of the tibia (or backward gliding of femur) is controlled by the anterior cruciate ligament (ACL). In addition, ACL controls the internal rotation and the hyper extension. PCL controls the external rotation.

In detailed mechanism, ACL is divided to antero-medial bundle (AMB) and postero-lateral bundle (PLB) in function and structure. AMB maintains stretching for full range of flexion-extension and intensely stretches at flexion, but PLB stretches at extension and becomes relax at flexion. PCL is divided to antero-lateral bundle (ALB) and postero-medial bundle (PMB). ALB stretches at flexion and PMB stretches at extension.

Medial collateral ligament (MCL) and lateral collateral ligament (LCL) also contribute to the stability of tibiofemoral joint. MCL controls the valgus and external rotation. LCL controls varus, external rotation and extension.

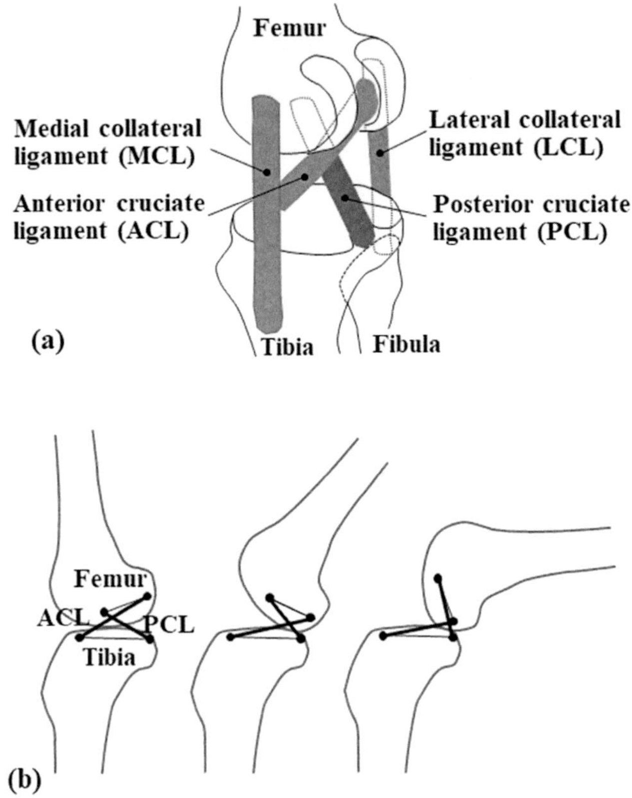

Figure 2.20 Role of ligaments to contribute to stability of knee joint. (a) Simplified schema of knee joint with ligaments (gap between femur and tibia is distended to illustrate the configuration of ligaments in a comprehensive way.) (b) Function of anterior and posterior cruciate ligaments in flexion and extension.
Modified from Makikawa, M. & Yoshida, M. (2008). *Biomechanics of human motion—Its hardware and software*. Corona Publishing Co., Ltd.

To establish the smooth movement of knee joint in wide range of flexion and extension, the combination of rolling and sliding motions is required. ACL tension contributes to sliding at flexed conditions, and PCL tension contributes to sliding at extension.

Motions in synovial joint are categorized as pure rolling, pure sliding, and rolling with sliding (Walker, 1977).

The motion patterns of rolling and sliding for a cylinder rolling along a plane are shown in Fig. 2.21. The sliding ratio is defined as follows:

$$s = \frac{l_1 - l_2}{l_1} \tag{2.33}$$

where l_1 is arc length of cylinder for A′B and l_2 is the rubbing distance for AB.

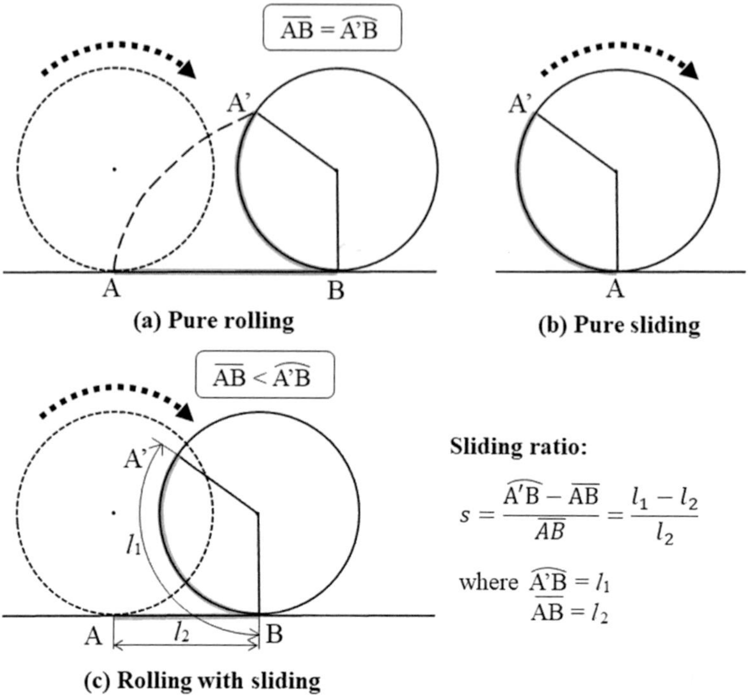

Figure 2.21 Motion types of rolling, sliding, and rolling with sliding. (a) pure rolling: $s = 0$. (b) pure sliding: $s = 1$. (c) rolling with sliding: $0 < s < 1$.

In pure rolling without sliding (the sliding ratio: $s = 0$), the initial contact point of cylinder exhibits the cycloid curve as a locus and maintains to be $l_1 = l_2$. In pure sliding, the contact point on the plane is the same and $l_2 = 0$, then $s = 1$. In rolling with sliding, $l_2 < l_1$.

In flexion-extension motion of knee joint, the rollback occurs in rolling with sliding. If the femur flexes in pure rolling in wide range of motion, the femur becomes dislocated. If the femur flexes in pure sliding in wide range of motion, the posterior part of femur impinges the posterior edge of tibia at high flexion. Therefore, the femur and tibia operate in rolling with sliding controlled with ligament tensions in wide range of motion as shown in Fig. 2.22.

As another feature of knee joint motion in flexion-extension, the screw home movement in rotation with flexion is observed as shown in Fig. 2.23 (Hirokawa, 2013). The radius of lateral condyle is larger than the medial condyle, therefore at roll back the femur rotates externally in flexion motion on the tibial plateau, or tibia rotates internally. This movement is confirmed from the feature that about 20° external direction of toe at extended position changes to the frontal direction at 90° flexion.

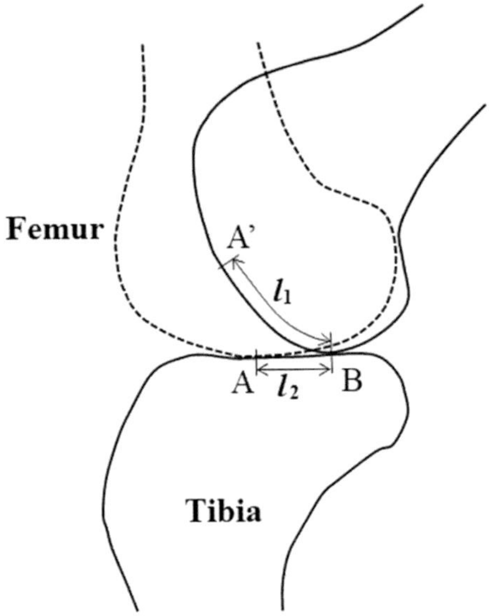

Figure 2.22 Rolling with sliding in tibiofemoral joint. At flexion, $l_1 > l_2$, $0 < s < 1$.

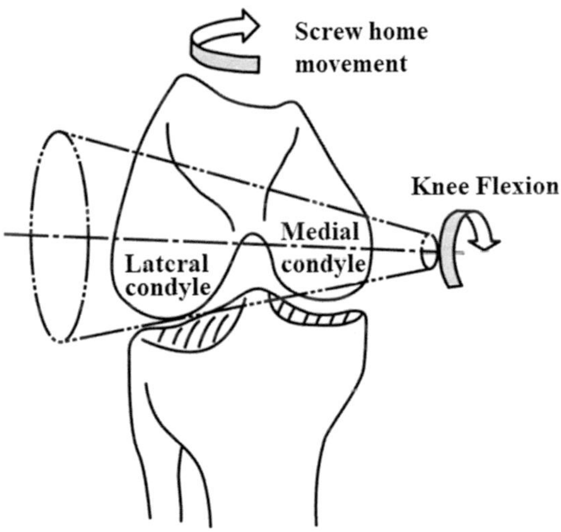

Figure 2.23 Screw home movement in knee joint. At rollback in flexion motion, the femur rotates externally on the tibial plateau.
Modified from Hirokawa, S. (2013). Biomechanics of natural and artificial joints. In *Frontiers of biomechanics* (pp. 109–166). Kyoritsu Shuppan Co., Ltd.

2.5 Importance of alignment in human joints

It has been indicated that the malalignment for synovial joint enhances the mechanical severity of rubbing surfaces. For example, the alignment of the lower limb can be evaluated with the anteroposterior projection of full-length standing weight-bearing radiograph as a primary tool (Luís & Varatojo, 2021). The mechanical axis of the lower limb, also called the Mikulicz line, is drawn by connecting a point in the center of the femoral head to a point in the center of the ankle. In normal mechanical axis of the lower limb, this line's physiological position runs, on average, 4 ± 2 mm medial to the center of the knee. The mechanical axis of the lower limb creates an approximate angle of 3° to the perpendicular axis of the body (Fig. 2.24a). Any deviation from this normal physiological range indicates either a valgus if the line runs lateral (Fig. 2.24b), or a varus if it runs medially (Fig. 2.24c). The value of the deviation is measured in millimeters and is named mechanical axis deviation (MAD).

In the presence of a frontal plane deformity, a nonphysiological load distribution occurs in the knee's medial or lateral compartment. The resulting mechanical overload originates progressive cartilage wearing and progressive degenerative disease.

In many cases of osteoarthritis, varus knees are found. As clinical treatments to correct the weight bearing axis and the lower limb's load distribution, osteotomy or total joint arthroplasty have been applied.

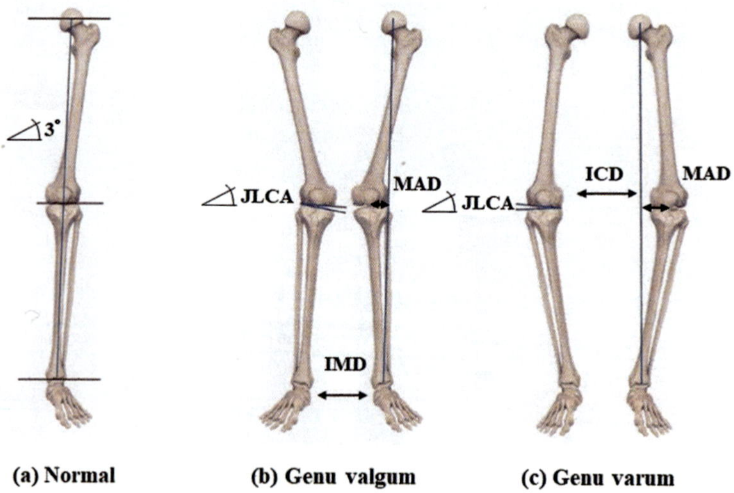

Figure 2.24 Normal mechanical axis of the lower limb and mechanical axis deviation in genu valgum and genu varum. (a) Normal mechanical axis of the lower limb, (b) Genu valgum (Genu valgus): anatomic tibiofemoral angle decreased (aTFA <173−175 degrees). The weight-bearing axis runs lateral to the 4 mm point and the distance between the medial malleoli (intermalleolar distance = IMD) is increased. The angle between the tibial plateau and the tangent to the femoral condyles (joint line convergence angle: ILCA) is opened medially. (c) Genu varum (Genu varus): anatomic tibiofemoral angle increased (aTFA >173−175°). The Mikulicz line runs medial to the 4 mm point and mechanical axis deviation medially increased (MAD >15 mm medially). The intercondylar distance (ICD) is increased. The joint forms an open angle laterally (in JLCA).
Modified with permission from Luís, N. M. & Varatojo, R. (2021). Radiological assessment of lower limb alignment. *EFORT Open Reviews*, 6(6), 487−494. https://doi.org/10.1302/2058-5241.6.210015.

2.6 Biomechanics in artificial joints

As effective medical treatment for patients with severely diseased joints in osteoarthritis, rheumatoid arthritis, and other arthritis with restriction in motion, deficiency in load support ability and severe pain, these joints have been replaced by the various artificial joints. Successful application of artificial joints as replacement arthroplasty could give the recovery in motion and load support abilities with reduction of severe pains to the patients. However, the revisions with alternate artificial joints have been applied in cases of joint loosening, severe wear, mechanical failure, dislocation, infection, and others. Artificial joints mimic human joints but have usually simplified geometric designs which are composed of artificial materials with different properties, therefore they have some limitations for effective use. To improve the performance and durability of artificial joints, biomechanical and biotribological evaluations are required. In this section, the fundamental features in biomechanics are mentioned for artificial hip and knee joints operating under high loading conditions.

2.6.1 Artificial hip joints

Hip replacement is an effective surgical procedure in which the diseased hip joint is replaced by hip prosthesis as a total hip replacement (total hip arthroplasty), hip resurfacing replacement (hip resurfacing arthroplasty), or a hip hemi replacement (hip hemiarthroplasty).

In total hip replacement (Fig. 2.25a), the femoral head and acetabulum are both removed and replaced with metal, ceramic, or polymeric components.

In hip resurfacing (Fig. 2.25b), the femoral head is not removed, but is instead trimmed and capped with a smooth metal covering. The damaged bone and cartilage

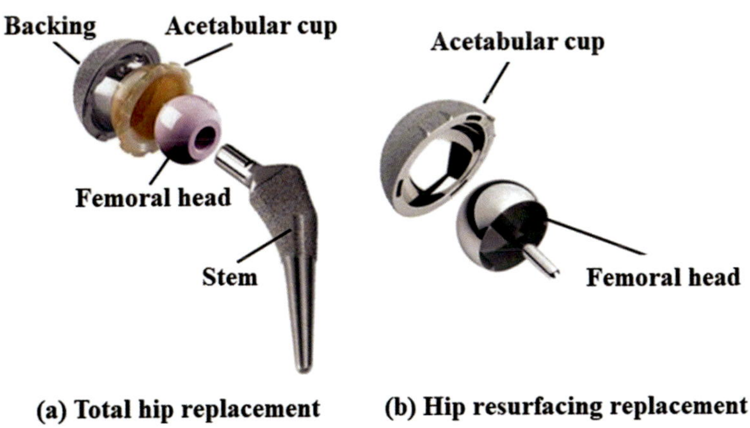

Figure 2.25 Total hip replacement and hip resurfacing replacement. (a) Total hip replacement, (b) Hip resurfacing replacement.
Modified from Mattei, L., Di Puccio, F., Piccigallo, B., & Ciulli, E. (2011). Lubrication and wear modeling of artificial hip joints: A review. *Tribology International*, *44*(5), 532–549. https://doi.org/10.1016/j.triboint.2010.06.010.

of acetabulum is removed and replaced with a metal shell. Hip resurfacing is recommended for younger, large-framed patients with strong and healthy bone. Hip resurfacing has the advantages such as easier revision and decreased risk of hip dislocation but disadvantages such as femoral neck fracture and metal ion risk. For adverse tissue reaction due to metal ions, hip resurfacing is now performed less frequently than in the past.

A hip hemiarthroplasty is an operation that is applied most commonly to treat a fractured hip. In hip hemiarthroplasty, only the femoral head is replaced by metallic or ceramic femoral head, where hard material is rubbed against nondamaged acetabular cartilage. As femoral head, unipolar or bipolar type is commonly used.

These hip replacements are fundamentally ball-in-socket joints with common or different materials and designs.

Total hip prostheses as ball-in-socket spherical bearings (Fig. 2.26a) are classified in two types according to the combination of material properties.

1. Hard-on-hard
2. Hard-on-soft

As hard materials, metal (Co-Cr-Mo alloy, stainless steel, Ti alloy) or ceramic (alumina, zirconia, zirconia toughened alumina) has been used. As a soft material, ultra-high molecular weight polyethylene (UHMWPE) has been widely used as acetabular cup.

The lubrication and wear phenomena in total hip replacement are described in Chapter 4, so the contact situation is focused in this section. In total hip replacement as a spherical bearing, the contact pressure between femoral head and acetabulum is controlled by the material properties and radial clearance.

The contact pressure in dry condition can be predicted from the following Hertzian contact formula in an equivalent ball-on-plane model (Fig. 2.26b).

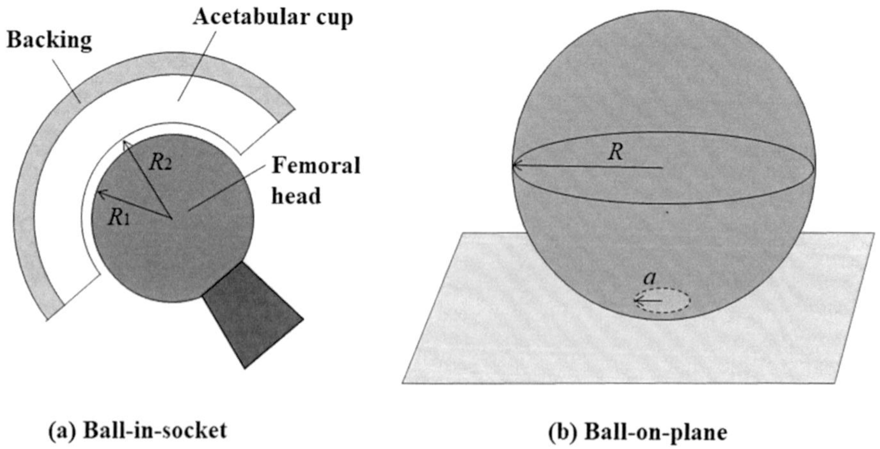

Figure 2.26 Simplified models of total hip prostheses. (a) Ball-in-socket and (b) Ball-on-plane models.

The pressure profile for p is parabolic for radial coordinate r;

$$p = p_{max}\sqrt{1 - \left(\frac{r}{a}\right)^2} \tag{2.34}$$

The maximum pressure: p_{max}

$$p_{max} = \frac{3W}{2\pi a^2} \tag{2.35}$$

where W is load.

Radius of contact area: a

$$a = \sqrt[3]{\frac{3WR}{2E'}} \tag{2.36}$$

The equivalent radius of the ball R in an equivalent ball-on-plane model is determined from the following equation:

$$R = \frac{R_1 R_2}{R_2 - R_1} \tag{2.37}$$

where R_1 and R_2 are the radii of the femoral head and the socket (acetabular cup), respectively. Radial clearance is defined as $R_2 - R_1$.

E': equivalent elastic modulus;

$$\frac{1}{E'} = \frac{1}{2}\left(\frac{1 - {\nu_1}^2}{E_1} + \frac{1 - {\nu_2}^2}{E_2}\right) \tag{2.38}$$

E_1: Young's modulus of femoral head.

E_2: Young's modulus of acetabular cup

ν_1: Poisson's ratio of femoral head,

ν_2: Poisson's ratio of acetabular cup.

Mechanical properties of materials are shown in Table 2.2.

Equivalent radius, equivalent elastic modulus, radius of contact area, and maximum contact pressure at 2.5 kN loading are predicted for ball-on-plane model as follows:

1. Metal (Co-Cr–Mo)-on-metal (Co-Cr–Mo) hip joint with $R_1 = 18$ mm and $R_2 = 18.08$ mm (radial clearance: $R_2 - R_1 = 0.08$ mm);

   $R = 4.07$ m, $E' = 253$ GPa, $a = 3.92$ mm ($< R_1 = 18$ mm), $p_{max} = 77.7$ MPa

2. Ceramic (alumina)-on-ceramic (alumina) hip joint with $R_1 = 14$ mm and $R_2 = 14.03$ mm (radial clearance: $R_2 - R_1 = 0.03$ mm);

   $R = 6.55$ m, $E' = 418$ GPa, $a = 3.89$ mm ($< R_1 = 14$ mm), $p_{max} = 78.9$ MPa

Table 2.2 Mechanical properties of materials.
Representative values of material properties are shown.

		Young's modulus	Poisson's ratio
Material category	**Material**	***E* GPa**	**ν**
Plastics	UHMWPE	1	0.4
Metal	Stainless steel	210	0.3
	Co-Cr–Mo alloy	230	
Ceramic	Alumina	380	0.3
	Zirconia	210	

Modified from Mattei, L., Di Puccio, F., Piccigallo, B., & Ciulli, E. (2011). Lubrication and wear modeling of artificial hip joints: A review. *Tribology International*, *44*(5), 532–549. https://doi.org/10.1016/j.triboint.2010.06.010.

3. Metal (Co-Cr–Mo)-on-UHMWPE hip joint with $R_1 = 14$ mm and $R_2 = 14.1$ mm (radial clearance: $R_2 - R_1 = 0.1$ mm);

$R = 1.97$ m, $E' = 2.37$ GPa, $a = 14.6$ mm ($> R_1 = 14$ mm)

It is noticed that in both metal-on-metal and ceramic-on-ceramic prostheses the radius of contact area is markedly smaller than the head radius R_1, whereas the estimated radius of contact area is larger than the head radius R_1 in metal-on-UHMWPE prosthesis at 2.5 kN loading, although the radius of contact area at lower load becomes lower than the head radius R_1(at 1.0 kN, $a = 10.8$ mm $< R_1, p_{max} = 4.09$ MPa). Therefore, the assumption of semi-infinite space, hence the use of an equivalent ball-on-plane model is justified only for hard-on-hard prostheses. For metal-on-UHMWPE at high load, it can be necessary to use a ball-in-socket model with a layered surface for taking into account the acetabular cup and the metal backing, where the elastic deformation is estimated with elasticity equation based the constrained column model (Jalali-Vahid et al., 2001a, 2001b).

2.6.2 Artificial knee joints

The knee joint is generally divided into three compartments of medial, lateral, and patellofemoral. Most patients with significant wear and damage in two or more of the compartments are usually treated with total knee replacement (total knee arthroplasty (TKA)). A minority of patients with osteoarthritis have wear primarily in one compartment, usually the medial, and may be candidates for unicompartmental knee replacement (unicompartmental knee arthroplasty (UKA)) as partial knee replacement with smaller incision.

A simplified schema for total knee replacement is shown in Fig. 2.27. In the tibiofemoral joint, the distal end of femur and the proximal end of tibia are cut to adjust the femoral and tibial components, which are fixed to bones using polymethylmethacrylate (PMMA) cement or cement-less techniques. Therefore, the cartilages and menisci are removed. As articulating surfaces of total knee prosthesis, metallic or ceramic femoral component is rubbed against UHMWPE tibial insert fixed on metallic tibial

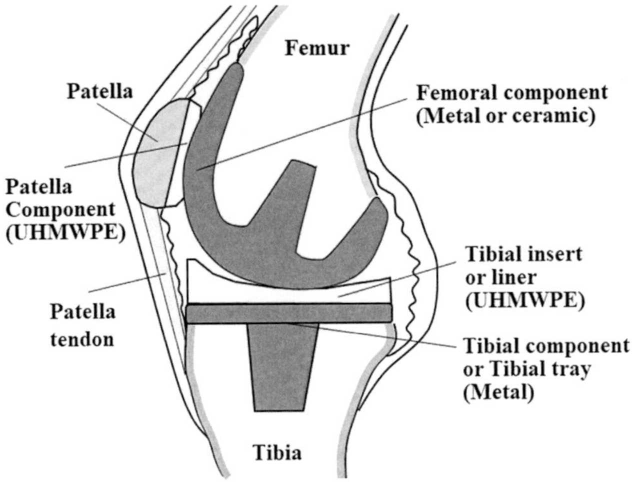

Figure 2.27 Simplified schema of total knee replacement. Tibiofemoral and patellofemoral articulating surfaces are replaced.

component. In patellofemoral joint, the articular surface of the patella is removed and replaced by a UHMWPE patella component cemented to the posterior surface of the patella. In other cases, the patella is replaced unaltered.

The anterior and posterior cruciate ligaments are removed or preserved depending on the selected knee prosthesis types, but the medial and lateral collateral ligaments are preserved.

In the total knee prosthesis as resurfacing joint, the metallic or ceramic femoral component mimic femoral natural shape but have usually simplified geometric design. The UHMWPE tibial insert plays the alternate role in replace of articular cartilage and meniscus, although UHMWPE has more than two order higher stiffness than articular cartilage and meniscus. As exhibited in Fig. 2.27, the geometrical congruity between femoral component and tibial insert is very low compared with conformal total hip prosthesis. Therefore, the contact pressure becomes high depending on the lowering of geometrical congruity compared with hard-on-soft hip prosthesis with small radial clearance.

The artificial knee joint is driven by muscular forces and the motion is controlled by articular surface geometry and ligament force.

From the viewpoint of articular surface geometry, the combination of femoral component and UHMWPE tibial insert in different congruity have been selected according to the design concept focusing stability and/or movability.

1. Flat type: movability-intensive design with flat tibial insert, insufficient stability and severe contact pressure
2. Laxity type: medium congruity design to compromise for stability and movability
3. Conformity type: stability-intensive conformal design with lower contact pressure and insufficient movability

From the viewpoint of the treatment of the ligament retaining or sacrificing, the total knee prostheses are categorized as the following three types:

1. Posterior stabilized (PS) type: Both anterior cruciate ligament (ACL) and posterior cruciate ligament (PCL) are removed (Fig. 2.28a). The PS implant uses a post that is built into the implant to accommodate for the loss of PCL. Post-cam mechanism is arranged to induce femoral rollback. In certain cases, the wear and breakage of the UHMWPE post have been observed and the optimization of post-cam design is required. Range of flexion-extension motion is enlarged but the movability in rotation is restricted.
2. Cruciate retaining (CR) type: ACL is removed but PCL is retained (Fig. 2.28b). The posterior cruciate ligament (PCL) improves the stability of the knee by preventing posterior subluxation of the tibia, reducing shear stress, increasing flexion and lever arm of the extensor mechanism by inducing femoral rollback upon flexion. The selection of flat tibial insert is suitable for femoral rollback but may enhance the contact pressure and wear. The abnormal kinematics and posterior instability by sacrificing of ACL should be reduced.
3. Bicruciate retaining (BCR) type: Both ACL and PCL are retained. The BCR knee retains the anterior and posterior cruciate ligaments to try to mimic the normal tension of ligaments of the knee which produces normal sagittal plane kinematics.

In fixed knee bearing mentioned above, the tibial insert is fixed to the tibial component. Therefore, the geometrical congruity between femoral and tibial insert articulating surfaces appears to control the stability and the movability. To satisfy both the stability and movability in knee prosthesis, the mobile bearings were proposed. A mobile bearing design allows for free motion of the polyethylene insert as anterior-posterior translation (Fig. 2.29a), internal-external rotation (Fig. 2.29b) or multidirectional movement of the tibial insert on the tibial tray. In the conformal design for femoral and tibial insert surfaces, sliding occurs between femoral component and tibial

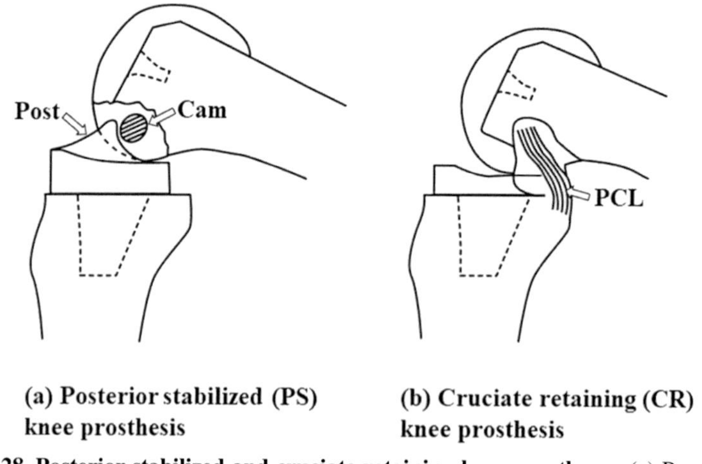

Figure 2.28 Posterior stabilized and cruciate retaining knee prostheses. (a) Posterior stabilized (PS) type, (b) Cruciate retaining (CR) type.
Modified from Hirokawa, S. (2013). *Biomechanics of natural and artificial joints* In Frontier of biomechanics (pp. 109–166). Kyoritsu Shuppan Co., Ltd.

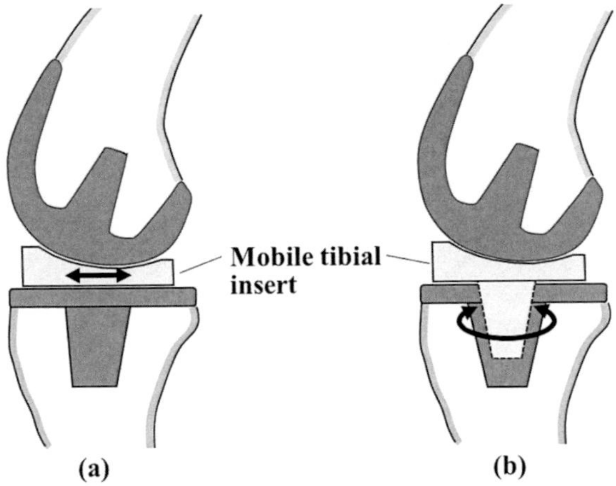

Figure 2.29 Simplified schema of mobile bearing knee. (a) Mobile bearing with anterior-posterior translation, (b) Mobile bearing with rotation.

insert at low contact pressure with good alignment, which is effective to suppress the delamination and severe wear. In certain cases with insufficient mobility, however, it should be noticed that considerable wear can occur at severe contact conditions. Undersurface wear should be prevented.

In cruciate retaining mobile knee with rollback at high flexion, a greater range of motion can be obtained.

To improve the clinical performance and the durability of knee prostheses with various designs, appropriate understanding of actual situations of knee prostheses in clinical use is required from the viewpoints of biomechanics and biotribology. In the following sections, the in vivo measurement of kinematics of total knee prostheses and the reproduction of physiological and active knee motion in the novel knee simulator are described.

2.6.3 In vivo measurement of kinematics of total knee prostheses by single-plane fluoroscopy

In clinical evaluation of total knee prostheses after arthroplasty, it is important to examine the adequacy between femoral and tibial components, the occurrence of edge loading and lifting between rubbing components, the range of motion of prosthesis components, and so on. In usual X-ray image where the UHMWPE tibial and patellar components are invisible, it was difficult to evaluate the actual contact situation, edge loading, and lifting between femoral and tibial components. Therefore, in vivo measuring methods of kinematics of total knee prostheses using single-plane fluoroscopy such as (a) pattern matching, (b) image matching, and (c) 2D-3D registration methods have been developed (Mahfouz et al., 2003; Hirokawa, 2011).

a. Pattern matching

To evaluate in vivo kinematics of total knee prosthesis, Banks and Hodge (1996) proposed the measurement method of three-dimensional total knee replacement

(TKR) kinematics using single plane fluoroscopy as an extended application of three-dimensional aircraft recognition algorithm using normalized Fourier descriptors developed by Wallace et al. (Wallace & Wintz, 1980). A library of contours of the CAD model of knee prosthesis in its various poses was prepared. Fourier descriptors were used for compact representation of library templates and contour extracted from single projective X-ray images of prosthesis.

Fig. 2.30 (Hirokawa et al., 2003) shows the coordinate conventions for perspective imaging for femoral component, where the object coordinates $\sum x-y-z$ and the screen coordinates $\sum X-Y-Z$ are set mutually parallel, and the Z and z axes are aligned perpendicular to the image plane. Principal distance as the distance between the projection center and the image plane is described as C_{pd}, and the distance between the objective center and image plane as Z_{lib}.

In Fig. 2.30, the position (x, y) of the femoral component can be estimated from the position (X, Y) at the image plane with affine transformation. The position z can be estimated on the basis of the projected size (area) of femoral component at the projection plane. The flexion/extension angle θ which rotates in plane about z-axis can be specified with Θ. The pose estimation of the valgus/varus angle ϕ and the internal/external rotation angle ϕ as out-of-plane rotations about x-axis and y-axis, respectively, should be performed by library matching.

To estimate the position and orientation of TKR, a template matching technique has been used. As shown in Fig. 2.31, Fourier descriptors are used to represent library templates and projection contours. If the silhouettes contour is continuously traced, it could be seen as a complex function and the discrete Fourier transform is applied.

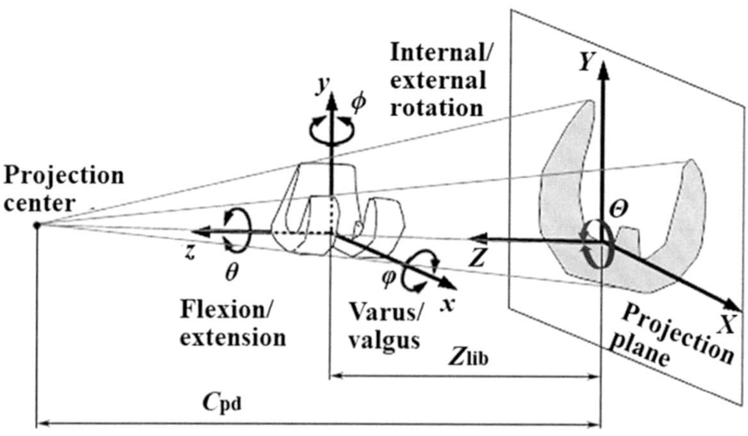

Figure 2.30 Coordinates of perspective transformation. C_{pd} is principal distance as the distance between the projection center and the image plane and Z_{lib} is the distance between the objective center and image plane.
Modified from Hirokawa, S., Ariyoshi, S., Takahashi, K., & Maruyama, K. (2003). Kinematic measurement of knee prosthesis from single-plane projection images. *JSME International Journal, Series C: Mechanical Systems, Machine Elements and Manufacturing*, *46*(4), 1368–1376. https://doi.org/10.1299/jsmec.46.1368/.

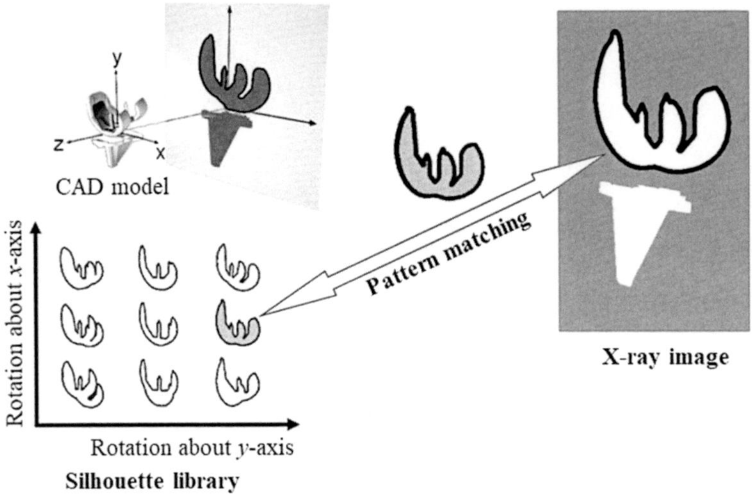

Figure 2.31 Principle of pattern matching method. Pattern matching is explored for the silhouette in X-ray image and the silhouette library.
Modified from Hirokawa, S. (2011). In vivo kinematic measurement of artificial knee joint. *Toraibarojisuto/Journal of Japanese Society of Tribologists*, *56*(12), 758–764.

Fig. 2.32 (Hirokawa et al., 2008) shows an example of visualization for the estimation of position and orientation of knee prosthesis represented with CAD model after application of pattern matching method for single-plane projection image. The position and orientation of femoral component and tibial tray estimated by pattern matching method for the silhouette in X-ray image (Fig. 2.32a) are represented with CAD model in Fig. 2.32b. Then the tibial insert is added and modified at neutral position to keep the relative position/orientation between femoral and tibial components (Fig. 2.32c). In Fig. 2.33, contact points are explored for femoral component and tibial insert.

In this estimation of a six-degree-of-freedom position/orientation for knee replacement motion from two-dimensional fluoroscopic images, the accuracy of the technique was assessed by Banks et al. (Banks & Hodge, 1996). In their computer modeling studies and controlled mechanical tests, the accuracy of approximately one degree for knee rotations and approximately 0.5 mm for sagittal plane translations was confirmed as a useful technology.

b. Image matching

Novel model-based method for the estimation of the three-dimensional (3-D) position and orientation (pose) of both the femoral and tibial knee prosthesis components based on the 3-D geometry of the components and a single plane projection view in a fluoroscopic image during activity was proposed by Zuffi et al. (1999). In this method, the best alignment of the component designs with the corresponding projection on the image plane was arranged by minimizing the Euclidean distance of the projection rays

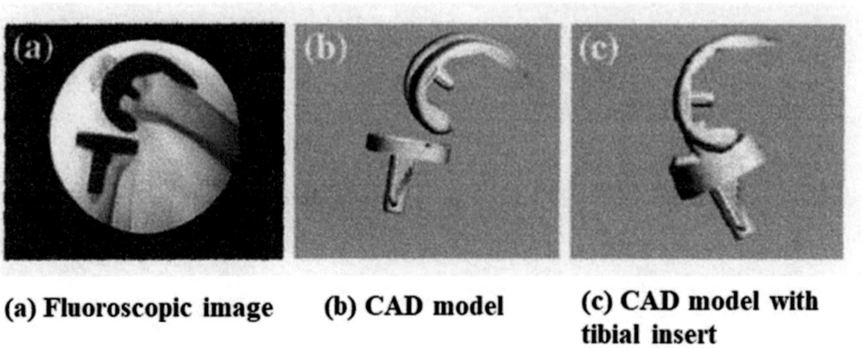

Figure 2.32 An example of pattern matching for the knee flexion−extension measurement. (a) Still picture captured from the fluoroscopy, (b) CAD model representation with the estimated position/orientation without the insert, (c) CAD model representation of the tibial rotations/translations relative to the femur.
From Hirokawa, S., Abrar Hossain, M., Kihara, Y., & Ariyoshi, S. (2008). A 3D kinematic estimation of knee prosthesis using X-ray projection images: Clinical assessment of the improved algorithm for fluoroscopy images. *Medical and Biological Engineering and Computing*, *46*(12), 1253−1262. https://doi.org/10.1007/s11517-008-0398-8.

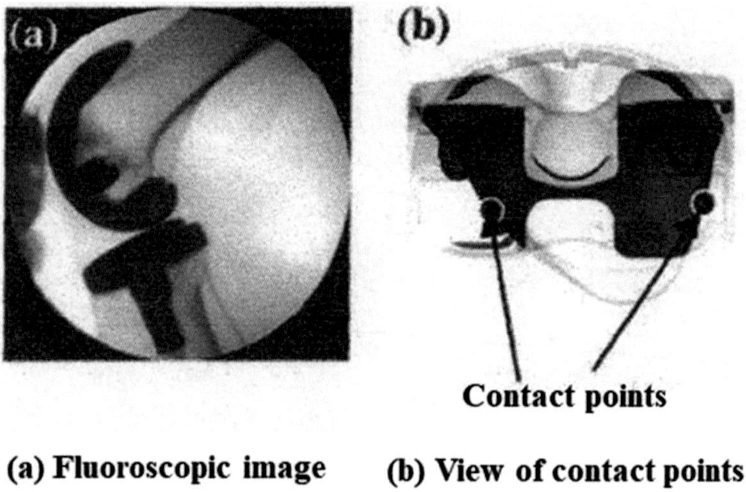

Figure 2.33 Estimation of contact points in knee prosthesis for walking-like motion. (a) Still picture, (b) calculated contact points' locations between the femur and the tibial insert.
From Hirokawa, S., Abrar Hossain, M., Kihara, Y., & Ariyoshi, S. (2008). A 3D kinematic estimation of knee prosthesis using X-ray projection images: Clinical assessment of the improved algorithm for fluoroscopy images. *Medical and Biological Engineering and Computing*, *46*(12), 1253−1262. https://doi.org/10.1007/s11517-008-0398-8.

from the object surface. On the accuracy, the computer simulation and static/dynamic in vitro tests using real knee prosthesis components in known and typical relative positions have shown that the accuracy with which component orientations and positions are estimated is better than 1.5 degrees and 1.5 mm, respectively. It is considered that the method is well suited for 3-D TKR kinematics analysis.

In the image matching method, several projection rays (3 rays for case in Fig. 2.34) are drawn from several points on the silhouette of X-ray image. Then, in the iterative estimation process for the pose of the CAD model of knee prosthesis by changing the rotation and translation, the summation of the squared Euclidean distance of the projection rays from the object surface: E is minimized to get the best alignment.

$$E = \sum_{i=1}^{n} d_i^2 \tag{2.39}$$

As a merit in the image matching method, the silhouette library necessary for the pattern matching method is not required. Even in cases of silhouette of X-ray image with partially blurry contour or partially shielded part, the appropriate estimation of the three-dimensional (3-D) position and orientation (pose) is available. The image matching method has the issues requiring deep thought on how to change in the rotation/translation and how to select the reference contour points.

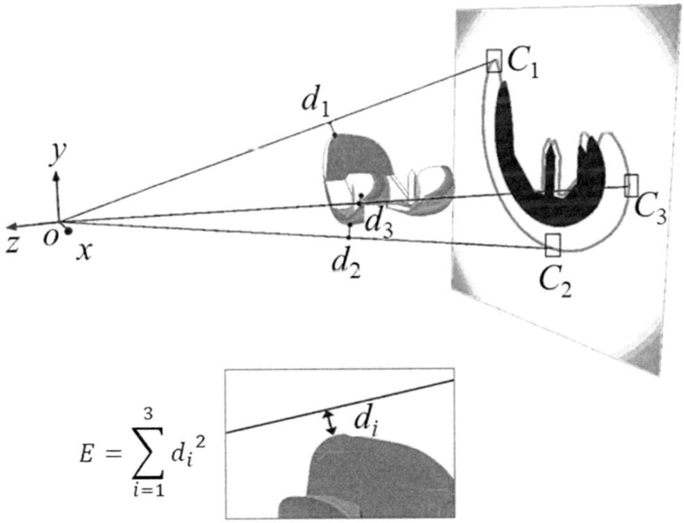

Figure 2.34 Principle of image matching method. The best alignment is obtained by minimization of the summation of squared Euclidean distance of the projection rays from the object surface.
From Hirokawa, S. (2011). In vivo kinematic measurement of artificial knee joint. *Toraibarojisuto/Journal of Japanese Society of Tribologists*, *56*(12), 758–764.

c. 2D-3D registration method

Mahfouz et al. (2003) developed a robust method for the registration of three-dimensional knee implant models to two-dimensional (single plane) X-ray fluoroscopy images.

In the 2D-3D registration method, the best 3D position and orientation is explored by maximizing the overlapping between the X-ray image of total knee prosthesis and the projection image of CAD knee prosthesis model. Fig. 2.35 shows the X-ray projection image of total knee prosthesis and the extracted silhouette contour image for femoral component. Fig. 2.36 shows the projection image of CAD model for the estimated position and orientation from Fig. 2.35 and the extracted silhouette contour image for the femoral component.

In Fig. 2.35a, X-ray projection image of femoral component is depicted as $G(x, y)$ where x and y are pixel coordinates and G corresponds to gray level intensity at (x, y). In Fig. 2.36a, $H(x, y)$ corresponds to gray level intensity. The overlapping between X-ray image of femoral component and projection image of CAD model can be evaluated based on the following cross-correlation function

$$I_1 = \sum G(x,y)H(x,y) / \sum H(x,y) \tag{2.40}$$

Here, I_1 corresponds to intensity matching score, and higher I_1 indicates higher overlapping.

In similar manner, the extracted silhouette contour images for femoral component for Fig. 2.35b and Fig. 2.36b are depicted as $J(x, y)$ and $K(x, y)$, respectively. The

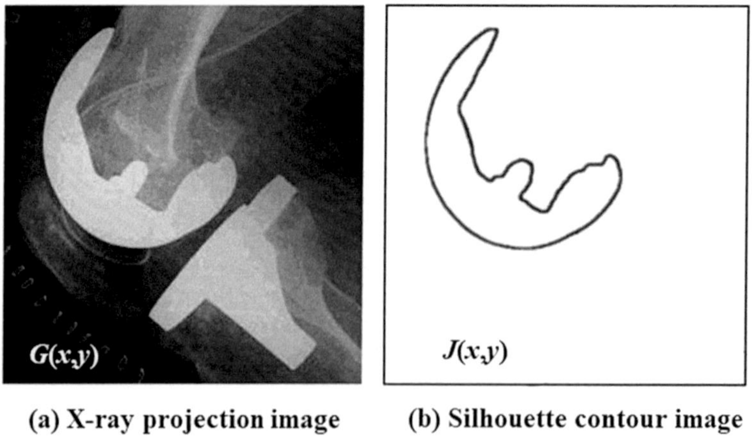

Figure 2.35 X-ray projection image of total knee prosthesis and extracted silhouette contour image. Here, the silhouette contour image of femoral component is extracted from X-ray projection image.
Modified from Hirokawa, S. (2011). In vivo kinematic measurement of artificial knee joint. *Toraibarojisuto/Journal of Japanese Society of Tribologists*, *56*(12), 758–764.

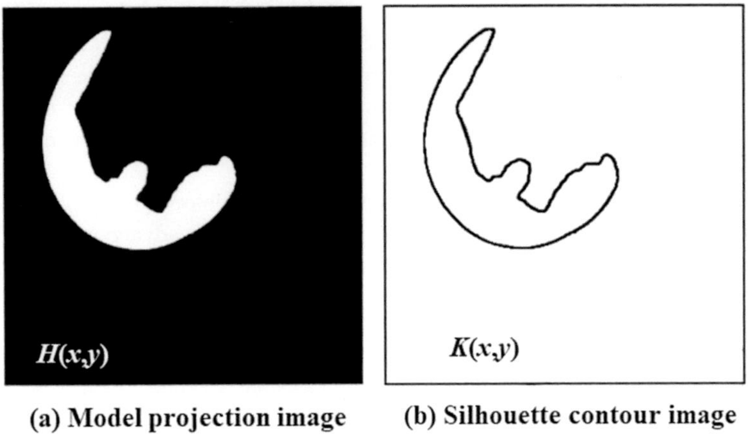

Figure 2.36 Projection image of CAD model and extracted silhouette contour image. The projection image of CAD model is arranged for the position and orientation estimated for the image in Fig. 2.35.
Modified from Hirokawa, S. (2011). In vivo kinematic measurement of artificial knee joint. *Toraibarojisuto/Journal of Japanese Society of Tribologists, 56*(12), 758–764.

overlapping between the extracted silhouette contour images can be evaluated based on the following cross-correlation function

$$I_2 = \sum J(x,y)K(x,y) / \sum K(x,y) \tag{2.41}$$

Here, I_2 corresponds to contour matching score.

Finally, the overlapping between X-ray projection image of total knee prosthesis and projection image of CAD model can be adjusted by maximizing the following I value with changing in rotation and translation of CAD model.

$$I = \omega_1 I_1 + \omega_2 I_2 \tag{2.42}$$

where ω_1 and ω_2 are weighting factors.

2.6.4 Reproduction of physiological and active knee motion in the novel knee simulator

Many kinds of joint simulators have been developed, but most of them may not appropriately reproduce actual physiological and active joint motion. The existing simulators for wear testing are useful for in vitro evaluation of wear in various kinds of total hip and knee joint prostheses, but usually driven using joint kinematics and loading profile data obtained from a normal gait (Hip: ISO-14242-3, 2009, Knee: ISO-14243-1, 2009) and may not always reflect in vivo situation. In simulators to measure kinematics and kinetics of lower limb using dissected limb, the passive way is applied for flexion (Zavatsky, 1997) or the muscular force of quadriceps is given by hydraulic action but with nonphysiological manner (Halloran et al., 2010). And, in most of knee simulators, it is difficult to reproduce deep knee flexion in squatting position or seiza (straight sitting, sitting in a kneeling position). In this section, it is focused for the reproduction of physiological and active knee motion in the novel knee simulator.

Hirokawa et al. (2015) developed the novel knee simulator that reproduces physiological and active knee motion by introducing the following advantages and innovative features for required properties as indicated in Fig. 2.37.

a. Active motion (Fig. 2.37a)

In human knee joints, knee kinematics and knee joint forces are "the outcome" caused by muscular forces and external loads applied to the knee joint. As for the simulators for the kinetics and kinematics measurement of the total knee arthroplasty (TKA), the knee is flexed in a "passive way" by applying constant muscular tension forces or other control mechanism. In the clinical evaluation of the replaced joint prosthesis in passive and manual flexion-extension test by medical doctor after arthroplasty, the replacement joint can show good contact condition and appropriate range of motion. However, in the actual flexion-extension motion by patient

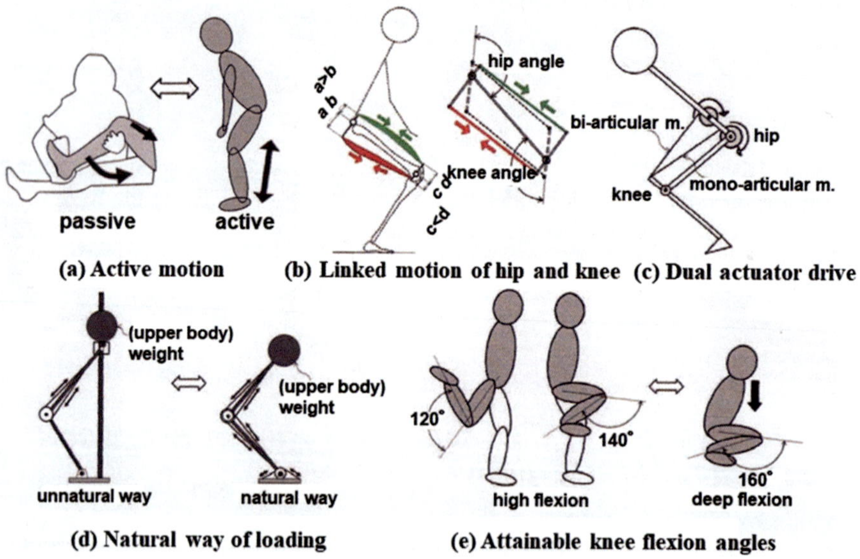

Figure 2.37 Required properties for the novel knee simulator. (a)–(e) Important features are illustrated to clarify the required properties.
Modified from Hirokawa, S., Murakami, T., Kiguchi, K., & Fukunaga, M. (2015). Development of the novel simulator that reproduces physiological and active knee motion. *Transactions of the JSME (in Japanese)*, *81*(824), 14-00488-14–00488. https://doi.org/10.1299/transjsme.14-00488.

on bed and walking training, the kinematics of replacement joint may be worsened. This being so, simulators should be driven by muscular forces and external loads, i.e., in an "active way." In the new simulator, it is driven directly by the muscles' tension forces as the motor-driven wire tugging system, thereby generating the active flexion-extension.

b. Linked motion of hip and knee (Fig. 2.37b)

In most of existing knee simulators, the effects of hip joint motion are not taken into account. For example, in human body, the cocontraction of a pair of rectus femoris and hamstrings as biarticular muscles for hip and knee joints acts to extend both of hip and knee joints in a synergistic way, irrespective of antagonist muscles.

At single contraction, the contraction of rectus femoris can flex the hip joint and extend the knee joint. In contrast, the contraction of hamstrings can extend the hip joint and flex the knee joint. Here, it is noticed for cocontraction of a pair of rectus femoris and hamstrings in Fig. 2.37b that $a > b$ as the moment arms of muscular forces around femoral head in hip, and $c < d$ as the moment arms of muscular forces around femoral condyle in knee. Therefore, if the same tension forces occur in rectus femoris and hamstrings at the cocontraction of both muscles, both hip and knee joints extend in the form of a closed link. That is, the extension of hip joint induces the extension of knee joint through the rectus femoris, and the extension of knee joint induces the extension of hip joint through the hamstrings. In this case with cocontraction, the rectus femoris as flexor muscle for hip can influence on the extension of hip joint, and the hamstrings as flexor muscle for knee can influence on the extension of knee joint. This is called Lombard's paradox (Lombard & Abbott, 1907).

In the new knee joint simulator, a hip joint is incorporated and the lower limb motion is achieved in a synergistic way between the hip and knee joints.

c. Dual actuator drive (Fig. 2.37c)

As described above, the biarticular muscles play an important role in synergistic motion of hip and knee joints. To attain various knee motions, however, the addition of the monoarticular muscles around knee joint is required. In the new simulator, dual actuator drive system with monoarticular and biarticular muscle mechanism is applied.

d. Natural way of loading (Fig. 2.37d)

In the external load system along the vertical rod in left figure of Fig. 2.37d, the upper body weight is not applied in a physiological way. The upper body weight should be applied on the upper link of inverted pendulum system as indicated in right figure.

e. Attainable knee flexion angle including deep flexion (Fig. 2.37e)

Few simulators are capable of knee flexion greater than about 130°. In the novel knee simulator, it is capable to make a complete deep knee flexion at 160°.

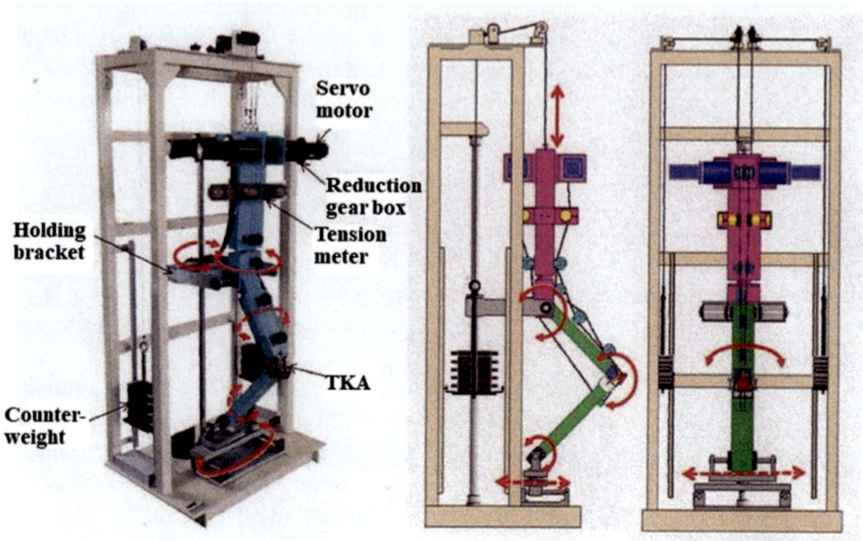

Figure 2.38 An overview photo and CAD pictures of the simulator. Solid line arrows indicate each link's freedom of motion, while broken line arrows indicate each link's play. From Hirokawa, S., Murakami, T., Kiguchi, K., & Fukunaga, M. (2015). Development of the novel simulator that reproduces physiological and active knee motion. *Transactions of the JSME (in Japanese)*, *81*(824), 14-00488-14−00488. https://doi.org/10.1299/transjsme.14-00488.

The constructed novel knee simulator and CAD pictures are shown in Fig. 2.38. The freedom of motion for each link and main devices/components are indicated. Ankle joint system and the holding bracket were arranged to prevent the falling of the simulator and to stabilize the posture. The size and weight for each link are arranged to correspond to the values for a human subject by adjusting with counterweight.

The wire-pull mechanism for substituting the lower limb muscles' pull force is shown in Fig. 2.39.

Here, biarticular muscles (B: rectus femoris and D: hamstrings) give joint moment for both hip and knee joints. Monoarticular muscles (A: vasti and C: Gluteus maximus) generate the joint moment for knee and hip, respectively.

Tibio-femoral and patella-femoral compartments including total knee prosthesis with springs for collateral ligaments are shown in Fig. 2.40. The sensor to measure tibio-femoral contact force was set between the tibial insert and tibial tray. The sensor to measure patello-femoral contact force was settled in patella block.

The quasi-static sit-to-stand motion test with legs in parallel was performed for total knee prosthesis of PS design in this simulator. The knee joint forces of tibial axis (tibiofemoral) contact force: F_n and the anterior-posterior tangential force: F_t in the range of deep flexion of 150 degrees are shown in Fig. 2.41. In comparison with numerical analysis by Hirokawa et al. (Hirokawa & Fukunaga, 2013) in

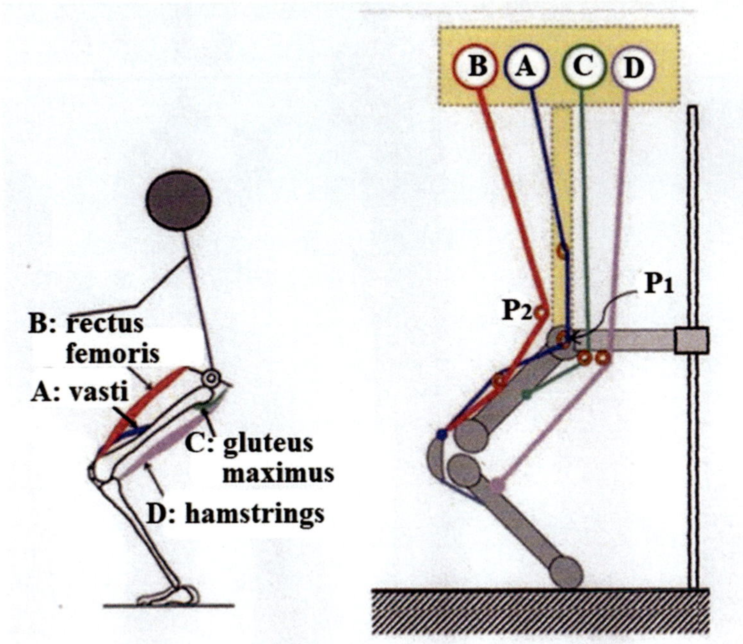

Figure 2.39 Wire-pull mechanism for substituting the lower limb muscles' pull force. As wire A passes through pulley P_1 which is located on the hip flexion axis, its pull force does not generate hip flexion moment, but knee extension moment. While wire B passes through pulley P_2 which is located apart from the hip flexion axis, and therefore its pull force generates hip flexion moment as well as knee extension moment. The same is true for wire C and D.
Modified from Hirokawa, S., Murakami, T., Kiguchi, K. & Fukunaga, M. (2015). Development of the novel simulator that reproduces physiological and active knee motion. *Transactions of the JSME (in Japanese)*, *81*(824), 14-00488-14–00488. https://doi.org/10.1299/transjsme.14-00488.

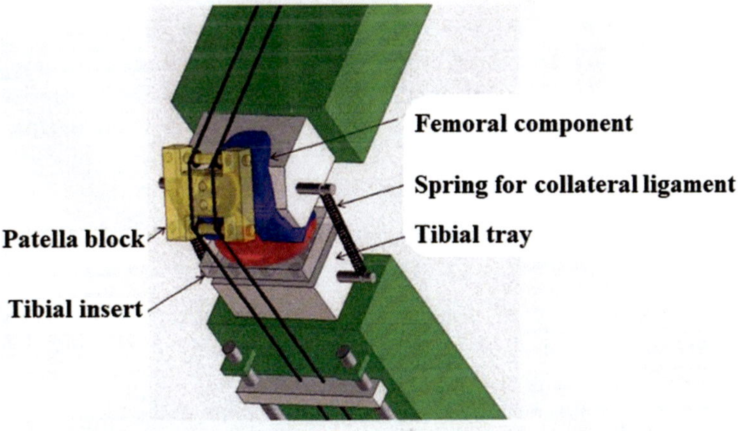

Figure 2.40 Schematic drawing of the tibio-femoral and patello-femoral compartments. The spring for collateral ligament was made so as to reproduce the mechanical property of the human collateral ligament.
From Hirokawa, S., Murakami, T., Kiguchi, K., & Fukunaga, M. (2015). Development of the novel simulator that reproduces physiological and active knee motion. *Transactions of the JSME (in Japanese)*, *81*(824), 14-00488-14–00488. https://doi.org/10.1299/transjsme.14-00488.

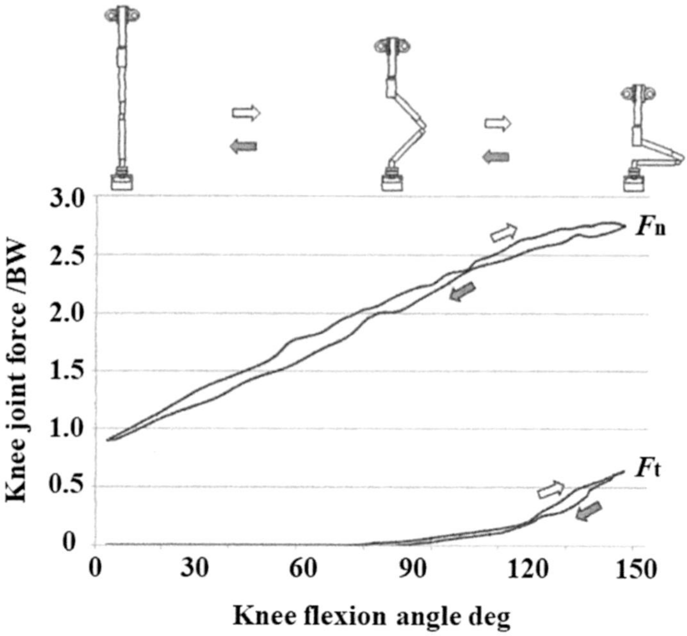

Figure 2.41 Variation of knee joint force as a function of knee angles during quasi static sit-to-stand motion of the simulator. F_n: The tibial axis directional component of tibiofemoral contact force, F_t: The vertical component to the tibial axis direction of tibiofemoral contact force.

From Hirokawa, S., Murakami, T., Kiguchi, K., & Fukunaga, M. (2015). Development of the novel simulator that reproduces physiological and active knee motion. *Transactions of the JSME (in Japanese)*, *81*(824), 14-00488-14−00488. https://doi.org/10.1299/transjsme.14-00488.

Fig. 2.42 for 2D musculoskeletal model with additional features including simulated muscle cocontraction and antagonistic muscle activations, patellofemoral mechanism (relation between the quadriceps and the patellar tendon forces as a function of knee angle) and the effect of the thigh-calf contact force during knee bending motion, the following common features were confirmed.

1. F_n of almost a body weight (1.0 × BW) at extension position,
2. F_n of about (2.0∼2.5) × BW at 120 degrees, F_t of about (0.2∼0.3) × BW at 120°.

In this simulator, therefore, joint contact forces are appropriately evaluated under the condition loaded by muscular forces and upper body weight while the loading system stabilizes the posture. The simulator that reproduces actual physiological and active joint motion is expected to contribute to the modeling analysis of total joint prostheses in various daily activities and the complementary validation for in vivo measurement using an instrumented knee prosthesis (Kutzner et al., 2010).

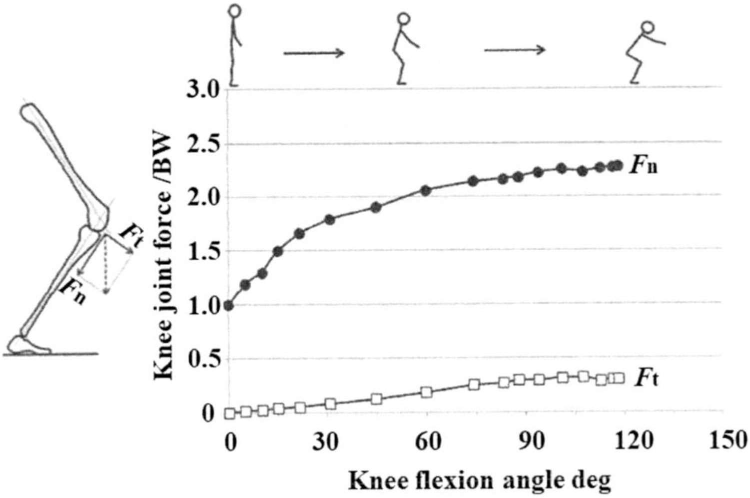

Figure 2.42 Variation of knee flexion angle and knee joint forces during knee bend motion. The graph are modified with those by Hirokawa and Fukunaga (2013).
From Hirokawa, S., Murakami, T., Kiguchi, K., & Fukunaga, M. (2015). Development of the novel simulator that reproduces physiological and active knee motion. *Transactions of the JSME (in Japanese), 81*(824), 14-00488-14–00488. https://doi.org/10.1299/transjsme.14-00488.

References

Banks, S. A., & Hodge, W. A. (1996). Accurate measurement of three-dimensional knee replacement kinematics using single-plane fluoroscopy. *IEEE Transactions on Biomedical Engineering, 43*(6), 638–649. https://doi.org/10.1109/10.495283

Davidovits, P. (2013). *Physics in biology and medicine* (4th ed.). Academic Press/Elsevier.

Dumbleton, J. H. (1981). *Tribology of natural and artificial joints. (Book).* Amsterdam, The Netherlands: ELSEVIER SCI. PUBLISHING CO., 1981.

Frankel, H., & Nordin, M. (1980). *Basic biomechanics of the skeletal system.* Lea and Febiger.

Fung, Y. C. (1968). Biomechanics – its scope, history, and some problems of continuum mechanics in physiology. *Applied Mechanics Reviews, 21*, 1–20.

Halloran, J. P., Clary, C. W., Maletsky, L. P., Taylor, M., Petrella, A. J., & Rullkoetter, P. J. (2010). Verification of predicted knee replacement kinematics during simulated gait in the Kansas knee simulator. *Journal of Biomechanical Engineering, 132*(8). https://doi.org/10.1115/1.4001678

Hirokawa, S. (2011). In vivo kinematic measurement of artificial knee joint. *Toraibarojisuto/Journal of Japanese Society of Tribologists (In Japanese), 56*(12), 758–764.

Hirokawa, S. (2013). Biomechanics of natural and artificial joints. In *Frontiers of biomechanics (In Japanese)* (pp. 109–166). Kyoritsu Shuppan Co., Ltd.

Hirokawa, S., Abrar Hossain, M., Kihara, Y., & Ariyoshi, S. (2008). A 3D kinematic estimation of knee prosthesis using X-ray projection images: Clinical assessment of the improved algorithm for fluoroscopy images. *Medical and Biological Engineering and Computing, 46*(12), 1253–1262. https://doi.org/10.1007/s11517-008-0398-8

Hirokawa, S., Ariyoshi, S., Takahashi, K., & Maruyama, K. (2003). Kinematic measurement of knee prosthesis from single-plane projection images. *JSME International Journal, Series C: Mechanical Systems, Machine Elements and Manufacturing, 46*(4), 1368–1376. https://doi.org/10.1299/jsmec.46.1368

Hirokawa, S., & Fukunaga, M. (2013). Knee joint forces when rising from kneeling positions. *Journal of Biomechanical Science and Engineering, 8*(1), 27–39. https://doi.org/10.1299/jbse.8.27

Hirokawa, S., Murakami, T., Kiguchi, K., & Fukunaga, M. (2015). Development of the novel simulator that reproduces physiological and active knee motion. *Transactions of the JSME (In Japanese), 81*(824). https://doi.org/10.1299/transjsme.14-00488, 14-00488-14–00488.

Jalali-Vahid, D., Jagatia, M., Jin, Z. M., & Dowson, D. (2001a). Prediction of lubricating film thickness in a ball-in-socket model with a soft lining representing human natural and artificial hip joints. *Proceedings of the Institution of Mechanical Engineers, J: Journal of Engineering Tribology, 215*(4), 363–372. https://doi.org/10.1243/1350650011543600

Jalali-Vahid, D., Jagatia, M., Jin, Z. M., & Dowson, D. (2001b). Prediction of lubricating film thickness in UHMWPE hip joint replacements. *Journal of Biomechanics, 34*(2), 261–266. https://doi.org/10.1016/S0021-9290(00)00181-0

Kapandji, A. I. (2019a). The upper limb. In *The physiology of the joints* (Vol. 1)Handspring Publishing Ltd. East Lothian.

Kapandji, A. I. (2019b). The lower limb. In *The physiology of the joints* (Vol. 2)Handspring Publishing Ltd. East Lothian.

Kutzner, I., Heinlein, B., Graichen, F., Bender, A., Rohlmann, A., Halder, A., Beier, A., & Bergmann, G. (2010). Loading of the knee joint during activities of daily living measured in vivo in five subjects. *Journal of Biomechanics, 43*(11), 2164–2173. https://doi.org/10.1016/j.jbiomech.2010.03.046

Lombard, W. P., & Abbott, F. M. (1907). The mechanical effects produced by the contraction of individual muscles of the thigh of the frog. *American Journal of Physiology-Legacy Content, 20*(1), 1–60. https://doi.org/10.1152/ajplegacy.1907.20.1.1

Luís, N. M., & Varatojo, R. (2021). Radiological assessment of lower limb alignment. *EFORT Open Reviews, 6*(6), 487–494. https://doi.org/10.1302/2058-5241.6.210015

Mahfouz, M. R., Hoff, W. A., Komistek, R. D., & Dennis, D. A. (2003). A robust method for registration of three-dimensional knee implant models to two-dimensional fluoroscopy images. *IEEE Transactions on Medical Imaging, 22*(12), 1561–1574. https://doi.org/10.1109/TMI.2003.820027

Makikawa, M., & Yoshida, M. (2008). *Biomechanics of human motion — its hardware and software (In Japanese)*. Corona Publishing Co., Ltd.

Martelli, S., & Pinskerova, V. (2002). The shapes of the tibial and femoral articular surfaces in relation to tibiofemoral movement. *Journal of Bone and Joint Surgery - Series B, 84*(4), 607–613. https://doi.org/10.1302/0301-620X.84B4.12149

Miyazaki, S. (2002). Walking motion. In Japanese Society for Medical Electronics and Biological Engineering (Ed.), *Biomechanics of Human Movement (In Japanese)* (pp. 102–125). Corona Publishing Co., Ltd.

Morrison, J. B. (1970). The mechanics of the knee joint in relation to normal walking. *Journal of Biomechanics, 3*(1), 51–61. https://doi.org/10.1016/0021-9290(70)90050-3

Paul, J. P. (1966–67). Forces transmitted by joints in the human body. *Proceedings of the Institution of Mechanical Engineers, 181*(Pt3J), 8–15.

Stansfield, B. W., Nicol, A. C., Paul, J. P., Kelly, I. G., Graichen, F., & Bergmann, G. (2003). Direct comparison of calculated hip joint contact forces with those measured using instrumented implants. An evaluation of a three-dimensional mathematical model of the lower limb. *Journal of Biomechanics, 36*(7), 929–936. https://doi.org/10.1016/S0021-9290(03)00072-1

Sugiura, Y. (1985). *Bone age of Japanese (In Japanese)*. Chugai Igakusya Press.

Walker, P. S. (1977). *Human joints and their artificial replacements*. Charles C Thomas publisher.

Wallace, T. P., & Wintz, P. A. (1980). An efficient three-dimensional aircraft recognition algorithm using normalized fourier descriptors. *Computer Graphics and Image Processing, 13*(2), 99–126. https://doi.org/10.1016/S0146-664X(80)80035-9

Wright, V. (1981). Joint anatomy and overall function. In D. Dowson, & V. Wright (Eds.), *Introduction to the biomechanics of joints and joint replacement* (pp. 1–7). Mechanical engineering publications Ltd.

Yamamoto, T., Suzuki, T., & Tasaki, K. (2005). *Modern anatomy and physiology (In Japanese)*. Nankodo Co., Ltd.

Zavatsky, A. B. (1997). A kinematic-freedom analysis of a flexed-knee-stance testing Rig. *Journal of Biomechanics, 30*(3), 277–280. https://doi.org/10.1016/S0021-9290(96)00142-X

Zuffi, S., Leardini, A., Catani, F., Fantozzi, S., & Cappello, A. (1999). A model-based method for the reconstruction of total knee replacement kinematics. *IEEE Transactions on Medical Imaging, 18*(10), 981–991. https://doi.org/10.1109/42.811310

Biotribology of natural joints

3

3.1 Importance of tribology in natural joints

Natural synovial joints or diarthroses are composed of bones covered with lubricious articular cartilage as soft layer and have a cavity containing the synovial fluid between bones. Natural articular cartilage is the compliant and hydrated tissue with high water content as the biphasic material composed of the extracellular matrix of proteoglycan reinforced by the collagen fibers and the chondrocytes. The synovial fluid is non-Newtonian viscoelastic hyaluronate solution containing various lubricating and nutritive constituents. The lubrication mechanism of the natural synovial joints with extremely low friction and minimal wear for longer life has been one of the most valuable and attractive subjects in the field of biotribology which was introduced as "those aspects of tribology concerned with biological systems" by Dowson and Wright (1973).

The healthy synovial joints maintain excellent load-carrying capacity and tribological properties with extremely low friction and minimum wear even under heavily loaded conditions in hip, knee and ankle joints. This superior lubricating performance appears to be actualized by not single lubrication mode but the synergistic combination of various modes from the fluid film lubrication without direct contact to boundary lubrication corresponding to the severity of rubbing conditions, as pointed out by Dowson (1966–67). At the symposium on "Lubrication and Wear in Living and Artificial Joints" (1967), he indicated that "the major lubrication mechanism would seem to be some form of elastohydrodynamic action determined by sliding or squeeze film action between porous surfaces with boundary lubrication providing the surface protection in cases of severe loading and little movement. The human joint seems to operate in the 'mixed' lubrication – sometimes elastohydrodynamic and sometimes boundary lubrication." Thus, the lubrication mechanisms are likely to have multimode systems, and the further investigations to elucidate the actual lubrication mechanisms in natural synovial joints have been continued. In this chapter, the predominant lubrication mechanisms are shown.

The importance of tribology in natural joints is considered from the viewpoints of friction, wear, and lubrication. The extremely low friction of 0.002–0.03 as friction coefficient in natural synovial joints contributes to the smooth movement of the musculoskeletal system with lower energy. The protection of the articular cartilage through various lubrication mechanisms to prevent the direct contacts between rubbing articular surfaces is effective to keep the function of natural joints for a whole life under healthy conditions. However, if the multimode lubrication mechanisms fail to sustain the protective role, some direct contacts between the rubbing articular cartilage surfaces may occur and then induce increases in friction and local surface failures such as excessive deformation, surface roughening, and wear. Raised friction enhances

Biotribology of Natural and Artificial Joints. https://doi.org/10.1016/B978-0-12-823669-7.00002-8

the rubbing severity of contact surfaces and bulk body of the cartilage and then the surface failure will proceed to worse level. The failures and diseases in natural joints can change the constituents and properties of the synovial fluid with reducing in viscosity. The articular cartilage has an inferior restoration ability and therefore the robust sustaining of lubrication mechanisms to protect articular cartilage become important, where both superior properties of the articular cartilage as rubbing surface and load-carrying material and the synovial fluid as lubricant containing nutritive constituents play fundamental roles.

In this chapter, first the structure and various properties of the articular cartilage as load-bearing material with low friction are described. The constituents and roles of the synovial fluid as lubricant are mentioned. Next, the various lubrication mechanisms from the fluid film lubrication, biphasic lubrication, hydration lubrication, boundary lubrication, and gel film lubrication are summarized from the viewpoint of adaptive multimode lubrication. Finally, wear and degradation mechanism in natural joints are discussed.

3.2 Articular cartilage

3.2.1 Structures and properties of articular cartilage

The types of cartilage as soft tissue composed of the cells and the extracellular matrix are divided as the hyaline cartilage, the elastic cartilage, and the fibrocartilage, which differ in composition and structure. The articular cartilage in natural synovial joint is the hyaline (glasslike) cartilage which has pearl-gray semi-translucent appearance and slippery surface. It is compliant but resilient and recovers its shape usually quickly when the load is removed although it shows some time-dependent viscoelastic behavior as biphasic material. In natural synovial joints, it covers the ends of bones and enables to move smoothly in joint cavity containing synovial fluid with superior lubricating ability. The thickness of cartilage layer changes depending on the joint type, size, and location. In case of adult human knee joint, the cartilage thickness is 2–5 mm or so.

The articular cartilage is compliant and hydrated tissue with high water content of 70%–80% as biphasic material which is composed of the extracellular matrix of proteoglycan reinforced by the collagen fibers (mainly type II) and the chondrocytes. It contains no blood vessels or nerves, unlike bone.

The schema of structure of the articular cartilage is shown in Fig. 3.1 (Mow et al., 1990), where the distribution of collagen fibers is depicted. The articular cartilage is usually divided as the three compliant layers, i.e., the superficial (tangential) zone, the middle zone, and the deep zone as non-calcified cartilage, and the stiff calcified cartilage zone. The differences in the orientation and distribution of the collagen fibers give remarkable influence on the stiffness and strength in each layer. The proteoglycan occupies the interfibrillar space and its concentration increases from minimum level in the surface zone to a maximum in the middle zone, then decreases toward the deep

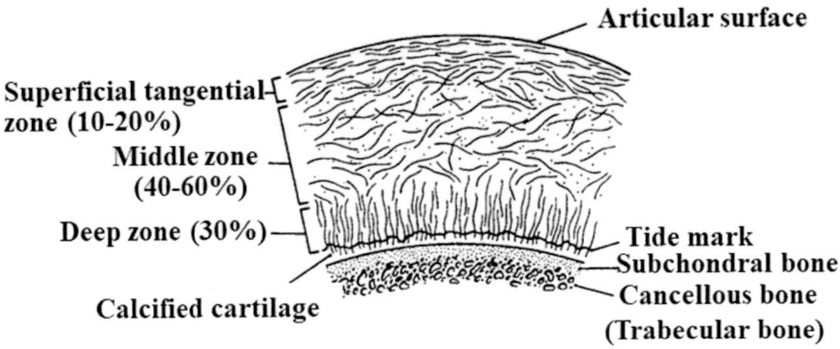

Figure 3.1 Schematic diagram of articular cartilage depicting the collagen fibril ultrastructure. The articular cartilage has been classified into four distinct zones or layers referred to as the superficial tangential zone, the middle or transitional zone, the deep or radial zone, and the zone of calcified cartilage.
Modified with permission from Mow, V. C., Hou, J. S., Owens, J. M., & Ratcliffe, A. (1990). Biphasic and quasilinear viscoelastic theories for hydrated soft tissues. In Ratcliffe, A., Woo, S.L.Y. & Mow, V.C.(Eds.) *Biomechanics of diarthroidal joint* (Vol. 1, pp. 215–260). Springer, New York, NY. https://doi.org/10.1007/978-1-4612-3448-7_8

zone. Within each zone, the regions for cellular organization for chondrocytes exist for only 5% or so of the tissue volume.

As shown in Fig. 3.1, the superficial layer consists primarily of fine densely packed collagen fibrils to the articular surface. In the middle zone, the collagen fibrils have much larger diameters and are more randomly arranged. In the deep zone, the collagen fiber bundles orient radially and insert across the tidemark to securely anchor the articular cartilage to the bone.

The cancellous bone (trabecular bone) is porous bone composed of trabecular bone tissue which is arranged in trabecular structure to bear the external loading. The appropriate deformation of the underlying cancellous bone contributes to reducing the contact stress in cartilage tissue. The subchondral bone is the layer of bone just below the cartilage and has an important influence on the deformation of cartilage layer and acts as a supplier of nutrients to cartilage.

To elucidate the detailed tribological mechanisms of friction, wear, and lubrication in synovial joints, it is important to clarify the structure and properties of the articular surfaces, in conjunction with the bulk structure and properties of cartilage. The uppermost acellular surface layer without the collagen fiber meshes appears to be discriminated from superficial tangential zone. This articular surface layer is sometimes also referred to as lamina splendens (MacConaill, 1951), surface amorphous layer (Kobayashi et al., 1995), uppermost superficial surface layer (Kumar et al., 2001), or articular surface (Crocket, 2014).

Higaki et al. (1998) examined the surface structure of articular cartilage of rat patella using a transmission electron microscope (TEM). In the TEM images (Fig. 3.2), a single continuous biomembrane structure was distinctly detected as the laminar structure with fibrous molecules where hydrophobic layer is positioned between two

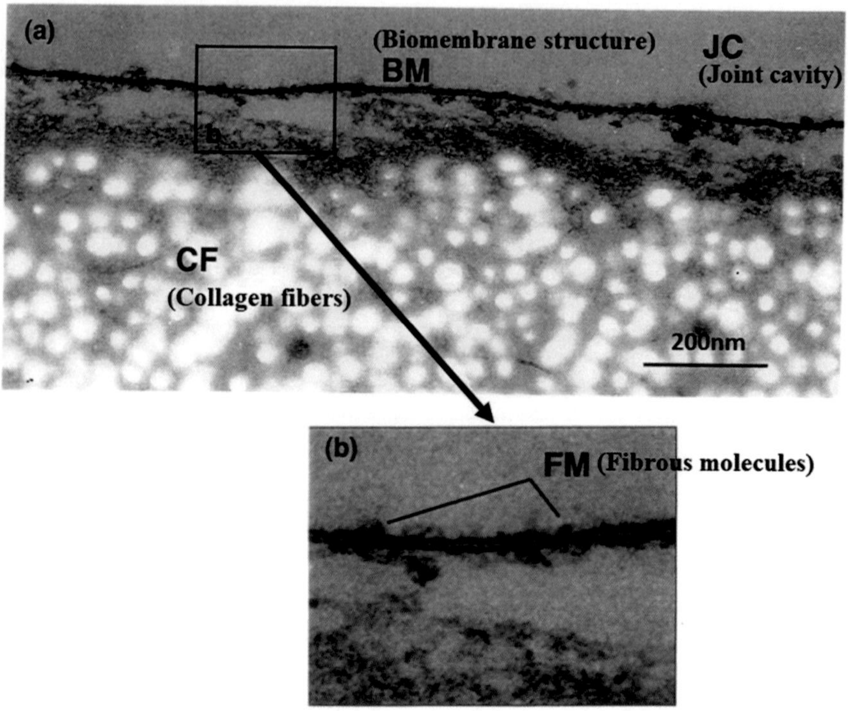

Figure 3.2 TEM image of transverse section of articular cartilage from rat patella fixed with osmium tetroxide. (a) The articular cartilage surfaces are covered by adsorbed films as a continuous membrane-structure with fibrous molecules. White circular parts correspond to the sections of collagen fibers. In the intermediate zone between the biomembrane and collagenous zone, the nonfibrillar layer is observed. (b) Enlarged image of biomembrane structure on underlying nonfibrillar layer.
From Higaki, H., Murakami, T., Nakanishi, Y., Miura, H., Mawatari, T., & Iwamoto, Y. (1998). The lubricating ability of biomembrane models with dipalmitoyl phosphatidylcholine and γ-globulin. *Proceedings of the Institution of Mechanical Engineers, Part H: Journal of Engineering in Medicine*, *212*(5), 337–346. https://doi.org/10.1243/0954411981534114

hydrophilic layers on the articular surface. This biomembrane structure appears to act as an adsorbed film. In addition, the intermediate nonfibrillar layer was confirmed between the biomembrane and the collagenous zone. As discussed later in Section 3.4.6, after removal of adsorbed films on cartilage surfaces, nonfibrillar gellike surface layer maintains low friction due to its low shearing resistance and protects the cartilage bulk tissue (Murakami et al., 1999). The surface structure covered with adsorbed films on underlying proteoglycan gel-like layer is the failsafe system in synovial joints.

The schema of extracellular matrix is shown in Fig. 3.3 (Mow et al., 1990), where the proteoglycan, the collagen fibers, and the interstitial fluid are indicated, but the chondrocyte is not shown. The chondrocyte is a specialized cell in cartilage and

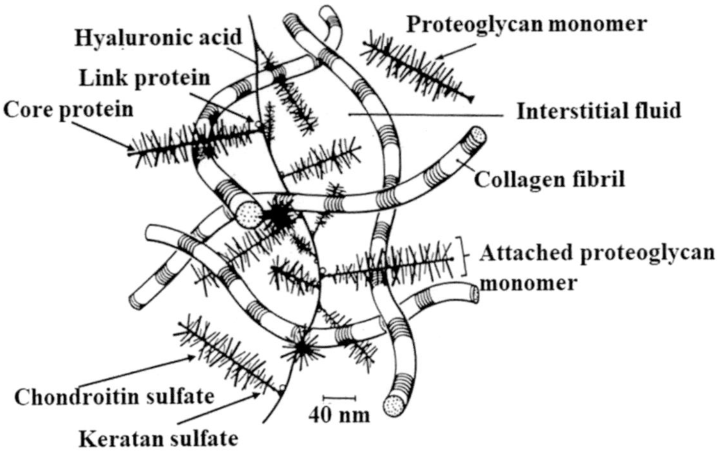

Figure 3.3 Extracellular matrix of articular cartilage. Schematic representation of the molecular organization of cartilage. Collagen and proteoglycan aggregate interact to form a porous solid matrix which is swollen with water.
Modified with permission from Mow, V. C., Hou, J. S., Owens, J. M., & Ratcliffe, A. (1990). Biphasic and quasilinear viscoelastic theories for hydrated soft tissues. In Ratcliffe, A., Woo, S.L.Y. & Mow, V.C.(Eds.) *Biomechanics of diarthroidal joint.* (Vol. 1, pp. 215–260). Springer, New York, NY. https://doi.org/10.1007/978-1-4612-3448-7_8

produces collagenous extracellular matrix. The main proteoglycan in cartilage is aggrecan which forms large aggregates with hyaluronic acid.

The proteoglycans are composed of glycosaminoglycan (GAG, or mucopolysaccharide) chains covalently linked to core protein (Alberts et al., 2021). Except for the hyaluronan (or hyaluronic acid) (Fig. 3.9), all GAGs such as chondroitin sulfate and keratin sulfate (Fig. 3.4) are covalently attached to protein as proteoglycan. The proteoglycans are composed of about 95% polysaccharide and about 5% protein. The aggregating proteoglycans or aggrecans are composed of monomers with chondroitin sulfate and keratan sulfate attached to the protein core. In most aggrecan molecules, link proteins connect these monomers to a hyaluronic acid chain. It consists of about 100 aggrecan monomers noncovalently bound through the N-terminal domain of the core protein to a single hyaluronan chain. A link protein binds both to the core protein of the proteoglycan and to the hyaluronan chain, thereby stabilizing the aggregate.

The proteoglycans of cartilage contribute significantly to the compressive and swelling properties. In articular cartilage, the normally water-soluble proteoglycan molecules are prevented by the collagen mesh from passing into solution. Proteoglycans carry a high density of negative charge groups (SO_4^-, COO^-) which are firmly attached to the GAG chains. Compression of the cartilage would bring these charge groups closer together thereby increasing the Donnan osmotic pressure and electrostatic repulsive forces. These internal swelling forces are balanced by tensile forces in the surrounding collagenous network. Therefore, the articular cartilage rich in proteoglycan have high water content but low permeability and high compressive stiffness.

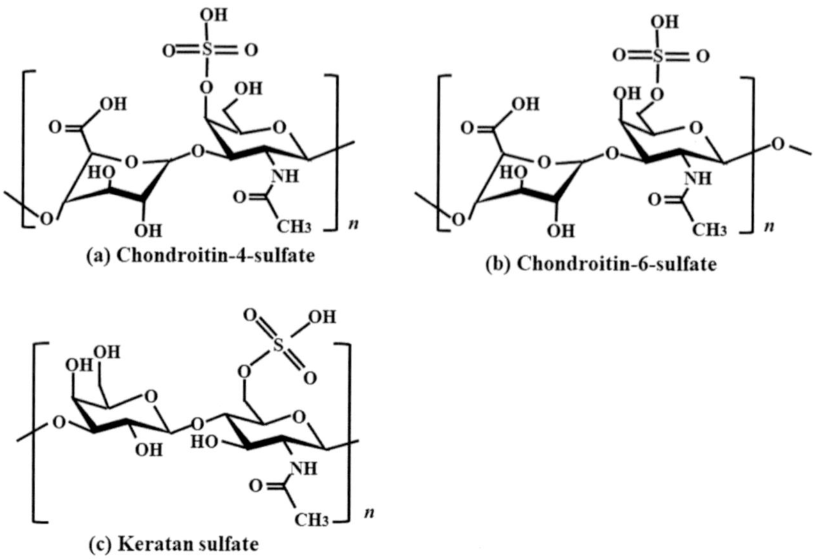

Figure 3.4 Chondroitin sulfate and keratan sulphate. Chondroitin sulfate is a sulfated glycosaminoglycan (GAG) composed of a chain of alternating sugars: D-glucuronic acid and *N*-acetyl-D-galactosamine. Chondroitin-4-sulfate and chondroitin-6-sulfate are also called chondroitin sulfate A and chondroitin sulfate C, respectively. Keratan sulfate is a sulfated glycosaminoglycan (GAG) composed of a chain of alternating sugars: D-galactose and N-acetyl-D-glucosamine.

Two-thirds of the dry mass of adult articular cartilage are polymeric collagen. The most important mechanical properties of collagen fibers are their tensile stiffness and strength. Type II collagen is the principle molecular component in mammals. But collagens III, VI, IX, X, XI, XII, and XIV all contribute to mature matrix (Eyre, 2002). In developing cartilage, the core fibrillary network is a cross-linked copolymer of collagens II, IX, and XI. Type IX collagen as a short nonfibrillar collagen binds covalently to type II collagen fibrils and may help link fibrils together or bind fibrils to other matrix molecules. Type XI collagen as a minor fibrillar collagen contributes to controlling the diameter of the type II collagen fibrils (Mow et al., 1990).

As described above, the solid phase of articular cartilage consists of collagen, proteoglycans, minor collagen types, and other noncollagenous proteins and glycoproteins. Articular cartilage is a biphasic material which is composed of solid and fluid phases. The fluid phase of cartilage consists of chiefly water and dissolved inorganic salts. Articular cartilage contains large amount of water at 65%–80% of weight of cartilage tissue (80% at the surface to 65% in the deep zone). The exudation of interstitial water from the cartilage tissue can be derived by applying a pressure gradient across the tissue or compressing the solid matrix. Here, the frictional resistance to interstitial fluid flow through the porous-permeable solid matrix is very high. Therefore, the permeability of articular cartilage is extremely low.

To evaluate friction, wear, lubrication mode, and contact situation in rubbing between the cartilage surfaces, it is required to get appropriate knowledge on surface topography/roughness. The articular cartilage is compliant material and the surface profiles measured with a fine diamond stylus, which would readily dig into the soft articular cartilage, may lead to erroneous indications of surface topography and damage to the surfaces. Desiccation and shrinkage of specimen may affect the real profile. To suppress these phenomena, the contact profilometry with stylus has been applied for tracing profiles from hard replicas of soft cartilage surfaces. In this method, however, the wet condition of cartilage surface may be still disturbed. Therefore, it is recommended to have surface images of articular cartilage immersed in liquid by atomic force microscopy (AFM) with fluid tapping mode as a noncontact mode (Fig. 3.5) (Murakami, Hayakawa, et al., 1997). Asperity height can be estimated as 1−2 μm in the trace for this sample. It should be noticed the vertical magnification is 5 times of horizontal one and actual profile has relatively gentle sloping.

3.2.2 Biphasic properties of articular cartilage

Articular cartilage as hydrated soft tissue is composed of a solid matrix phase and an interstitial fluid phase and exhibits the complicated creep and stress-relaxation behaviors with exudation of interstitial fluid and penetration. Therefore, it is difficult to describe the creep and stress-relaxation of articular cartilage by the viscoelastic models which combine springs and dashpots in Maxwell model, Voigt or Kelvin model, three-elements model, and so on. Mow et al. (1980) proposed the biphasic theory as the mixture theory in which it is assumed that the solid matrix and the interstitial fluid of the tissue are immiscible and each phase is intrinsically incompressible. The interstitial fluid may be moved through the tissue by a pressure gradient or the compression

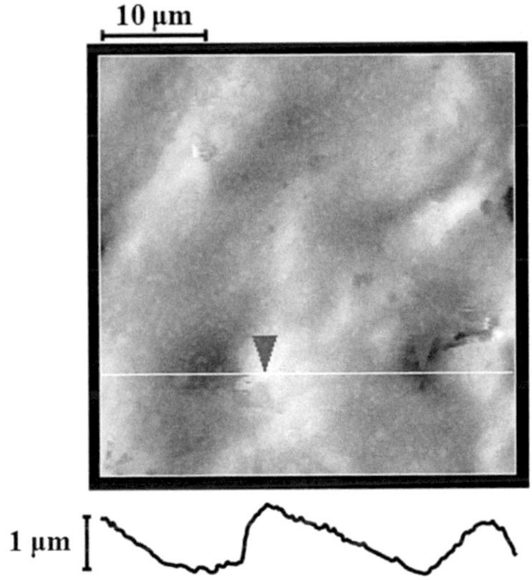

Figure 3.5 AFM image of porcine knee femoral articular cartilage immersed in saline (fluid tapping mode). Upper trace corresponds to the white line in AFM image.
Modified from Murakami, T., Hayakawa, Y., Higaki, H., & Sawae, Y. (1997). Tribological behavior of sliding pairs of articular cartilage and bioceramics. In *Proceedings of International Conference on New Frontiers in Biomechanical Engineering* (pp. 233−236).

of the solid matrix. Viscous dissipation in the tissue is dominated by the friction drag of interstitial fluid flow through the porous-permeable collagen-proteoglycan solid matrix.

The fundamental formulae for the biphasic theory for hydrated soft tissues such as articular cartilage and meniscus are shown (Mow et al., 1990).

For the incompressible solid and fluid phases, the continuity equation as the balance of mass for the tissue is described as below:

$$\mathrm{div}\left(\phi^{s}\boldsymbol{v}^{s}+\phi^{f}\boldsymbol{v}^{f}\right)=0 \tag{3.1}$$

where ϕ^{s} and ϕ^{f} are the volume fractions of the solid phase and the fluid phase, and $\boldsymbol{v}^{s}$ and $\boldsymbol{v}^{f}$ are their velocities, respectively. In the binary phases,

$$\phi^{s}+\phi^{f}=1 \tag{3.2}$$

The equations of motion for solid phase and the fluid phase for quasi-static situations are:

$$\mathrm{div}\ \boldsymbol{\sigma}^{s}+\boldsymbol{\pi}=0 \tag{3.3}$$

$$\mathrm{div}\ \boldsymbol{\sigma}^{f}-\boldsymbol{\pi}=0 \tag{3.4}$$

where $\boldsymbol{\sigma}^{s}$ and $\boldsymbol{\sigma}^{f}$ are the stress tensors acting on the solid phase and fluid phase, respectively, and $\boldsymbol{\pi}$ is the action-and-reaction interaction force between the two phases or the momentum supply from one phase to the other. For infinitesimal strains, it is assumed that the solid phase is a linearly elastic solid and the fluid phase is a Newtonian viscous fluid. Under these conditions, the constitutive relations for the stresses and the momentum supply are given by:

$$\boldsymbol{\sigma}^{s}=-\phi^{s}p\boldsymbol{I}+\lambda_{s}\,\mathrm{tr}(\boldsymbol{E})\boldsymbol{I}+2\,\mu_{s}\,\boldsymbol{E} \tag{3.5}$$

$$\boldsymbol{\sigma}^{f}=-\phi^{f}p\boldsymbol{I}-\frac{2}{3}\mu_{a}\left(\mathrm{div}\ \boldsymbol{v}^{f}\right)\boldsymbol{I}+2\mu_{a}D \tag{3.6}$$

$$\boldsymbol{\pi}=p\nabla\phi^{f}+K\left(\boldsymbol{v}^{f}-\boldsymbol{v}^{s}\right) \tag{3.7}$$

where λ_{s} and μ_{s} are Lamè constants for the solid phase, μ_{a} is the apparent viscosity of the fluid phase in cartilage, $\boldsymbol{E}$ is the infinitesimal strain tensor for the solid matrix, $\boldsymbol{D}$ is the rate of the deformation tensor for the interstitial fluid, and K is the coefficient of the diffusive drag caused by the relative motion between the two phases.

If the constitutive Eqs. (3.5)−(3.7) are inserted into the equations of motion (3.3) and (3.4), the next equations are obtained.

$$-\phi^{\mathrm{s}}\nabla p+(\lambda_{\mathrm{s}}+\mu_{\mathrm{s}})\nabla(\operatorname{div}\boldsymbol{u})+\mu_{\mathrm{s}}\nabla^2\boldsymbol{u}+K\left(\boldsymbol{v}^{\mathrm{f}}-\boldsymbol{v}^{\mathrm{s}}\right)=0 \tag{3.8}$$

$$-\phi^{\mathrm{f}}\nabla p+\frac{1}{3}\mu_{\mathrm{a}}\nabla\left(\operatorname{div}\boldsymbol{v}^{\mathrm{f}}\right)+\mu_{\mathrm{a}}\nabla^2\boldsymbol{v}^{\mathrm{f}}+K\left(\boldsymbol{v}^{\mathrm{s}}-\boldsymbol{v}^{\mathrm{f}}\right)=0 \tag{3.9}$$

where $\boldsymbol{u}$ is the displacement vector of the solid matrix. For infinitesimal deformation the velocity of solid matrix is given by:

$$\boldsymbol{v}^{\mathrm{s}}=\frac{\partial\boldsymbol{u}}{\partial t} \tag{3.10}$$

The permeability coefficient k for tissue with the constant permeability is related to K:

$$k=\frac{\phi f^2}{K} \tag{3.11}$$

The permeability k for the strain-dependent permeability case, the following formula is applied:

$$k=k_0\,\mathrm{e}^{M\mathrm{tr}(\boldsymbol{E})} \tag{3.12}$$

where M is a material constant, and $tr(\boldsymbol{E})$ is the dilatation of the solid matrix.

The creep deformation and stress-relaxation of articular cartilage accompanied with movement of the interstitial fluid can be evaluated on the basis of the biphasic theory mentioned above.

Here, the effect of interstitial fluid viscosity is negligible, i.e., $\mu_{\mathrm{a}}=0$ in Eqs. (3.6) and (3.9).

3.2.3 Chondrocytes

Chondrocytes are the cells specialized in healthy articular cartilage. The cartilaginous matrix composed of collagen fibers and proteoglycans are produced and maintained by the chondrocytes.

Fig. 3.6 shows microphotographs of cross-section for articular cartilage with chondrocytes treated by hematoxylin-eosin (HE) staining. Fig. 3.7 shows morphological changes of chondrocytes at different zones observed by confocal laser scanning microscopy in which the viable chondrocytes in porcine femoral condyle are stained with Calcein-AM at 37°C for 30 min (Sakai et al., 2005). The morphology and distribution of chondrocytes correspond to the distribution of collagen fibers (Fig. 3.1).

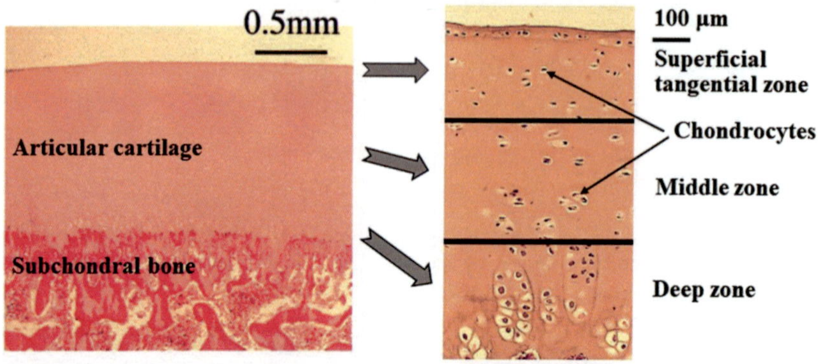

Figure 3.6 Photomicrograph of cross-section for articular cartilage of porcine shoulder joint (HE staining). The morphology and distribution of chondrocytes are different in superficial tangential, middle, and deep zones.
Modified from Murakami, T. (2007). 2·2·2 Musculoskeletal system. In: *JSME mechanical engineers' handbook, design β8:Bioengineering, JSME* (pp. 48–57).

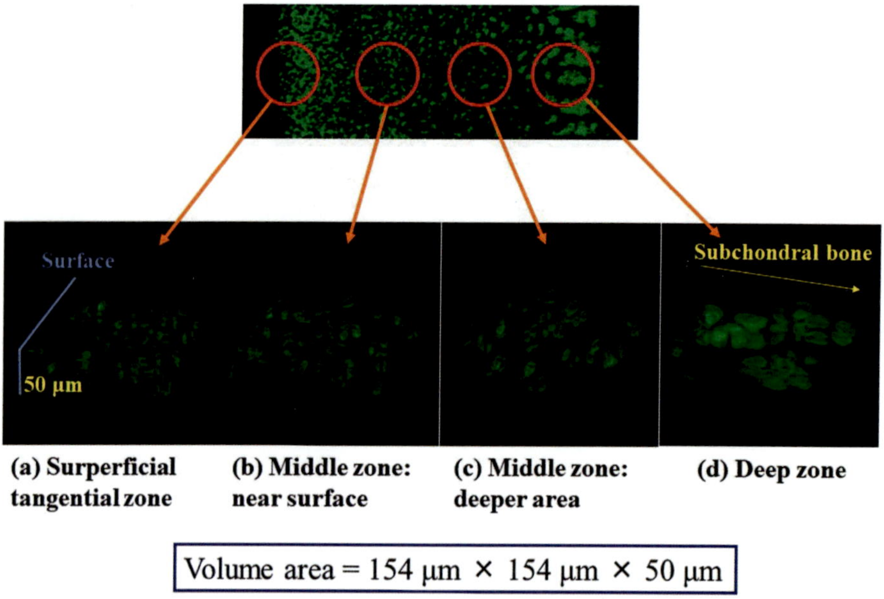

Figure 3.7 Morphological changes of chondrocytes in different zones in articular cartilage of porcine femoral chondyle observed by the confocal laser scanning microscopy. The 3D images of chondrocytes located in rectangular volume of 154 × 154 × 50 μm in each layer are shown, where the rendering with contour enhancement clarifies the difference in cell morphology.
Modified from Sakai, N., Sawae, Y., Murakami, T., Kurohara, Y., & Okamoto, M. (2005). The development of micro testing machine equipped with microscope and visualization method for investigation of cartilaginous tissue. *Japanese Journal of Clinical Biomechanics*, *26*, 59–63.

Most of chondrocytes in cartilage matrix have rounded or polygonal structure, but near articular surfaces the chondrocyte is flattened or discoid. Flattened cells are oriented parallel to the surface along the collagen fibers in the superficial zone. In the middle zone, chondrocytes are large and more rounded and have a random distribution, in which the collagen fibers also are more randomly arranged. In the deep zones, chondrocytes form columns that are oriented perpendicular to the subchondral bone, along with the collagen fibers. Different responses by chondrocytes may be exhibited depending on their position within the different layers under compressive or shearing loading conditions.

3.3 Synovial fluid

In the synovial joints, opposing articular cartilage surfaces are rubbed with extremely low friction and low wear under lubricated conditions with synovial fluid preserved within the joint cavity which is enclosed with joint capsule (Fig. 3.8). The synovial fluid normally functions as a biological lubricant as well as a biochemical pool through which nutrients and regulatory cytokines traverse (Blewis et al., 2007). The actual volume and composition of synovial fluid varies from joint to joint. The synovial fluid is supplied as the dialysate from the blood capillaries to which hyaluronic acid (hyaluronan) (Ogston and Stanier, 1953), lubricin or proteoglycan 4 (PRG4) (Swann et al., 1985), and surface-active phospholipids (Hills, 1989) are added. Synovial fluid as an ultrafiltrate of blood plasma contains many of globular proteins such as albumin and globulin, but the total protein level is less than that in blood plasma, about one-thirds. In Table 3.1 (Sasada et al., 1988), the composition of healthy human synovial fluid is shown in comparison with that of human blood plasma. The fibrinogen appears to be absent in synovial fluid. In normal synovial fluid, hyaluronic acid at 0.25–0.50 g/

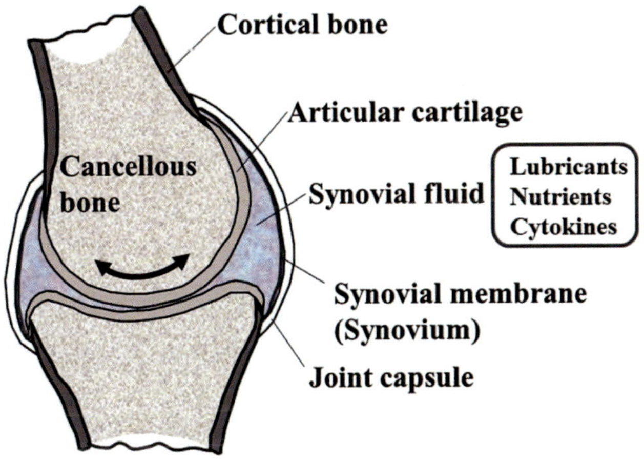

Figure 3.8 Simplified model of synovial joint. Constituents of synovial fluid play roles as lubricants, nutrients, and cytokines.

Table 3.1 Comparison of compositions of human synovial fluid and blood plasma.

	Synovial fluid	**Blood plasma**
Total proteins	1–3 g/dL	6–8 g/dL
Albumin	55%–70%	50%–65%
α_1-Globulin	6%–8%	3%–5%
α_2-Globulin	5%–7%	7%–13%
β-Globulin	8%–10%	8%–14%
γ-Globulin	10%–14%	12%–22%
Hyaluronic acid	0.25%–0.50% (0.25–0.5 g/dL)	–
Glucose	70–110 mg/dL	70–110 mg/dL

Modified from Sasada, T., Tsukamoto, Y., & Mabuchi, K. (1988). *Biotribology – Friction and lubrication of synovial joints.*

dL, lubricin (PRG4) at 0.005–0.035 g/dL, and surface-active phospholipids at 0.01 g/dL are included. Hyaluronic acid is viscous agent to control the lubricating fluid film formation. Lubricin (PRG4) and phospholipids are important lubricating constituents in synovial fluid. Synoviocytes in synovium secrete hyaluronic acid and lubricin and are major sources of surface-active phospholipids. Chondrocytes in the superficial layer of articular cartilage also secrete lubricin. Here, pH of synovial fluid is 7.30–7.40, while pH of blood plasma is 7.38–7.44 (arterial blood) or 7.36–7.43 (venous blood).

The viscosity, concentration, and molecular weight of hyaluronic acid and concentration of proteins for normal synovial fluid are shown in Table 3.2. It is noticed that the decrease in viscosity with age appears to be derived from the decrease in concentration of hyaluronic acid, while the molecular weight maintains the similar level. However, in synovial joints of patients with osteoarthritis, traumatic arthritis or rheumatoid arthritis, the viscosity remarkably decreases due to both lowering of the concentration and the molecular weight of hyaluronic acid.

The synovial fluid is a viscous/viscoelastic fluid in which molecular weight and concentration of hyaluronic acid play important role in non-Newtonian viscous behaviors. Hyaluronic acid or hyaluronan is a major constituent of synovial fluid and a hydrophilic polymer of disaccharides composed of D-glucuronic acid and *N*-acetyl-D-glucosamine (Fig. 3.9). The shearing rate has a significant effect on the viscous property of synovial fluids (Fig. 3.10) (Murakami, Higaki, et al., 1997). At low shearing rate for squeeze action under loading process where articular cartilage surfaces are approaching, high viscosity of synovial fluid suppresses the close approaching. For wedge action under walking conditions, the shearing rate in synovial joints of lower limb is expected to be ranged at 10^5 to 10^6 s^{-1} and the viscosity becomes very low, which suppresses the viscous resistance.

As shown in Fig. 3.10, an increase in temperature reduces viscosity, but the increased pressure (up to 10 MPa) has little or no effect on the viscosity (Cook, 1974). Furthermore, it is noted that the healthy human synovial fluid is significantly viscous than

Table 3.2 Viscosity, concentration and molecular weight of hyaluronic acid and concentration of proteins for normal synovial fluid.
Decrease in viscosity with age corresponds to decrease in concentration of hyaluronic acid.

Age	N	Viscosity	Concentration of hyaluronic acid	Molecular weight of hyaluronic acid	Concentration of proteins
		mPa·s	mg/mL		mg/mL
		At 3.84 s^{-1}			
17–19	7	2809	4.85	3.90×10^6	24.7
20–29	9	2409	3.91	4.09	27.2
30–39	6	2397	3.7	4.09	20.6
40–49	5	1997	2.77	4.01	27.4
50–59	4	1564	2.72	3.87	39.6
60–69	2	1357	3.74	4.08	26.4
70–79	5	1178	2.68	4.00	32.1
Average		1992	3.65	3.96	26.7

Modified from Sasada, T., Tsukamoto, Y., & Mabuchi, K. (1988). *Biotribology – Friction and lubrication of synovial joints*. Sangyo Tosho Publishing Co., Ltd.

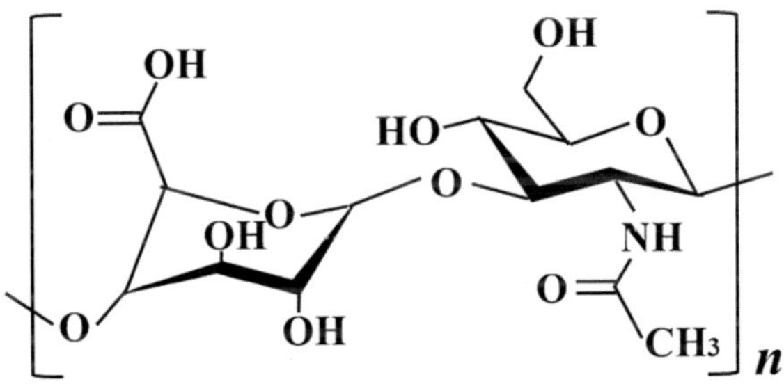

Figure 3.9 Structure of hyaluronic acid. Hyaluronic acid is a polymer of disaccharides composed of D-glucuronic acid and N-acetyl-D-glucosamine.

that of porcine synovial fluid, since the concentration of hyaluronic acid in human synovial fluid (0.25–0.5 g/dL) is higher than porcine synovial fluid (0.12 g/dL) although not so large difference in molecular weight is confirmed as $(3.9–4.1) \times 10^6$ for human to 4.77×10^6 for porcine. In diseased human joints with osteoarthritis or rheumatoid arthritis, the viscosity of synovial fluid is decreased, probably due to reduction of concentration and/or molecular weight of hyaluronic acid. The injections of hyaluronic acid as medical treatment have been purported to affect the biology in order to restore the viscosity and the protective functions of the synovial fluid.

In cases where the evaluation of lubrication modes in natural and artificial joints, the appropriate simulated synovial fluids are required. Fig. 3.11 shows the viscous

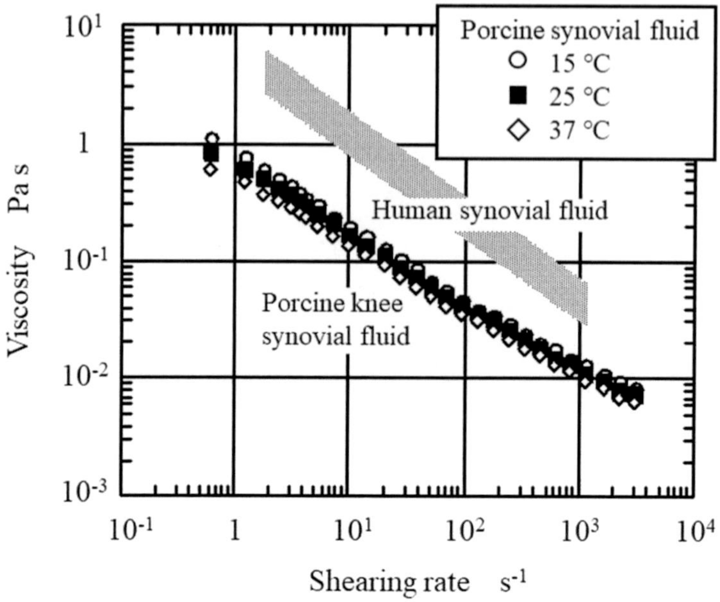

Figure 3.10 Viscous properties of synovial fluids. The viscosity of healthy human synovial fluids shows remarkable non-Newtonian behavior. Measured values for viscosity of porcine knee synovial fluid in cone-plate viscometer at different temperatures are plotted, which exhibit lower level than human synovial fluids.
Modified from Murakami, T., Higaki, H., Ando, H., & Nakanishi, Y. (1997). Viscous property of synovial fluids and related solutions and its role in joint lubrication. *Transactions of the Japan Society of Mechanical Engineers, Part C, 63*(607), 750−756. https://doi.org/10.1299/kikaic.63.750

properties of hyaluronate (HA) solutions with or without proteins. It is noted that HA solution containing albumin alone has significantly lower viscosity than HA solution, which may be caused by the interaction between HA and albumin molecules. On the contrary, the addition of γ-globulin alone or both albumin and γ-globulin has little effect on viscosity of these HA solutions. The synovial fluid is HA solution containing various constituents such as serum proteins, glycoproteins, lipids, and so on (Table 3.1), and is expected to maintain the similar viscous property to HA solution.

3.4 Lubrication mechanism in natural joints

3.4.1 Diversity of lubrication modes and low friction properties

The healthy natural synovial joints have excellent load-bearing characteristics with extremely low friction, high load-carrying capacity, good shock absorption capability, smooth movement with stability and minimum wear (long durability) even in heavily

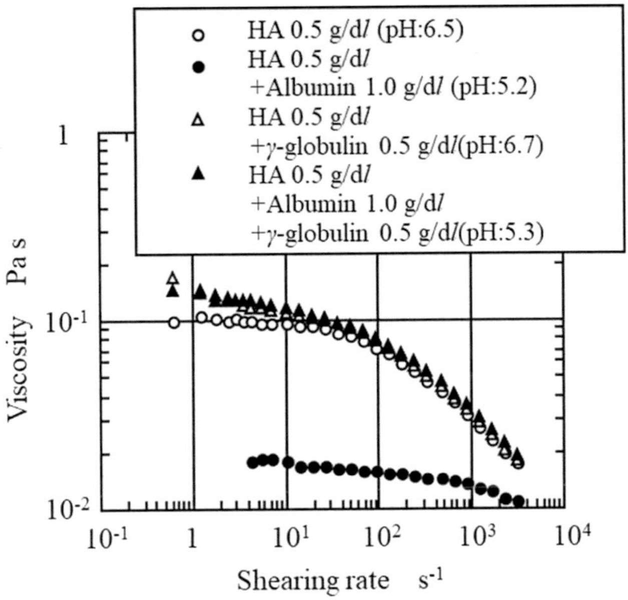

Figure 3.11 Viscous properties of hyaluronate solutions with or without proteins. HA solution containing albumin alone has significantly lower viscosity than HA solution. The addition of γ-globulin alone or both albumin and γ-globulin has little effect on viscous property.
Modified from Murakami, T., Higaki, H., Ando, H., & Nakanishi, Y. (1997). Viscous property of synovial fluids and related solutions and its role in joint lubrication. *Transactions of the Japan Society of Mechanical Engineers, Part C, 63*(607), 750–756. https://doi.org/10.1299/kikaic.63.750

loaded hip, knee, and ankle joints. These superior functions of natural synovial joints are likely to be maintained by the ingenious and robust lubrication mechanisms. As previously mentioned on friction and lubrication for lubricated rubbing pairs in Chapter 1, the differences of lubrication modes depending on the severity of operating conditions can remarkably change the friction level as indicated by Stribeck curve (Fig. 1.8 and Fig. 3.12). With increasing of load, lowering of rubbing speed, and lowering of lubricant viscosity, the fluid film becomes thinner, which enhances the occurrence of direct contact between rubbing surfaces. Thus, the lubrication modes and corresponding friction levels change depending on the relative ratio of fluid film thickness to surface roughness. In popular industrial rubbing pairs, friction at thin film condition with considerable direct contacts between rubbing surfaces covered with adsorbed films in boundary lubrication regime shows high level, but friction at thick fluid film condition without direct contact between rubbing surfaces in fluid film lubrication regime exhibits lower one caused by small fluid shearing resistance. In mixed lubrication regime, the transition of friction level is observed. On the contrary, friction levels in healthy natural synovial joints in aqueous-based lubrication appear to sustain the extremely lower range in wider operating conditions as shown

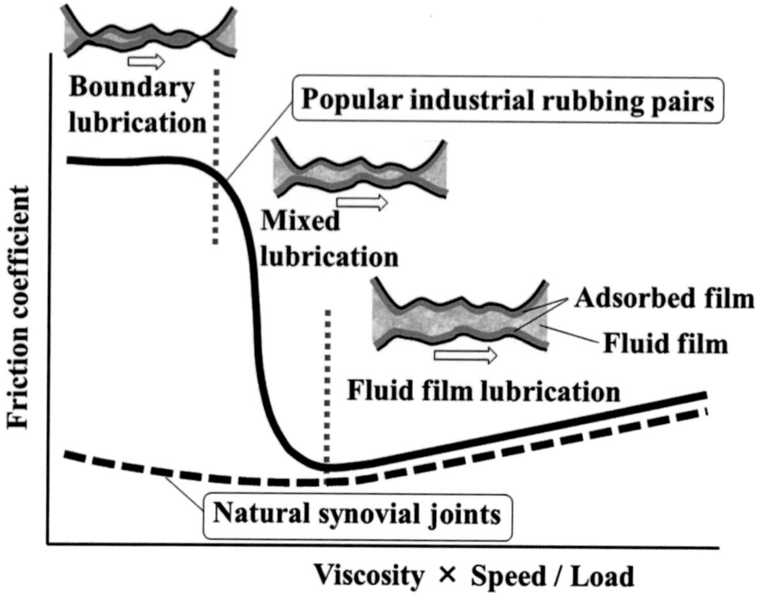

Figure 3.12 Stribeck curves for popular industrial rubbing pairs and natural synovial joints. For popular industrial rubbing pairs, friction remarkably changes depending on the corresponding lubrication modes from fluid film lubrication to boundary lubrication. On the contrary, in natural synovial joints friction maintains lower level for wider operating conditions, where the effective lubrication mechanisms are discussed in this chapter.

in Fig. 3.12. Distinct differences in their frictional behaviors suggest that the effective lubrication mechanisms are different between popular industrial rubbing pairs and natural synovial joints. How the various lubrication modes become effective in natural synovial joints are discussed in this chapter.

It has been well known that the synovial joints show extremely low friction in various daily activities, and different types of studies have been performed to elucidate the lubrication mechanism in natural synovial joints.

Reynolds who established the fundamental theory of fluid lubrication mechanism did not examine the hydrodynamic pressurization and friction in natural synovial joints, but suggested the importance of self-acting hydrodynamic lubrication by squeeze action as common phenomena in both steam engines and animal synovial joints in his paper (Reynolds, 1886) as follows:

> *The only other self-acting system of lubrication is that of reciprocating joints with alternate pressure on and separation (drawing the oil back or a fresh supply) of the surfaces. This plays an important part in certain machines, as in the steam engine, and is as fundamental to animal mechanics as the lubricating action of the journal is to mechanical contrivances.*

After many years since Reynolds' suggestion of the importance of the squeeze film action in animal joint lubrication, the investigations which recognized the importance of the hydrodynamic pressurization as the wedge film action in natural synovial joints were reported from the viewpoint of fluid film lubrication by MacConaill (1932) and Jones (1936). MacConaill indicated in his paper with special reference to the knee and inferior radioulnar joints that intraarticular fibrocartilages act to bring about the formation of convergent wedge-shaped films of synovia for the weight transmitting parts of joints in movement. Here, these wedges are narrowed in the direction of motion. Jones evaluated the frictional behaviors of a human amputated proximal interphalangeal finger joint as a fulcrum in a pendulum machine and noticed that the decay in swing amplitude was exponential. He indicated that this damping is caused by viscous damping in fluid film lubrication. In addition, he mentioned that the pressure-film is dependent for its formation upon (a) viscosity of lubricant, (b) speed of surfaces, and (c) eccentricity of surfaces. Furthermore, he described the occurrence of some form of solid friction under insufficient speed and/or eccentricity, and friction increase due to viscosity rising at lower temperature.

To these prevailing concepts of the hydrodynamic lubrication in natural synovial joints, Charnley (1959, pp. 12–19), (Charnley, 1960) contested on the basis of the experimental results for pendulum tests of cadaveric human ankle joints as high congruity joint without ligament and capsule which showed linear decay of a pendulum and pointed out the predominance of boundary lubrication in synovial joints. Before experimental examination of frictional behaviors of animal joints, he indicated the unlikely situation of hydrodynamic lubrication caused by (1) difficulty in the existence of wedge-shaped spaces due to the large deformation of resilient cartilage, (2) difficulty in the wedge-shaped fluid film formation in reciprocating motion, and (3) difficulty in the hydrodynamic fluid film formation with slow-moving surfaces under heavy loads. Furthermore, he gave comments on Jones' friction tests as the addition of resistance by retained ligaments and capsule at large amplitudes and the less resistance with rolling motion at low amplitudes in proximal interphalangeal finger joint, which has imperfect geometrical congruity and some rolling-sliding motion can occur in flexion-extension.

His viewpoint of boundary lubrication mechanism in natural joints was reflected to the selection of rubbing materials such as polytetrafluoroethylene (PTFE) and ultra-high molecular weight polyethylene (UHMWPE) at developing process of artificial joints, as described in detail in Chapter 4.

As an alternative lubrication mechanism of boundary lubrication for low friction (friction coefficient <0.01 to 0.02) in natural synovial joints, McCutchen (1959) proposed the weeping lubrication, or self-pressurized hydrostatic lubrication, on the basis of time-depending frictional behaviors of sliding pairs for the cut surface of closed cell sponge against glass plate and sausage casing on sponge lubricated with soapy water. He indicated the effectiveness of self-pressurized hydrostatic lubrication associated with deformation of microporous cartilage at intermittent loading on natural joints in daily activities. Lewis and McCutchen (1959) exhibited the experimental evidence for weeping lubrication in mammalian joints by the detection of fluid exuded from articular cartilage specimen at contact when the pressure is applied. The possibility of fluid exudation from cartilage surface in actual operating joint is discussed later in this chapter.

After that, it was shown in friction test for cartilage against glass plate at constant load that initial low friction and rise of both friction and deformation and lowering in friction after resoaking in lubricant at unloading (McCutchen, 1962). It was proved in his formulation that the liquid pressure in the cartilage matrix is less than that outside. Therefore, enough liquid can be imbibed in cartilage with fluid flow driven by the pressure gradient to reduce friction in appropriate resoaking time. Here, the mechanism how to weep from cartilage matrix in rubbing cartilage surfaces at conjunction in actual immersed state was not discussed.

McCutchen indicated the importance of weeping lubrication and suggested that the boundary lubrication with synovial fluid is a supplemental lubrication mechanism in natural synovial joints (McCutchen, 1966-67).

As mentioned in Chapter 1, fluid film lubrication is categorized as the hydrostatic lubrication and the hydrodynamic lubrication. Weeping lubrication is one of the hydrostatic lubrications. In the hydrodynamic lubrication, the pressurization is mainly brought by the wedge action and squeeze action. In natural synovial joints, the normal loading changes during walking, running and other daily activities, and it is expected that the squeeze film lubrication can play a predominant role at particularly slow moving or little movement activities. When rubbing surfaces are elastically deformed, the load-supporting area is enlarged and then contact pressure is reduced, and the fluid film formation is improved as known as elastohydrodynamic lubrication (EHL).

Dintenfass (1963) pointed out the importance of the viscoelastic properties of synovial fluid and the resiliency and time-dependent and stress-dependent elasticity of articular cartilage in natural joint lubrication. Tanner (1966) indicated the effect of elastic deflection of the cartilages in joint lubrication.

Fein Fein (1966-67) showed the effectiveness of a squeeze film of synovial fluid between compliant (cartilage) surfaces based on the analysis and experiments of the squeeze films at loading phase. The hydrodynamic replenishment through entrainment of fluid at the moving phase was mentioned.

Thus, the articular cartilage materials as rubbing surfaces of natural joints are compliant, and elastically deformed at loading which can extend the load support area in rubbing surfaces and facilitate to form thicker fluid film as soft-elastohydrodynamic lubrication (soft-EHL) in natural synovial joints.

As mentioned previously, various lubrication mechanisms had been proposed for the superior lubrication performances of natural synovial joints in various daily activities, but the full consensus on the actual lubrication mechanism had not been obtained. On this subject, Dowson (1966-67) indicated that "the major lubrication mechanism would seem to be some form of elastohydrodynamic action determined by sliding or squeeze film action between porous surfaces with boundary lubrication providing the surface protection in cases of severe loading and little movement." He pointed out that this superior lubricating performance appears to be actualized by not single lubrication mode but the synergistic combination of various modes from the fluid film lubrication to boundary lubrication corresponding to the severity of rubbing conditions.

On the changes in properties of lubricants, Maroudas (1966-67) suggested the gel formation of concentrated synovial fluid collected on the cartilage surface, and Walker et al. (1968) proposed "boosted lubrication" mechanism with fluid entrapment and enrichment, in which squeeze-film action leads to a concentration of hyaluronic acid-protein complex in the lubricant as a result of diffusion of water and low molecular weight substances through the porous cartilage and the restricted gap between the approaching cartilage surfaces. The effectiveness of boosted lubrication in squeeze film action was supported with the analysis and calculations (Dowson et al., 1970).

The detailed features of lubrication mode in daily activities were evaluated in the following investigations.

For example, it was shown that the lubrication modes may change for flexion-extension motion during normal walking motion composed of the loading stance phase with peak loads of about three to four times body weight and the swing phase with medium flexion at light load.

Unsworth et al. (1975) suggested on the basis of pendulum tests of human hip joints that the lubrication modes change from the elastohydrodynamic lubrication mode with squeeze fluid film action at heel strike and entraining action during load-bearing stance phase followed by short time boundary lubrication after squeeze action at toe off to the fluid film lubrication during swing phase. Thereafter, Roberts et al. (1982) measured the frictional torque of cadaveric hip joints in hip function simulator for walking motion and observed that friction at high load was maintained as low due to squeeze film effect even with very low viscosity condition while the transition from mixed lubrication to full fluid film lubrication was observed at low load at viscosity range of 0.025–0.1 Pa s for synovial fluid.

For evaluation of the possibility of full fluid film lubrication in natural joint, the estimation of fluid film thickness is required. Medley et al. (1984) estimated the changes in fluid film thickness and friction coefficient for the ankle joint (composed of tibial and talar surfaces) model as the partial journal bearing with compliant surface layer, i.e., cylinder-on-plane model during walking.

In their computational analysis of soft-EHL considering elastic deformation of soft cartilage layer, the lubricant flow was described by the Reynolds equation with both wedge and squeeze terms retained. In their analysis, the minimum film thickness throughout walking cycle remained reasonably constant at about 0.7 μm for typical simulated condition. This film thickness was not quite larger than surface roughness of articular cartilage. This result suggests the difficulty of full fluid film lubrication in ankle joint during walking, although the fluid film lubrication would be actualized if the rubbing cartilage surfaces have smooth surface. The detailed results are described in the next section.

On the possibility of fluid film lubrication during walking, Dowson and Jin (1986) applied the micro-elastohydrodynamic lubrication (micro-EHL) concept to ankle synovial joint as simplified model covered with soft elastic layer of sinusoidal surface roughness during walking, where they found that the surface roughness itself causes local pressure perturbations to occur which flatten the initial surface asperities. Thus, it was shown that higher ratio of the minimum fluid film thickness to the reduced surface roughness in loaded conjunction zone supports the possibility of full fluid film lubrication. In their analysis, however, the flow in articular cartilage is excluded.

The fluid flow between articular cartilage and joint cavity was evaluated by Mansour and Mow (1977) from the viewpoint of biphasic lubrication. They indicated that the interstitial fluid is exuded in front and near the leading half of loading conjunction zone and imbibed behind and near the trailing half.

In a numerical analysis of deformation and lubrication in a hip joint under step load by Ikeuchi et al. (1989), Ikeuchi (1995), the flows in both the articular cartilage and the gap between cartilages were considered. In their analysis, the local contact points appear in a circumferential zone while high pressure fluid was trapped at the center of loading zone just after step loading, and an outward flow to the gap from the cartilage at contact point was caused by a significant pressure gradient near the surface accompanied with creep deformation. The squeeze film behavior with creep deformation and outward flow from cartilage contribute to the improvement in the conformity and the lubrication. These behaviors were confirmed by experiment (Ikeuchi, 1995; Ikeuchi, Oka, & Gi, 1989). Furthermore, it was confirmed that no creep deformation is detected when the cartilage is pressed by fluid alone, which suggests that the cartilage can be simplified as a nonporous elastic material under full fluid film lubrication condition.

On boundary lubrication mechanism, Radin et al. (1970), Swann et al. (1981) clarified the molecular structure of lubricating glycoprotein as predominant boundary lubricant, and Hills (1989) indicated the effective lubrication by phospholipids.

As described above, in natural synovial joints where both the articular cartilage as biphasic compliant bearing material and the synovial fluid containing effective lubricative constituents play the important roles in sustaining low friction and minimal wear, various lubrication modes are likely to function depending on the severity of operating conditions in various daily activities. This superior lubrication mechanism was called "adaptive multimode lubrication" based on summarized review of related studies by Murakami (1990). This ingenious lubrication mechanism in natural synovial joints have been investigated from the various viewpoints of the adaptive multimode lubrication mechanism (Murakami, 2012; Murakami et al., 1998, 2011, 2014).

In the following sections, the important features in each lubrication mechanism are discussed.

3.4.2 Fluid film lubrication

In fluid film lubrication, in both hydrostatic and hydrodynamic lubrication mechanisms, the pressurized fluid supports the applied load and prevents the direct contact between rubbing surfaces. The viscous resistance due to fluid shearing in fluid film lubrication is extremely minute compared with the shearing resistance of solid-solid contacts in non-fluid film lubrication. Therefore, friction is maintained as very low level in fluid film lubrication. In addition, the prevention of direct contacts between rubbing surfaces by the fluid load support can contribute to the minimization of surface failures such as wear and rupture, and thus bring about longer durability. Establishment of fluid film lubrication is a desirable lubrication mode for natural synovial joints. In natural joints as rubbing pair of compliant articular cartilage surfaces, soft-elastohydrodynamic lubrication (soft-EHL) is likely to enhance fluid film formation.

For lubricated rubbing pair, the possibility of full fluid film lubrication is generally evaluated by comparison of the minimum fluid film thickness and surface roughness. For most cases, the film parameter (lambda ratio): Λ as the ratio of the minimum fluid film thickness: $h_{\min}$ to the composite surface roughness as root mean square: σ is applied as the critical parameter to evaluate the effectiveness of the EHL for rubbing of rough surfaces (Johnson et al., 1972; Tallian, 1967).

$\Lambda \geq 3$; full fluid film lubrication.

$1 \leq \Lambda < 3$; mixed lubrication.

$\Lambda < 1$; boundary lubrication or mixed lubrication with significant contacts

$$\text{where } \Lambda = \frac{h_{\min}}{\sigma} \tag{3.13}$$

$$\sigma = \sqrt{\mathrm{Sq}_1^2 + \mathrm{Sq}_2^2} \tag{3.14}$$

Sq_1, Sq_2: Root mean square roughness.

As mentioned above, the soft-EHL mechanism is likely to be essential to sustain the fluid film lubrication in natural synovial joints with compliant articular cartilage surfaces lubricated with synovial fluid.

The theoretical analysis for two-dimensional model for ankle joint was conducted by Medley et al. (1984). They represented the ankle joint as a partial bearing with compliant cartilage surface layer as a cylinder-on-plane configuration (Fig. 3.13).

The lubrication flow with pressurization was described by the Reynolds equation with both wedge and squeeze terms retained. In the right-hand side of the following equation, the first and second terms correspond to wedge and squeeze terms, respectively.

$$\frac{\partial}{\partial x}\left(h^3 \frac{\partial p}{\partial x}\right) = 12\eta_0\left(u\frac{\partial h}{\partial x} + \frac{\partial h}{\partial t}\right) \tag{3.15}$$

where.

h: Film thickness

$$h = h_0 + \psi(x, t) \tag{3.16}$$

h_0: Minimum film thickness,

$\psi(x,t)$: Surface profile function including the elastic deformation caused by the film pressure

p: Film pressure

u: Entrainment velocity, $U_1/2$

η_0: Lubricant viscosity under ambient conditions

x: Spatial coordinate

t: Time.

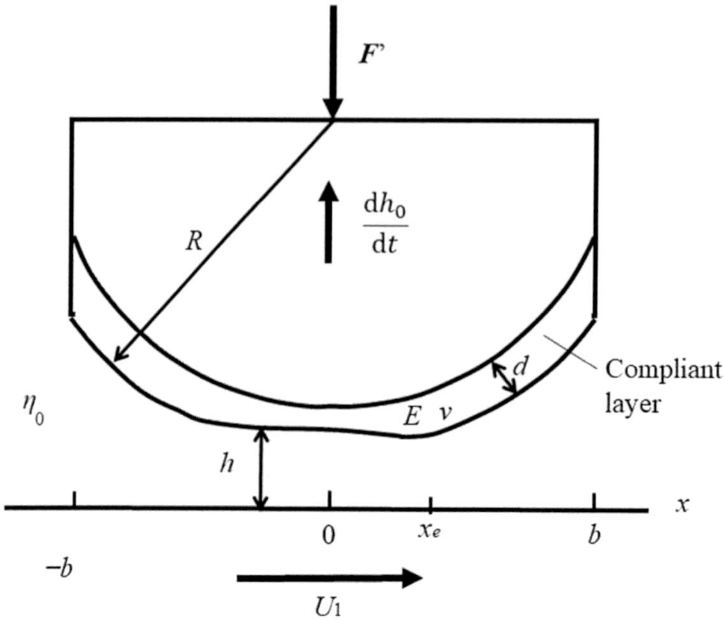

Figure 3.13 An equivalent bearing model representing the human ankle joint. Cylinder-on-plane model, R: Reduced radius of curvature, d: Thickness of compliant layer, F': Applied load per unit width, h: Film thickness, b: Half width of equivalent bearing, x_e: Cavitation boundary, U_1: Surface velocity, η_0: Lubricant viscosity under ambient conditions, E: Elastic modulus of compliant layer, ν: Poisson's ratio.
From Medley, J. B., Dowson, D., & Wright, V. (1984). Transient elastohydrodynamic lubrication models for the human ankle joint. *Engineering in Medicine*, *13*(3), 137−151. https://doi.org/10.1243/EMED_JOUR_1984_013_035_02

The loads and entrainment velocities for the analysis are shown in Fig. 3.14a, where F'_A: time averaged load per unit width for one cycle, and u_A: time averaged entrainment velocity for one cycle.

$$u = \frac{\pi u_A}{2}\left|\cos\left(\frac{2\pi t}{t_p}\right)\right| \tag{3.17}$$

where

t_p: Period of cycle.

Changes in dimensionless minimum film thickness: $H_0 = h_0/R$ versus dimensionless time: $T = t/t_p$ under cyclic steady state conditions for normal walking, running-1 and running-2 cases are shown in Fig. 3.14b.

The values of the input parameters for walking condition are as follows:

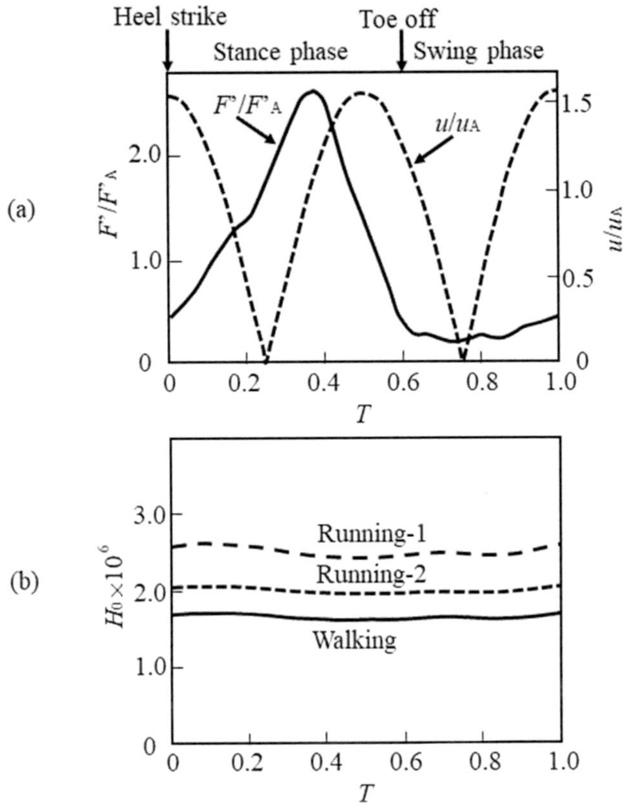

Figure 3.14 The load and entrainment velocity and dimensionless minimum film thickness. (a) Dimensionless load and entrainment velocity for one cycle. (b) Dimensionless minimum film thickness for walking and running conditions.
Modified from Medley, J. B., Dowson, D., & Wright, V. (1984). Transient elastohydrodynamic lubrication models for the human ankle joint. *Engineering in Medicine*, *13*(3), 137–151. https://doi.org/10.1243/EMED_JOUR_1984_013_035_02

$R = 0.35$ m, $d = 2.4$ mm, $t_p = 1.0$ s, $u_A = 19.1$ mm/s, $F'_A = 33.7$ kN/m, $\eta_0 = 0.01$ Pa s, $E' = 42.7$ MPa, $\nu = 0.5$.where E': Reduced elastic modulus.

$$E' = 2E/\left(1 - \nu^2\right) \tag{3.18}$$

On the input parameters for running-1 condition, cycle period and velocity conditions ($t_P = 0.5$ s, $u_A = 38.2$ mm/s) were different from walking condition. For running-2 conditions, furthermore, load was changed as $F'_A = 67.5$ kN/m. Actual running seems to exist between running-1 and running-2.

It is noticed in Fig. 3.14 that the percentage changes in minimum film thickness throughout the cycle are small for both walking and running conditions, although the values of film thickness at running conditions are thicker than walking condition. Calculated values for minimum film thickness h_0 during the one cycle for walking, running-1, and running-2 are 0.566 μm, 0.852 μm, and 0.690 μm, respectively. These results mean that the full fluid film lubrication can play a main lubricating role for completely smooth rubbing cartilage surfaces, but for actual cartilage surfaces with significant surface roughness with height larger than 1 μm (Fig. 3.5), considerable local solid-solid contact is likely to occur.

In this study, the prediction of minimum film thickness for the simplified model of plane inclined surface geometry was compared with the full dynamic solution. It is noticed that both values of the minimum film thickness showed similar ones although pressure and film profile were a little different.

In related numerical analyses (Murakami et al., 1998; Sawae et al., 1996), the changes in fluid film formation in simplified knee joint model neglecting the existence of menisci were evaluated by applying the plane inclined (PI) surface model proposed by Medley et al. (1984), Gladstone and Medley (1990), where the compliant surface was assumed to be elastically deformed as a PI surface. Fig. 3.15 shows the changes in minimum film thickness in simplified natural knee joint model composed of a cylindrical femoral condyle of 30 mm radius with a soft layer of 4 mm thickness and a tibial flat surface, lubricated with Newtonian lubricants with different viscosity. In this case, the fluid film thickness gradually decreases with the suppression by squeeze action during the high loading stance phase and then recovers to a thicker level mainly due to a wedge action producing an entraining effect during the light loading swing phase. Peak friction coefficients estimated as fluid shearing resistance at maximum speed around 7% walking phase were 0.0014 and 0.0004 for 0.01 Pa·s and 0.001 Pa s, respectively. The distinct changes in film thickness during walking were confirmed but the values of minimum film thickness were not fully thicker than surface roughness of cartilage, which suggests the difficulty of full fluid film lubrication. As further evaluation, the consideration from the viewpoint of micro-elastohydrodynamic lubrication is required.

To elucidate the actual lubrication mode in loading conjunction in natural synovial joint, Dowson and Jin (1986) focused the changes in surface profile of the initially rough cartilage surfaces as they pass through the loaded region. In a cylinder-near-a-plane model for ankle joint, the combination of a cylinder with compliant layer and a perfectly smooth rigid solid plane was used. The surface roughness was represented as transverse, sinusoidal waves of amplitude a and wavelength λ (Fig. 3.16). Therefore, the local film thickness h is given by

$$h = h_0 + \frac{x^2}{2R} + a\cos\left(\frac{2\pi x}{\lambda} + \phi\right) + \delta \quad (3.19)$$

where δ is the local elastic deformation of compliant layer surface, which is estimated by summing the deformation for smooth surface δ_s and one for surface roughness δ_r.

$$\delta = \delta_s + \delta_r = C_s p_s + C_r \Delta p \quad (3.20)$$

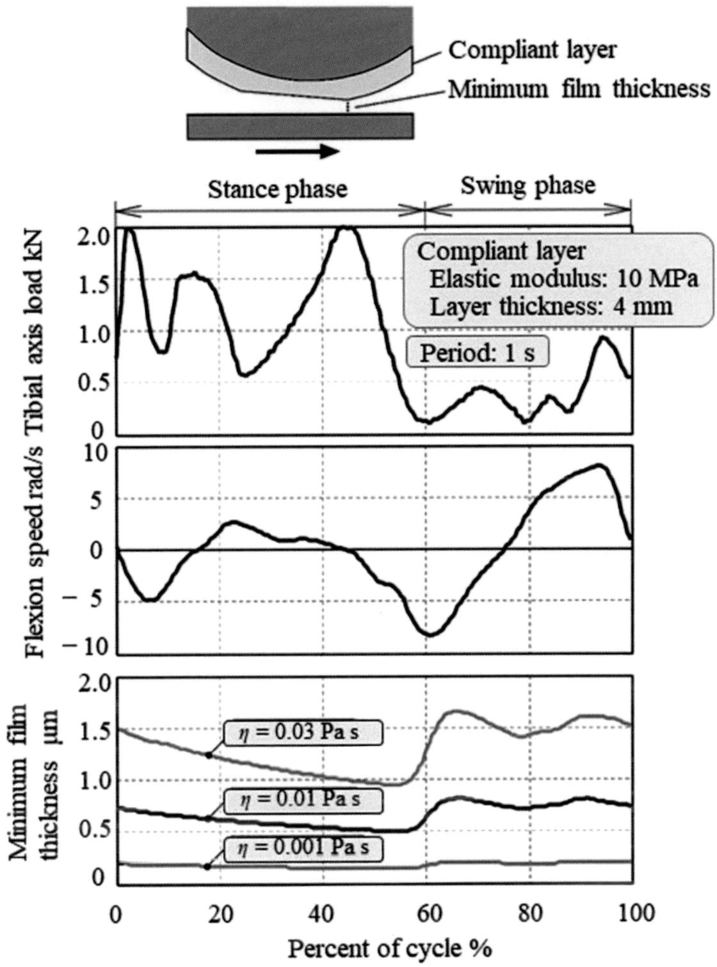

Figure 3.15 Changes in minimum film thickness in knee joint model during walking. It is noticed that the fluid film attains the minimum around toe off and recovers during swing phase. Modified from Murakami, T. (2007). 2·2·2 Musculoskeletal system. *JSME mechanical engineers' handbook, design β8:Bioengineering, JSME* (pp. 48–57).

where C_s and C_r represent smooth and rough cartilage elastic compliances, respectively.

The hydrodynamic pressure p is written in combination of the dry contact pressure p_s for a smooth surface and a small perturbation Δp.

$$p = p_s + \Delta p \tag{3.21}$$

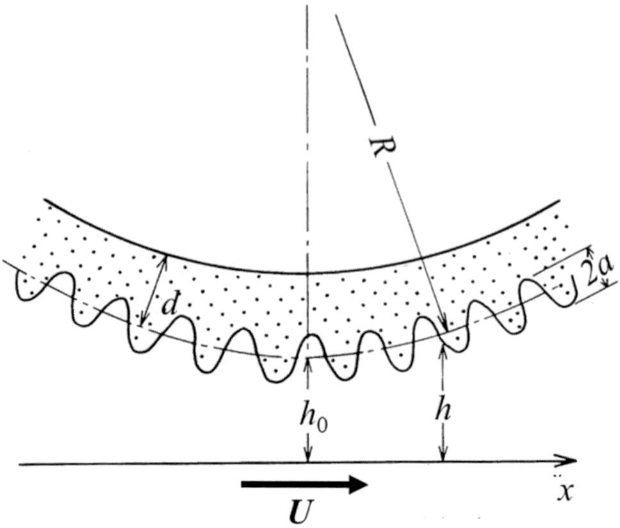

Figure 3.16 Micro-elastohydrodynamic model of a synovial joint with rough compliant layer on rigid cylindrical backing adjacent to a rigid sliding plane. The surface roughness of compliant layer is represented as transverse, sinusoidal waves.
Modified from Dowson, D., & Jin, Z.-M. (1986). Micro-elastohydrodynamic lubrication of synovial joints. *Engineering in Medicine, 15*(2), 63–65. https://doi.org/10.1243/EMED_JOUR_1986_015_019_02

Here, δ_s was calculated by adopting the constrained column model (Medley et al., 1984) because the mean thickness of the undeformed elastic layer d is generally small compared with the width of the loaded conjunction. The boundary condition for Reynolds equations is shown below:

$p = 0$ at inlet; $p = \frac{dp}{dx} = 0$ at outlet.

A micro-elastohydrodynamic lubrication solution for the pressure distribution, film shape, and deformed roughness in the ankle joint in steady walking for the representative conditions is shown in Fig. 3.17. It is noteworthy that the appropriate fluid film is formed to prevent the direct contact between rubbing surfaces in loaded conjunction area by smoothing the initially rough cartilage surfaces, which is achieved by modest perturbation to the smooth surface pressure profiles as the micro-elastohydrodynamic action.

In this case, the composite roughness as root mean square of undeformed surface σ is about 0.71 μm and the predicted minimum film thickness h_{min} is 0.68 μm, and then the film parameter is estimated as $h_{min}/\sigma = 0.96 \fallingdotseq 1$. Therefore, it is assessed in usual elastohydrodynamic analysis that considerable direct contact between rubbing surfaces can occur. However, it is indicated from the viewpoint of micro-elastohydrodynamic lubrication that the full fluid film lubrication is sustained with large smoothing of the initial cartilage surface roughness caused by micro-elastohydrodynamic action in natural synovial joints in steady walking (Fig. 3.17).

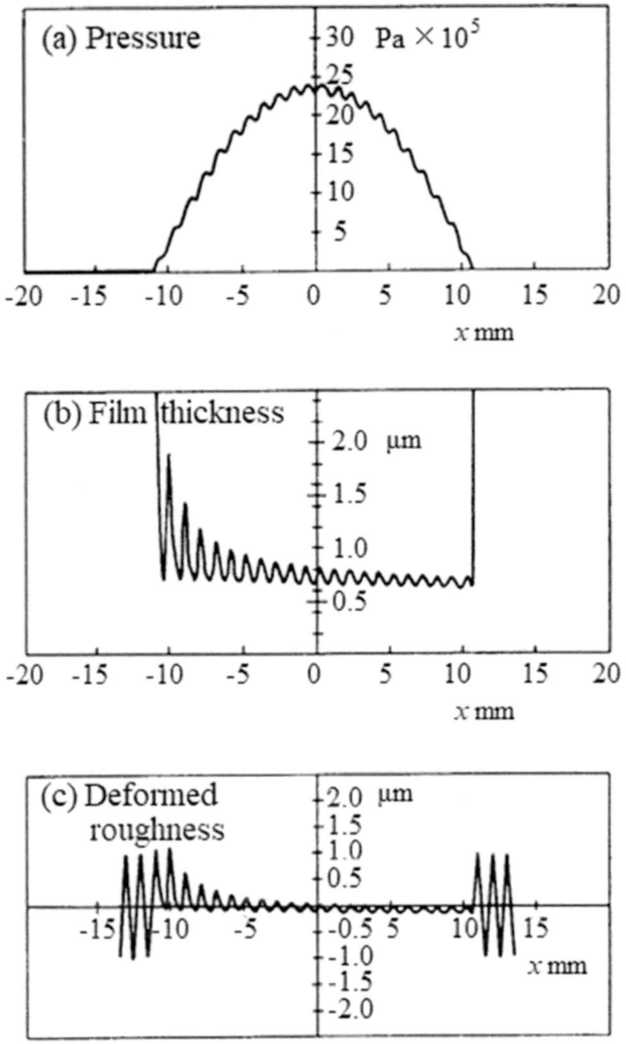

Figure 3.17 Micro-elastohydrodynamic solution for a synovial joint. Initial roughness amplitude: $a = 1$ μm, Wavelength: $\lambda = 1$ mm, Equivalent radius: $R = 0.35$ m, Mean total layer thickness: $d = 2.4$ mm, Elastic modulus: $E = 16$ MPa s, Poisson's ratio: $\nu = 0.4$, Lubricant viscosity: $\eta = 0.01$ Pa s, Sliding velocity: $U = 0.019$ m/s, Load per unit length: $F' = 33.7$ kN/m. (a) Pressure, (b) Film profile, (c) Deformed roughness.
Modified from Dowson, D., & Jin, Z.-M. (1986). Micro-elastohydrodynamic lubrication of synovial joints. *Engineering in Medicine*, *15*(2), 63–65. https://doi.org/10.1243/EMED_JOUR_1986_015_019_02

As discussed above, micro-elastohydrodynamic lubrication mechanism is likely to extend the fluid film lubrication mode in natural synovial joints in daily activities such as walking and other motion. However, some direct contacts may occur for thin film conditions at slower movement. In the next section, the effectiveness of the biphasic lubrication is described.

3.4.3 Biphasic lubrication

3.4.3.1 Features of biphasic lubrication

The concept of biphasic lubrication is originated from a new proposal for superior low friction in natural synovial joints even at operating conditions without sufficient hydrodynamic effects of an intervening lubricant fluid film. The articular cartilage has high water content from about 70% to 80% in tissue composed of type II collagen, proteoglycan, and chondrocytes, and thus exhibits a time-dependent biphasic behavior due to the simultaneous coexistence of solid and liquid phases (Mow et al., 1980, 1990). Furthermore, it is noted that the lubrication mode depends on the extent of exudation and rehydration in various activities. The load support by interstitial fluid pressure in biphasic cartilage controls to reduce friction and deformation.

Mansour and Mow (1977) indicated from the viewpoint of the biphasic lubrication that the normal healthy articular cartilage is capable of generating its own film of fluid lubricant at medium sliding speed of 25.4 mm/s to 76.2 mm/s. It was noticed for healthy cartilage with low permeability that the movement of interstitial fluid at the surface is circulatory, i.e., it is exuded in front and near the leading half of loading conjunction zone and imbibed behind and near the trailing half. In contrast, it was suggested that pathological cartilage with porous surface of high permeability and low stiffness cannot maintain fluid film formation. In their paper, the time-dependent frictional behavior with local direct contact and fluid film formation between cartilage surfaces at slow speed were not discussed.

After about two decades, Forster and Fisher (1996, 1999) examined the influence of loading time, unloading, and lubricant on the time-dependent friction of biphasic articular cartilage in sliding tests at low speed of 4 mm/s in mixed lubrication regime. They confirmed that the load carriage by the fluid phase becomes an important factor influencing friction, and the friction force was proportional to the load carried by the solid phase.

The theoretical formulation of a boundary friction model for biphasic articular cartilage was proposed and validated by Ateshian et al. (Ateshian, 1997, 2009; Ateshian et al., 1998; Krishnan et al., 2004). The frictional model is constructed on the basis of the biphasic theory for soft hydrated tissues as a mixture considering each volume ratio of solid and fluid phases proposed by Mow et al. (1980). In usual biphasic models, it is assumed that the interstitial fluid flow follows Darcy's law as shown below:

$$Q = k\frac{\mathrm{d}p}{\mathrm{d}h} \tag{3.22}$$

where Q is volume flow rate per unit cross-sectional area, k is permeability, p is pressure, and h is distance.

Thus, the pressure differences in the fluid phase cause flow. In biphasic materials, the permeability and the pressure gradient control the interstitial fluid pressurization, fluid flow, and deformation.

Furthermore, it was indicated that the Peclet (Pe) number as the ratio of convective velocity of sliding and the diffusive velocity of interstitial fluid flow for rubbing condition controls the effectiveness of interstitial fluid load support (Ateshian, 2009).

Here, only solid-to-solid contact is assumed to result in significant friction forces, while other interaction as solid-to-fluid is assumed to be negligible. In FE analysis, the effective friction coefficient μ_{eff} can be estimated for the friction coefficient μ_{eq} for solid-to-solid contact as equilibrium coefficient of friction achieved when the interstitial fluid pressure has subsided, using the following formula (Ateshian, 2009).

$$\mu_{eff} = \mu_{eq}\left(1 - (1-\phi)\frac{W^{p}}{W}\right) \tag{3.23}$$

where:

μ_{eq}: friction coefficient at equilibrium state when interstitial fluid pressure fully has subsided.

W: total load support.

W^{p}: interstitial fluid load support,

ϕ: fraction of contact area of solid phase, the solid-to-solid contact area fraction in microscopic view.

The value of ϕ was set as 0 under an assumption that the matrix was under completely mixed conditions and the direct contact solid surface in an intact state was small enough for nonsmooth cartilage surface.

3.4.3.2 Biphasic finite element analysis for articular cartilage during reciprocating motions

By applying the biphasic finite element (FE) analyses for articular cartilage in reciprocating motion as described by Sakai et al. (2012, 2016) and Murakami (2012), it was examined that the keeping of high ratio of load support by interstitial fluid pressurization due to very low permeability of cartilage tissue is capable of reducing friction even at slower speed (<10 mm/s) conditions.

The differences in biphasic behaviors were examined for two kinds of two-dimensional models (Fig. 3.18) (Murakami, 2012) in migrating and stationary loading conditions for cartilage specimens as follows:

a. Migrating contact (on-off loading) for the reciprocating biphasic cartilage flat plate specimen of 1.5 mm thickness loaded by cylindrical impermeable rigid specimen, and
b. Stationary contact (continuous loading) for the biphasic cartilage cylindrical specimen of 1.5 mm thickness layer against the reciprocating flat rigid plate.

Two-dimensional biphasic FE analyses were conducted using commercial package ABAQUS (6.8−4) (Pawaskar et al., 2007; Wu et al., 1997). The biphasic tissue was modeled by CPE4RP (four-node bilinear displacement and pore pressure, reduced

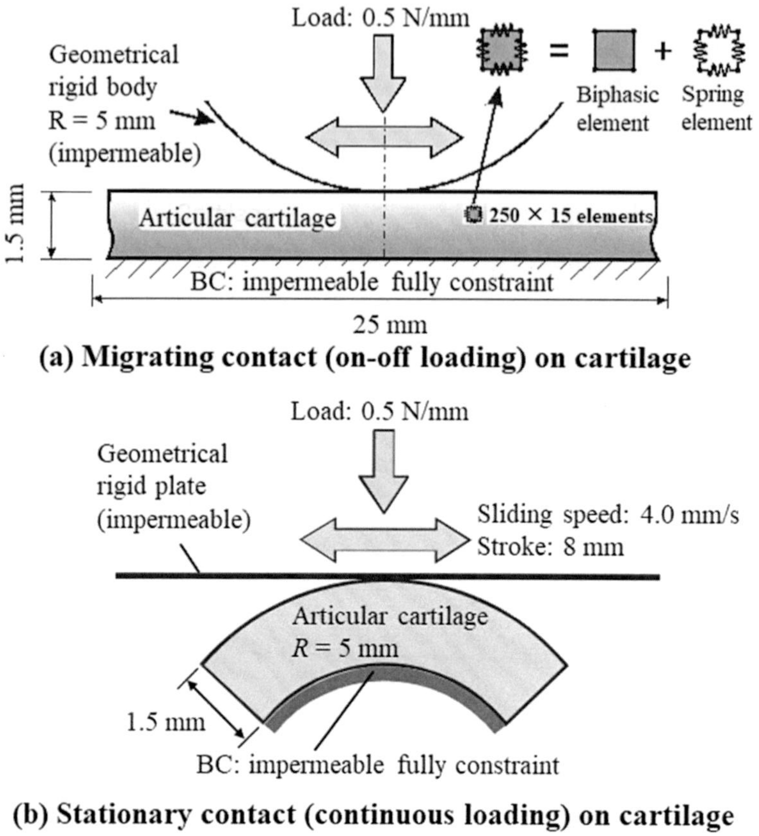

Figure 3.18 Finite element models for flat and cylindrical cartilage specimens. (a) In migrating contact on cartilage, the recovery of deformation with rehydration can occur during offloading phase. (b) In stationary contact on cartilage, the time-depending exudation of interstitial fluid can occur continuously in reciprocating motion.
Modified from Murakami, T. (2012). Importance of adaptive multimode lubrication mechanism in natural and artificial joints. *Proceedings of the Institution of Mechanical Engineers, Part J: Journal of Engineering Tribology*, *226*(10), 827−837. https://doi.org/10.1177/1350650112451377

integration with hourglass control) elements and the mesh size was chosen as 0.1 mm square. The horizontal and vertical fibrils were represented by spring element SPRINGA (axial spring between two nodes, whose line of action is the line joining the two nodes) of the software, in which the spring elements were configured to generate reaction force only in the tensile direction. The stiffness K of the spring elements was simplified to the uniform value over the tissue and both in horizontal and vertical direction for the first FE analysis.

The biphasic FE analysis was carried out by the following properties.

1. Inhomogeneous depth-dependent apparent Young's modulus of solid phase (Hosoda et al., 2008; Murakami et al., 2004):

total apparent Young's modulus: $E_0 = 0.83$ MPa,depth-depending Young's modulus at depth x:

$$E(x) = (\varepsilon_0 / \varepsilon)E_0 \tag{3.24}$$

local strain at depth x:

$$\varepsilon(x) = 0.462\, e^{-6.53x} + 0.0284 \tag{3.25}$$

Poisson's ratio was assumed as 0.125 (Wilson et al., 2004).

2. Strain-dependent permeability (compaction effect) where e is current void ratio (Jurvelin et al., 2003; Sakai et al., 2015; Wilson et al., 2004):

$$k = k_{min} + k_0\, exp\left(M \frac{e - e_0}{1 + e_0}\right) \tag{3.26}$$

initial permeability: $k_0 = 58.9 \times 10^{-15}$ m^4/N s
minimum permeability: $k_{min} = 5.0 \times 10^{-15}$ m^4/N s
compaction effect of permeability: $M = 22$
initial void ratio: $e_0 = 4.0$ (80% interstitial fluid)

k_{min} was introduced mainly for acceptable curve fitting for the creep behavior in experimental cylindrical indentation test (Sakai et al., 2012, 2015).

3. Collagen reinforcement in tensile strain for full cartilage thickness (Li et al., 1999):

Apparent stress component by spring reinforcement σ_{sp}:

$$\sigma_{sp} = K(\varepsilon)\varepsilon \tag{3.27}$$

spring stiffness component $K_0 = 17.5$ MPa for $K(\varepsilon) = K_0\varepsilon$.

The bottoms of the cartilage models were fixed and impermeable, where no flow was allowed through the bottom surfaces. Material properties were specified by curve fitting comparing FE calculation with experimental time-dependent reaction force of the cylindrical indenter. The other surfaces were not fixed and basically permeable except for the contact region. The surface seepage in the contact area was controlled by a user subroutine, in which the FLOW function regulated the surface seepage coefficient as zero in the contact area.

The friction coefficient for solid-to-solid contact μ_{eq} between the geometrical rigid indenter and the solid phase was set to 0.2.

In simulation of reciprocating test, the load of 0.5 N/mm was applied by rigid cylindrical or flat plate indenter with a ramp time of 1 s and then the load was held constant for further 508 s. Initial horizontal position of the indenter was at the center of the cartilage tissue surface. The reciprocation of rigid cylinder or flat plate was started at 4 mm/s for sliding speed and stroke of 8 mm immediately after loading at constant load of 0.5 N/mm and continued for 508 s, 127 cycles at period of 4 s.

The changes in interstitial fluid pressure and Mises stress of solid phase in simulated reciprocating tests based on biphasic FE analyses are shown for the migrating contact (on-off loading) to articular cartilage and for the stationary contact (continuous loading) to articular cartilage in Figs. 3.19 and 3.20 (Murakami, 2012), respectively.

It is noticed under migrating contact condition in Fig. 3.19 that the interstitial fluid pressure is high at start and the high level of pressure is maintained even after 508 s, 127 reciprocating cycles. And the Mises stress is low at start and the stress distribution is a little changed but the stress level is not so much changed even after 508 s, 127 cycles. This fact suggests that most of loading is supported by interstitial fluid pressure in reciprocating sliding with sufficient stroke length (Pawaskar et al., 2007). As pointed by Sakai et al. (2012), the model including the spring reinforcement exuded the interstitial fluid in the forward surface of the indenter, whereas the fluid flow in the backward region was drawn into the cartilage tissue.

On the contrary, the stationary contact to cartilage in reciprocating motion had large influence on time-dependent biphasic behaviors. It is remarkably noted under continuous loading to articular cartilage in Fig. 3.20 that the interstitial fluid pressure at start is considerably high and the Mises stress at start is low, but after 508 s, 127

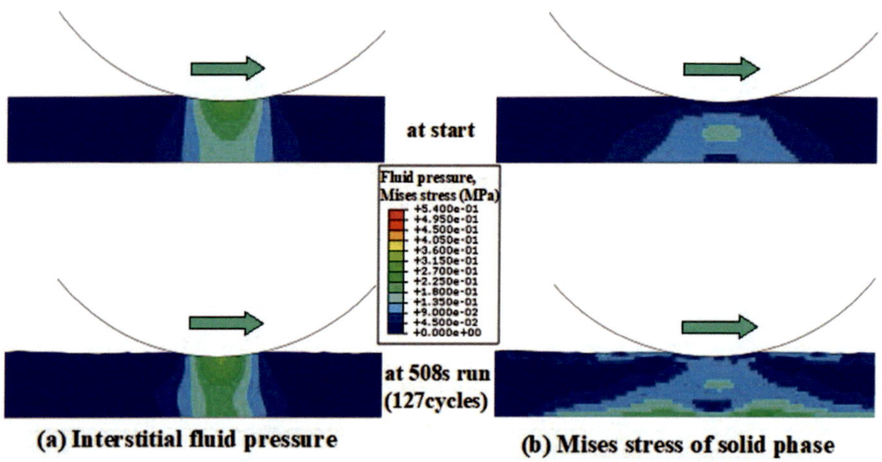

Figure 3.19 Changes in interstitial fluid pressure and Mises stress of solid phase in reciprocating test at migrating contact to cartilage. (a) High interstitial fluid pressure occurs at start and is maintained at 508 s run. (b) Mises stress at 508 s run is slightly changed from at start.
Modified from Murakami, T. (2012). Importance of adaptive multimode lubrication mechanism in natural and artificial joints. *Proceedings of the Institution of Mechanical Engineers, Part J: Journal of Engineering Tribology, 226*(10), 827–837. https://doi.org/10.1177/1350650112451377

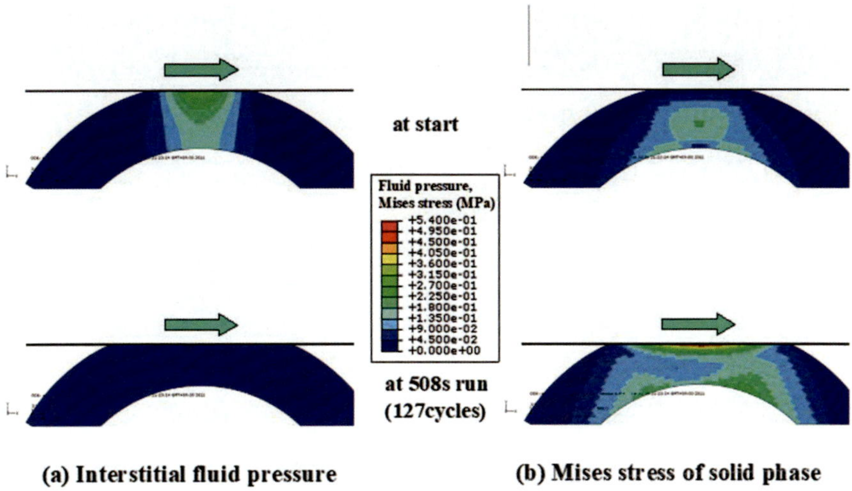

Figure 3.20 Changes in interstitial fluid pressure and Mises stress of solid phase in reciprocating test at stationary contact to cartilage. (a) High interstitial fluid pressure occurs at start but subsides at 508 s run. (b) Mises stress at 508 s run increases particularly in surface zone compared with that at start.
Modified from Murakami, T. (2012). Importance of adaptive multimode lubrication mechanism in natural and artificial joints. *Proceedings of the Institution of Mechanical Engineers, Part J: Journal of Engineering Tribology*, *226*(10), 827–837. https://doi.org/10.1177/1350650112451377

reciprocating cycles, the fluid pressure almost disappears and high Mises stress concentrates in cartilage surface zone. Furthermore, it is worth noting that the contact area increases from initial contact area, while the contact area under migrating contact condition maintains almost initial area in size.

To compare the changes in load supports by fluid pressure and vertical stress in solid phase, the supporting forces by fluid pressure and the vertical stress in solid phase were estimated and the percentages of fluid load support were calculated (Fig. 3.21) (Murakami, 2012). It is confirmed that the fluid load support maintains from 90% at start to 83% after 508 s in reciprocation motion under migrating contact condition, but under stationary contact condition the load support by interstitial fluid pressure decreases from 91% at start to 27% after 508 s as time-depending phenomenon. The maintaining of fluid pressure under migrating (on-off loading) is considered to be due to recovery of deformation with rehydration during off-loading phase, as indicated by Sakai et al. (2012).

In this biphasic FE analysis, the friction coefficient μ_{eq} for solid-on-solid contact is assumed as 0.2. The time-depending changes in friction coefficient μ_{eff} can be estimated by using formula (3.23) by Ateshian et al. (Ateshian, 2009). The frictional behaviors calculated in FE analyses are shown in Fig. 3.22 (Murakami, 2012). The friction for the migrating (on-off loading) condition shows a little increase but is not so much changed. In contrast, the friction for the stationary (continuous loading)

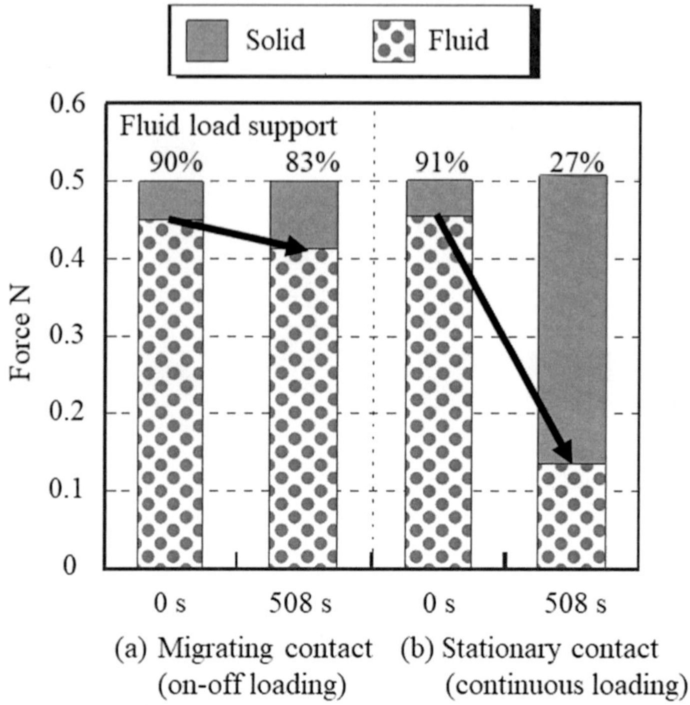

Figure 3.21 Changes in load support by interstitial fluid pressure and stress in solid phase for migrating contact and stationary contact to cartilage in reciprocating motion.
Percentages of load support by fluid pressure are shown with graph. Fluid load support at migrating contact is decreased only as 7% after 508 s run, but at stationary contact is decreased as 64% after 508 s run.
Modified from Murakami, T. (2012). Importance of adaptive multimode lubrication mechanism in natural and artificial joints. *Proceedings of the Institution of Mechanical Engineers, Part J: Journal of Engineering Tribology*, *226*(10), 827–837. https://doi.org/10.1177/1350650112451377

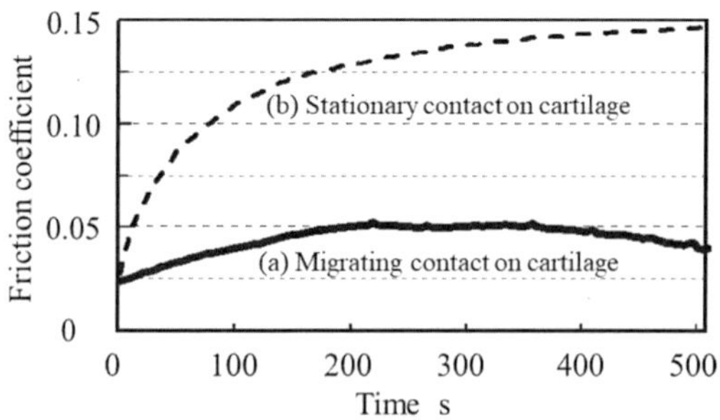

Figure 3.22 Friction behaviors based on biphasic analyses for migrating contact and stationary contact on cartilage in reciprocating motion. The friction for the migrating contact is suppressed at lower level, but for stationary contact friction gradually increases.
Modified from Murakami, T. (2012). Importance of adaptive multimode lubrication mechanism in natural and artificial joints. *Proceedings of the Institution of Mechanical Engineers, Part J: Journal of Engineering Tribology, 226*(10), 827–837.

gradually increases from initial low value and approaches to equivalent friction coefficient, although the friction level depends on the assumed value of 0.2 for solid-to-solid friction coefficient. The gradual increase in friction under continuous loading corresponds to the previous studies (Ateshian, 1997; Ateshian et al., 1998).

The time-dependent frictional behaviors with distributions of interstitial fluid pressure after 508 s, 127 reciprocating cycles for two kinds of reciprocating tests are shown in Fig. 3.22, where friction coefficient was calculated as the ratio of horizontal force divided by normal force. Immediately after starting, both conditions show similar low friction level less than 0.025, where the fluid phase load supports are 90% for migrating contact condition and 91% for stationary contact condition, both of which correspond to low friction level. However, friction at stationary contact condition gradually increased to high friction. High friction for stationary contact at 508 s correspond to the lowering of fluid phase load support as 27%, which means that the solid phase load support is 73%, and the coefficient of friction estimated from Eq. (3.23) as $0.2 \times 0.73 = 0.146$ approximately equals to calculated value in biphasic FE analysis (Fig. 3.22). In contrast, for migrating contact condition at 508 s, the fluid load support is 83%, and solid phase load partitioning is 17% which suggests that friction coefficient due to surface resistance is estimated as 0.034. In this case, the plowing resistance produces additional part of total friction in Fig. 3.22.

Thus, it was confirmed for migrating contact condition that the keeping of high ratio of load support by interstitial fluid pressurization in cartilage tissue with biphasic properties including uniform collagen fiber-reinforced mechanism is capable to reduce friction.

The first model described above had uniform reinforcement by collagen fiber network irrespective of anisotropic collagenous structure in actual articular cartilage (Fig. 3.1). Therefore, in the next example, the effect of reinforcement in the superficial tangential cartilage layer on mechanical behavior of cartilage in migrating contact condition was examined as the second FE analysis by using modified model proposed by Sakai et al. (2016). The side size of FE mesh for square was modified as 0.05 mm and the superficial tangential zone (STZ) was defined as 0.05 mm top surface layer, where the anisotropic (horizontal) fiber stiffness of the STZ was stiffened by 4.74 times larger value than the homogeneous isotropic stiffness of the base model. Furthermore, the tangential fluid permeability was modified by 1.090 times larger while the vertical one was modified by 0.818 times, to consider the structural permeability alteration by the fiber arrangement. A little severe reciprocating conditions for sliding speed of 2 mm/s and stroke of 4 mm were applied, at which *Pe* number for contact length of 2.1 mm was estimated as 86 at a lower limit to maintain the sufficient performance of the biphasic lubrication in a migrating contact condition.

The effectiveness of fiber reinforcement with anisotropic fiber stiffness in the superficial tangential layer (STZ model) was compared with the standard model with isotropic fiber network as shown in Fig. 3.23. Friction in STZ model maintains lower friction than that in standard one within 300 s running although both values get closer at 300 s. The frictional force in migrating contact condition include not only the shearing resistance at solid-to-solid contact in conjunction zone but also plowing resistance, and therefore the fluid load support contributes to the lowering surface friction.

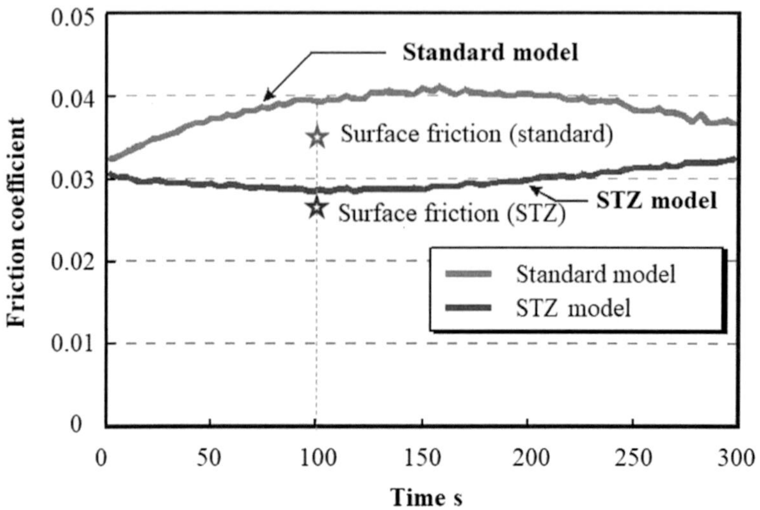

Figure 3.23 Frictional behaviors of standard and STZ models in migrating contact condition. It is shown that the fiber reinforcement with anisotropic fiber stiffness in the superficial tangential layer (STZ model) reduces friction. The values of surface friction excluding plowing term are plotted as star marks.
Modified from Sakai, N., Hashimoto, C., Yarimitsu, S., Sawae, Y., Komori, M., & Murakami, T. (2016). A functional effect of the superficial mechanical properties of articular cartilage as a load bearing system in a sliding condition. *Biosurface and Biotribology*, *2*(1), 26–39. https://doi.org/10.1016/j.bsbt.2016.02.004

In biphasic analysis the changes in load partitioning were examined. The fluid load support ratios of STZ model changed from 91.9% at starting to 86.7% at 100 s and 84.4% at 300 s, whereas those for the standard model changed from 90.5% at starting to 82.5% at 100 s and 84.3% at 300 s. The estimated friction coefficients (as star marks in Fig. 3.23) due to surface shearing resistance at 100 s are 0.0266 for STZ model and 0.035 for standard model, respectively, and these behaviors correspond to changes in total friction. It is noticed that the plowing effect is lower in STZ model than the standard model. In this case, the reinforcement by stiffer anisotropic fiber in superficial tangential zone appears to suppress not only the surface resistance but the plowing resistance.

The interstitial fluid pressures for both models at 100 s in migrating conditions are shown in Fig. 3.24. It is noted that the width of high-pressure region for STZ model is a little wider than that for the standard model in the superficial layer. Furthermore, it is noticed that the STZ model generated the negative pressure in front of the rigid indenter sliding to right direction. This negative pressure induce to intake fluid into the surface would be effective in rehydration of biphasic cartilage. In addition, it is confirmed that the von Mises stress level in superficial zone for the STZ model is maintained lower than that for the standard model. As described above, the fiber-

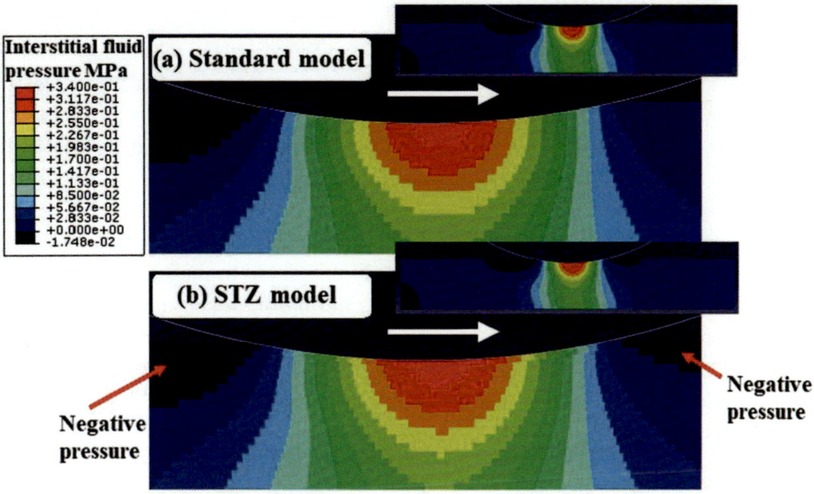

Figure 3.24 Interstitial fluid pressure for standard and STZ models in migrating contact condition. Small/wide size and enlarged size are depicted for contour of interstitial fluid pressure.
Modified from Sakai, N., Hashimoto, C., Yarimitsu, S., Sawae, Y., Komori, M., & Murakami, T. (2016). A functional effect of the superficial mechanical properties of articular cartilage as a load bearing system in a sliding condition. *Biosurface and Biotribology*, *2*(1), 26–39. https://doi.org/10.1016/j.bsbt.2016.02.004

reinforcement with anisotropic fiber stiffness in STZ layer contributes to sustaining higher level in interstitial fluid pressurization.

On the effect of biphasic properties of superficial layer on biphasic lubrication, Fujie et al. indicated that the anisotropic permeability (10 times lower in tangential direction than in normal direction) improved friction in analytical study using a fiber–reinforced poroelastic biphasic model (Fujie & Imade, 2015). Furthermore, they suggested that the collagen-induced residual stress plays an important role in enhancing the biphasic lubrication property of articular cartilage (Fujie et al., 2018).

As discussed in this section, biphasic lubrication appears to play an important role in reducing direct contact between cartilage surfaces with lowering of friction. Particularly, in the migrating contact condition with on-off loading, the biphasic lubrication sustains to be effective with recovery of deformation with rehydration during off-loading phase. Friction level is controlled with the surface lubricity of rubbing surfaces, and the role of other lubrication mechanisms such as the hydration and the boundary lubrication become important. In the following sections, the attractive features on hydration lubrication, boundary lubrication, and gel film lubrication are described.

3.4.4 Hydration lubrication

Under severer operating conditions at start up after standing for a long time and/or at slow movement in heavy loading where the effectiveness of biphasic lubrication can be reduced and some local direct contacts can occur between rubbing cartilage surfaces, the roles of alternate lubrication mechanisms such as hydration lubrication, boundary lubrication, and gel-film lubrication are likely to prevail at direct contact zones. In this section, the effectiveness of hydration lubrication is discussed.

Ikeuchi (2007) gave the simple definition for the hydration lubrication as "water lubrication of surface layer with hydrophilic polymers fixed to the surface." Therefore, the boundary lubrication is not included in this definition because the surface-active agent may move and the films may be repaired by the surface-active agent existing in solution.

The hydration lubrication has the common properties as follows:

1. When two bodies with hydrated surface layer such as surface gel or polymer brush approach, pressure rises in the surface layer and a repulsive force arises between them and most of the applied load is supported by water.
2. Under sliding motion, the water molecules are sheared with low stress, whereas the polymers are fixed to the surface.

If adequate water molecules are attracted by hydrophilic polymers fixed to the surface, the low friction would be sustained between hydrated layers at low velocity or even under static conditions.

Klein (2013) indicated as the detailed mechanism that a hydrated charge between sliding surfaces can sustain a large normal load because of the reluctance of the hydration water to be squeeze out, but under shear this hydration layer responds in a fluid manner because of its rapid relaxation. This combination of sustaining a large normal load together with a fluid response to shear between hydrated surfaces has been termed the hydration lubrication mechanism.

Raviv and Klein (2002) measured the frictional forces between atomically smooth mica surfaces in high concentration salt solutions in the hydration repulsion regime using a surface force balance (SFB) and revealed the hydration lubrication mechanism. The effective friction coefficient was not greater than 0.0002 (Klein, 2013).

The water molecule is overall neutral but possesses a large electric dipole by virtue of the residual charges on the H and O atoms (Fig. 3.25). Water molecules in the vicinity of a charge or polar group in aqueous media can surround it to form a hydration shell because of their large electric dipoles. A large energy is required to break up the shells. And individual water molecules in each shell can exchange positions with surrounding bulk water or with water in adjacent shells. The time scale for that exchange can be a nanosecond.

Fig. 3.26 shows two mica surfaces across a thin (1 nm) layer of aqueous solution (the ion sizes correspond to Na^+ ions and their first hydration shell). At sliding velocity V_s of the top surface, as long as the shear rate (V_s/D) is lower than the relaxation rate $\omega_{relaxation} = (1/\tau_{exchange})$ of the hydration shell, the hydrated ions would respond to shear in a liquidlike fashion. In their experiment (Raviv & Klein, 2002),

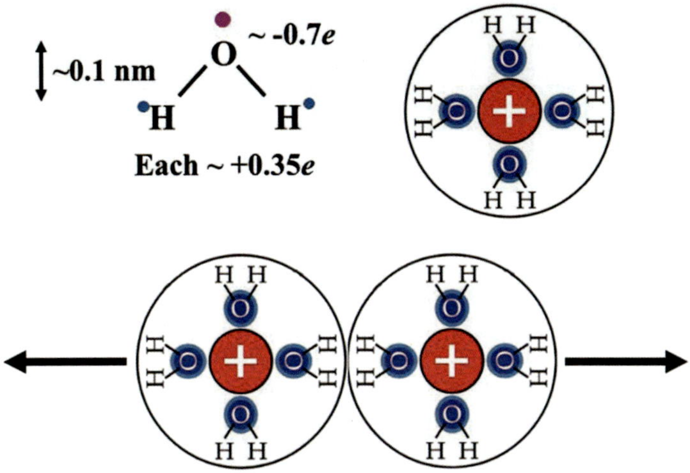

Figure 3.25 Large dipole of water and hydration shell about charge. The water molecules are overall neutral but possess large electric dipoles and form hydration shells about charges in aqueous media. Bottom: Adjacent charges experience a repulsive interaction of steric origin when their hydration shells overlap.
Modified from Klein, J. (2013). Hydration lubrication. *Friction*, *1*(1), 1–23. https://doi.org/10.1007/s40544-013-0001-7

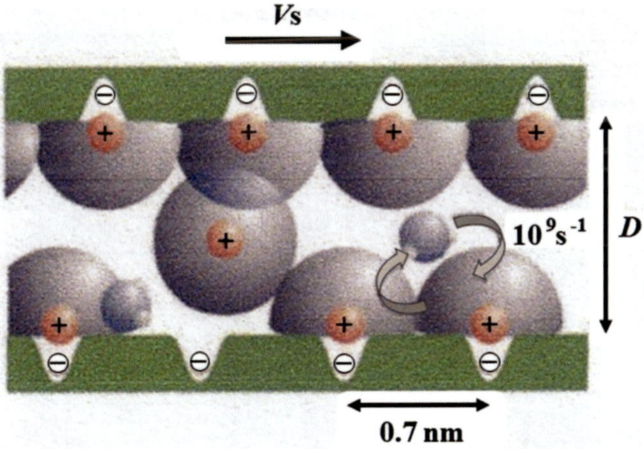

Figure 3.26 Mechanism of hydration lubrication between charged mica surfaces across trapped hydrated Na^+ ions. A large load can be supported by the hydration repulsion, but as long as the shear rate (V_s/D) is less than the relaxation/exchange rate $\omega_{relaxation} = 10^9\ s^{-1}$, the confined liquid behaves in a fluid manner.
Modified from Klein, J. (2013). Hydration lubrication. *Friction*, *1*(1), 1–23. https://doi.org/10.1007/s40544-013-0001-7

$V_s = 1000$ nm/s and $D = 1$ nm, then the shear rate: $V_s/D = 10^3\ s^{-1}$ is lower than the relaxation rate: $\omega_{relaxation} = 10^9\ s^{-1}$. At the same time, the hydration repulsion acts to keep surfaces act even up to high pressures. The combination of sustaining a high normal pressure and behaving as a fluid under shear brings about the very low friction arising from the hydration lubrication mechanism. The estimated friction coefficient was 0.0003 as an upper limit, which is very low similar to experimental upper limit as 0.0002.

One of examples for hydration lubrication is observed in phospholipid bilayer on cartilage surfaces (Fig. 3.27). The close-packed array of opposing hydrated head groups of phospholipids, comprising a negative charge: PO_4^-, next to a positive charge: $N^+(CH_3)_3$ reside at the slip plane and suppress frictional dissipation with hydration lubrication mechanism. The combination of the head group and hydrophobic CH_2 tails keeps the lipid bilayers robust enough to withstand high pressure.

Next, the superiority of hydration lubrication in the charged polymer brushes to neutral polymer brushes or adsorbed charged polymers is mentioned in terms of the structure and hydration of charged brushes (Fig. 3.28) (Klein, 2013). The charged brushes provide considerably better lubrication up to higher compression than either neutral brushes or adsorbed charged polymers. Low friction between sliding surfaces with charged polymer brushes is derived by not only the charged chains themselves but also the mobile counterions trapped within the brush layers. Up to moderate compressions, the polymer brushes, for charged and neutral brushes, interpenetrate only weakly because of configurational entropy effects. The load is borne by the osmotic pressure of the polymer segments. The addition of the osmotic pressure of the trapped counterions provides little lubricating effect. At the highest compressions, however, when the surface separation becomes comparable with or smaller than the radius of gyration of the polymer chains, it is expected that the entropic effect suppressing the interpenetration becomes negligible. At these high compressions, the lubrication by the hydration layers on the charged monomers becomes the dominant lubricating mode via hydration lubrication mechanism. Since frictional dissipation takes place

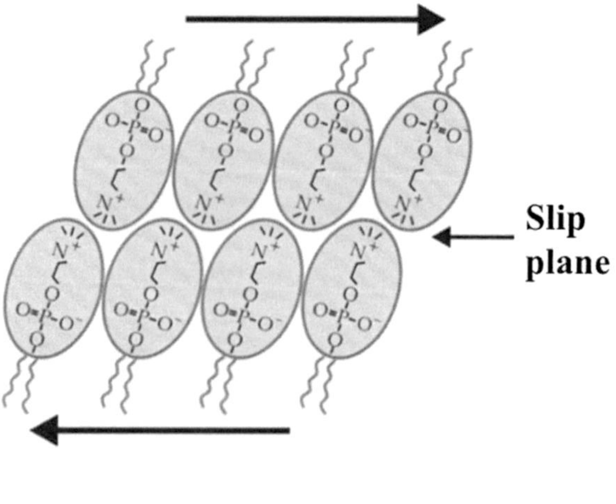

Figure 3.27 Hydrated lubrication in phospholipid bilayer Low friction is maintained with hydration lubrication at high pressure on cartilage surfaces. Reproduced from Jahn, S., & Klein, J. (2018). Lubrication of articular cartilage. *Physics Today*, *71*(4), 48–54, with the permission of AIP Publishing. https://doi.org/10.1063/PT.3.3898

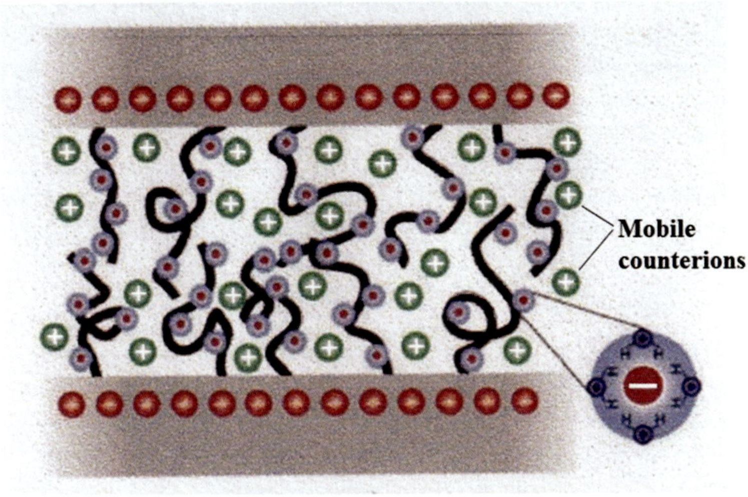

Figure 3.28 Hydration lubrication by charged polymer brushes. Up to moderate compression, there is little interpenetration between brushes, resulting in an unentangled overlapped layer and so little viscous dissipation. At higher compressions, the hydration layers on the monomers act via the hydration lubrication mechanism. The mobile counterions augment the osmotic pressure of the polymer segments and help to support the load. Modified from Klein, J. (2013). Hydration lubrication. *Friction*, *1*(1), 1–23. https://doi.org/10.1007/s40544-013-0001-7

across the sheared interpenetration zone rather than at a sharp interface, the hydration layers reduce the friction between the compressed monomers via the hydration lubrication mechanism. The mobile counterions augment the osmotic pressure of the polymer segments and contribute to support load. In neutral polymer brushes, both the lubricating effect of the hydration shells and the additional load-bearing osmotic pressure due to trapped counterions are unavailable. In adsorbed charged polymers, these effects are applied but suppressed by energy dissipation due to the bridging which occurs when a chain from one adsorbed layer adsorbs also to the opposing surface.

For example, very low friction coefficient 0.004 is measured for sliding surfaces with poly[2-(methacryloyloxy)ethyl phosphocholine (pMPC) brushes in water at high pressure of about 8 MPa (Chen et al., 2009). The phosphocholine group is zwitterionic as overall neutral but containing both negative (PO_4^-) and positive ($N^+(CH_3)_3$) charged groups and very highly hydrated, with up to 15 or more water molecules in the primary hydration shell. In this case pMPC polymers are overall neutral and there are no mobile counterions. Therefore, the additional load support by the osmotic pressure as for charged brushes at high compression (Fig. 3.28) is unavailable. However, friction is significantly reduced by hydration lubrication mechanism with the hydration shells surrounding the phosphocholine monomers.

In natural synovial joints, the hydration lubrication appears to play an important role in reducing friction and wear for cartilage-cartilage rubbing lubricated with synovial fluid even at slow movement under high contact pressure (up to 10 MPa or more) in lubrication regime of nonfluid film lubrication. The hydration shells surrounding charged, zwitterionic, or polar groups are tenaciously attached, and so able to support large normal stresses without being squeezed out, and at the same time exhibit fluidlike behaviors, so that their sliding can occur at very low shear stresses at the desired slip interfaces. The hydration lubrication mechanism becomes effective in not only charged polymer brushes (Fig. 3.28) and zwitterionic polymer brushes but also phospholipid bilayers in boundary lubrication agents (Fig. 3.27).

3.4.5 Boundary lubrication

In thin film lubrication under severer operating conditions at slow movement in heavy loading and/or at start up after standing for a long time where the adequate lubricating fluid film cannot be formed and the effectiveness of biphasic lubrication will be reduced, significant local direct contacts occur between rubbing cartilage surfaces. To reduce friction and protect rubbing surfaces, the alternate lubrication mechanisms such as hydration lubrication, boundary lubrication, and gel-film lubrication are expected to be effective at direct contact zones. In this section, the important features of boundary lubrication mechanism are mentioned.

In boundary lubrication regime, the adsorbed films such as glycoproteins (lubricin or proteoglycan 4 (PRG4), Fig. 3.29) (Radin et al., 1970; Swann, Hendren, et al., 1981; Swann, Slayter, et al., 1981; Swann et al., 1985; Zappone et al., 2007) and phospholipids (Fig. 3.30) (Hills, 1989, 2000; Hills & Butler, 1984; Schwarz & Hills, 1998) are likely to play significant roles in friction reduction and protection of surfaces.

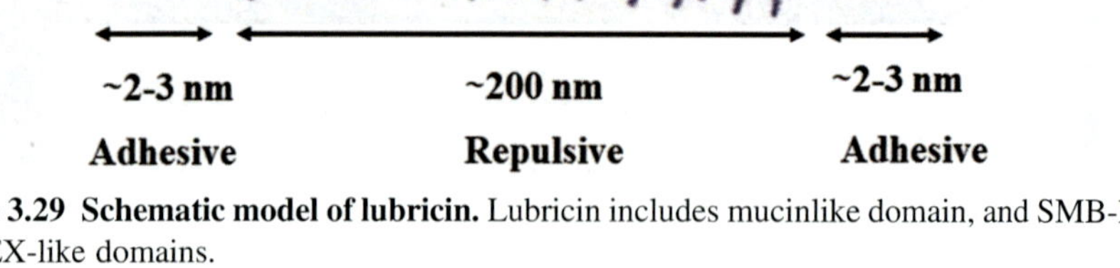

Figure 3.29 Schematic model of lubricin. Lubricin includes mucinlike domain, and SMB-like and PEX-like domains.
Peprinted with permission from Banquy, X., Burdyńska, J., Lee, D. W., Matyjaszewski, K., & Israelachvili, J. (2014). Bioinspired bottle-brush polymer exhibits low friction and amontons-like behavior. *Journal of the American Chemical Society*, *136*(17), 6199–6202. Copyright 2014 American Chemical Society. https://doi.org/10.1021/ja501770y

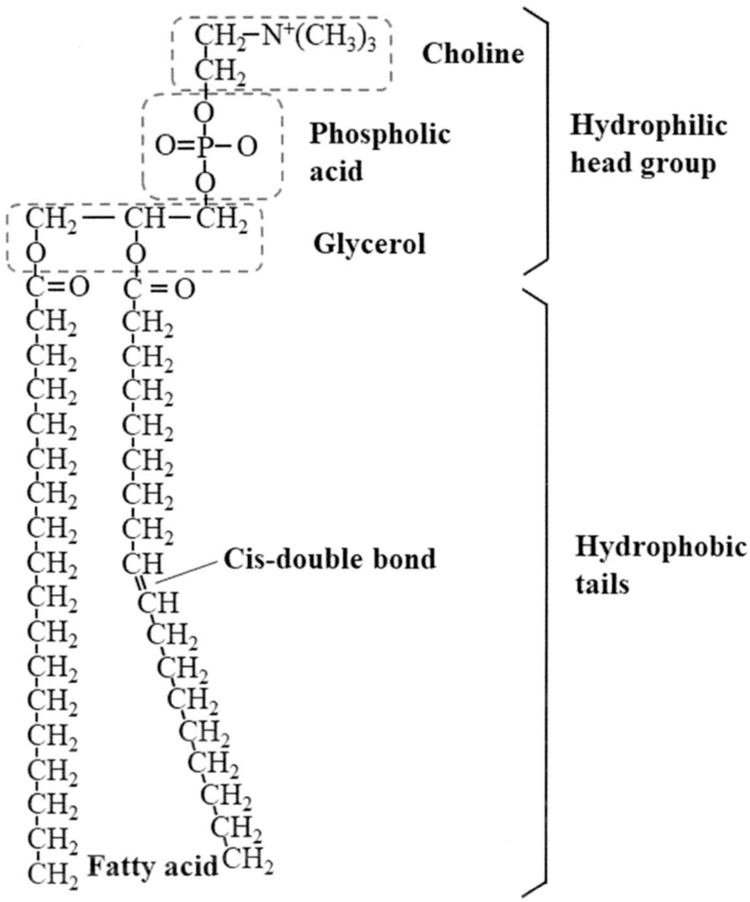

Figure 3.30 Chemical structure of phospholipid (Dipalmtoyl phosphatidylcholine, DPPC). This phospholipid is composed of hydrophilic head and hydrophobic tails.

The lubricin is composed of hydrophilic central mucinlike domain and both side parts as somatomedin-B (SMB)-like and homeopexin (PEX)-like domains. The lubricin with plenty of hydrophilic parts appears to form various adsorbed configurations such as a brushlike adsorbed film or a brushlike film partly anchored within the superficial zone of articular cartilage. In contrast, DPPC has the hydrophilic head group and the hydrophobic tails, and is likely to form monolayer, bilayer, multilayer, or combined structure with other constituents in synovial joints.

Serum proteins are divided into albumin, α_1-globulin, α_2-globulin, β-globulin, and γ-globulin via electrophoresis (Table 3.1). Serum proteins had been considered to have weak lubricating ability on cartilage surfaces, but the significant lubricating performance of γ-globulin with low friction and minimal wear in HA solutions was found by Higaki and Murakami (1996).

The γ-globulin forms thicker adsorbed film on cartilage surfaces than albumin in static immersion and in rubbing condition. Here, it should be considered that even albumin exhibited considerable adsorption after appropriate rubbing process compared with after static immersion for 1 h (Fig. 3.31) (Murakami et al., 2009). Severe rubbing removes the adsorbed protein films, but appropriate rubbing action to control denaturation and unfolding of proteins appears to enhance the adsorption of protein molecules on cartilage surface, as confirmed with in situ observation. Protein adsorption in coexistence of albumin and γ-globulin seems to be complicated, but the layer structure of both proteins at appropriate combination in HA solution showed low friction and minimal wear (Nakashima et al., 2005). Hyaluronic acid plays a main role for viscous property of synovial fluid but contributes to protection of articular cartilage surface with other synovia constituents.

In the examination of intact articular cartilage of rat patella with TEM by Higaki et al. (1998), a single continuous biomembrane structure was distinctly detected as the laminar structure with fibrous molecules as shown in Fig. 3.2. This bilayer biomembrane structure appears to act as an adsorbed film. In addition, the intermediate nonfibrillar layer was confirmed as underlying barrier layer between the biomembrane and the collagenous zone.

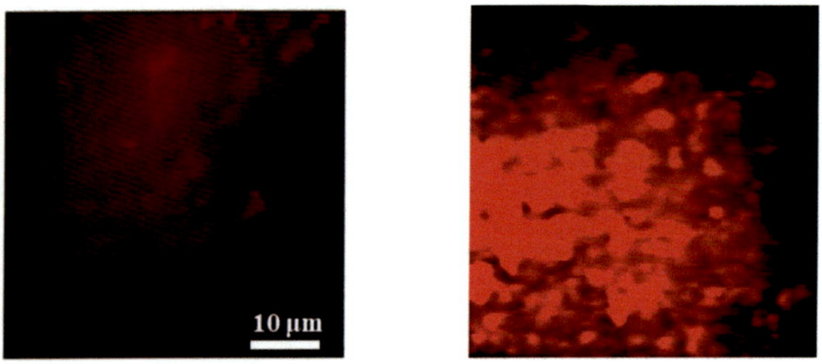

(a) After immersion in albumin solution for 1h

(b) After reciprocation of 1800 cycles (1h)

Figure 3.31 In-situ observation of protein molecules adsorbed on cartilage surface in albumin solution. Reciprocating test was carried out for rubbing pair of elliptical articular cartilage specimen and a cover glass in the reciprocating tester at 0.735 N (apparent contact pressure: 0.094 MPa) and the stroke of 10 mm in saline solution of 0.7 mass% albumin. The sliding speed was arranged as 0.2 mm/s for in situ observation and 10 mm/s at a period of 2 s during rubbing without observation. Albumin was stained with Rhodamine- B-Isothiocyanate. Modified from Murakami, T., Nakashima, K., Sawae, Y., Sakai, N., & Hosoda, N. (2009). Roles of adsorbed film and gel layer in hydration lubrication for articular cartilage. *Proceedings of the Institution of Mechanical Engineers, Part J: Journal of Engineering Tribology*, *223*(3), 287–295. https://doi.org/10.1243/13506501JET536

The molecular lubrication including polymer brush play roles as hydration lubrication as described in Section 3.4.4, but the adsorbed films are expected to become effective as the first solid-like barrier to maintain low friction and protect the surface.

Next, the effectiveness of synovia constituents as boundary lubricants is examined.

3.4.5.1 Influence of synovia constituents on friction in boundary lubrication

To evaluate of concentration of hyaluronic acid on frictional behaviors under constant load, the pendulum tests were performed for porcine shoulder joints (Higaki et al., 1998; Murakami et al., 1998). The shoulder joint specimens were prepared by articulating parts excised from pigs which were six-month-old and about 100 kg in body weight. The ellipsoidal surfaces of porcine shoulder joint composed of humerus head and glenoid cavity (cup) were used as the fulcrum and adjusted so that the swing motion was along the major radius. After placing of 0.5 cm^3 of lubricant into the concave cup part, the swinging of the pendulum was observed immediately after loading of 100 N (apparent mean contact pressure: $p_a \fallingdotseq 0.25$ MPa) and 1 kN ($p_a \fallingdotseq 1.8$ MPa) at room temperature of 20°C. The damping behavior of the pendulum at an initial angle of 0.1 rad was monitored. The periods at 100 N and 1 kN are 1.0 and 1.5 s, respectively. Friction coefficient was estimated from the linear damping of amplitude ranging between 0.08 and 0.03 rad.

At first, the pendulum tests were conducted using intact joints lubricated with saline solutions of sodium hyaluronate (HA, molecular weight: $8.8 \times 10^5 - 9.6 \times 10^5$, Seikagaku Corporation, Japan) of different concentrations (0.1−2.0 g/dL). The non-Newtonian viscosity of lubricants was measured by the cone-plate viscometer. The measured values of HA viscosity increased with increasing concentration.

After the first tests, the joints were rinsed with a saline solution containing polyoxyethylene p-t-octylphenyl ether (Triton X-100, 10 g/dL, Wako pure Chemical Industries, Japan) as a detergent, which is a highly surface-active nonionic surfactant, to remove the adsorbed film on the articular cartilage surfaces. The removal of adsorbed films mostly composed of amphiphilic molecules such as phospholipids, proteins, and biomembrane molecules from the cartilage surfaces was verified by significant reduction of contact angles for a droplet of distilled water on the dried cartilage surfaces. The changes from hydrophobic to hydrophilic surface (contact angle: from 82.8 ± 7.1 degree to 37.8 ± 6.7 degrees, N = 30) were confirmed by this detergent treatment.

Fig. 3.32 shows changes in friction of porcine shoulder joints at 100 N loading in pendulum tests from intact to after detergent treatment with removal of adsorbed film on cartilage surfaces. It is noticed in the first test for intact cartilage specimens that measured friction values exhibit a typical Stribeck curve, where it is suggested that high viscosity (high HA concentration) area corresponds to fluid film lubrication regime, and low viscosity (low HA concentration) area corresponds to boundary or mixed lubrication regime. After detergent treatment in the second tests, friction is increased, particularly friction levels at 0.1 g/dL and 0.2 g/dL HA concentration in boundary or mixed lubrication regime are significantly increased. This indicates that

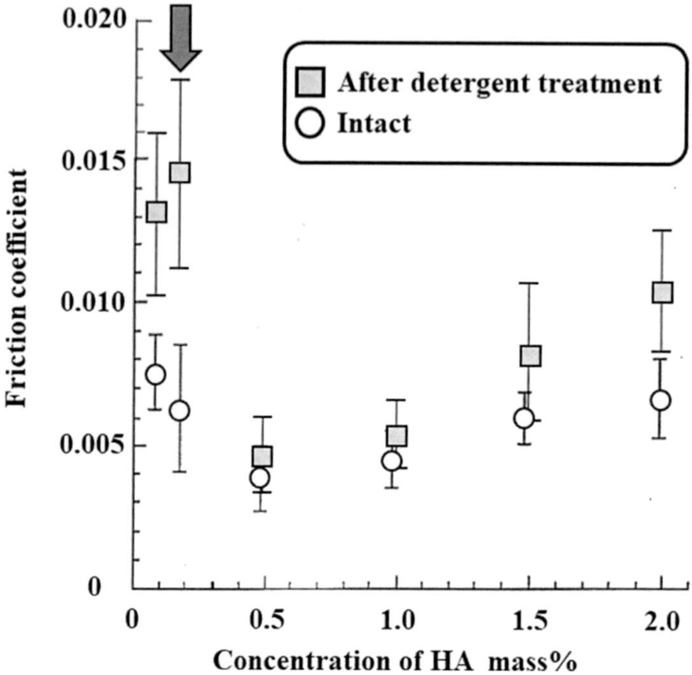

Figure 3.32 Changes in friction of porcine shoulder joints at 100 N loading in pendulum tests from intact to after detergent treatment with removal of adsorbed film on cartilage surfaces. Experimental values of friction coefficient for intact joints and after detergent treatments are plotted against concentration of HA. Friction increases after detergent treatment, and particularly friction at 0.2 mass% at boundary or mixed lubrication regime (indicated by arrow mark) remarkably increases. $N = 7$, except $N = 30$ for 0.2 g/dL, Error bars indicate standard deviation (SD).

Modified from Murakami, T., Higaki, H., Sawae, Y., Ohtsuki, N., Moriyama, S., & Nakanishi, Y. (1998). Adaptive multimode lubrication in natural synovial joints and artificial joints. *Proceedings of the Institution of Mechanical Engineers, Part H: Journal of Engineering in Medicine*, *212*(1), 23–35. https://doi.org/10.1243/0954411981533791

the removal of adsorbed films on cartilage surfaces brings about severer contact in boundary or mixed lubrication regime.

In contrast, the friction coefficient in the first test for intact cartilage at high load of 1 kN showed approximately constant low values around 0.005 for a wide range of HA concentrations or viscosities. This low friction at high load appears to be brought about by an elastohydrodynamic effect due to improved geometrical conformity of shoulder joints. In the shoulder joint, the convex articular surfaces have a slightly larger radius than the concave surface, similar to the hip joint. During light loading, direct contact seems to occur at the peripheral thin film zone and entrap a fluid film in the central zone. At high loading, the articular surfaces elastically deform to improve geometrical conformity, resulting in enhancement of the fluid film effect, particularly squeeze film effect. Similar phenomenon was observed in the hip joint by Roberts et al. (1982). The high friction in lower viscosity at 100 N is considered as the feature in boundary or mixed lubrication regime.

In the second tests after removal of adsorbed films, a significant increase in friction is noticed in the low-load and low-viscosity boundary or mixed lubrication regime. Consequently, the third test condition to evaluate the lubricating effect of the addition of synovia constituents on frictional behaviors was selected in the boundary or mixed lubrication regime, i.e., immediately after loading of 100 N, lubricated with HA solution of 0.2 g/dL (arrow mark in Fig. 3.32). The average value of friction coefficient at 0.2 g/dL HA in the first test of 0.0065 increased to 0.015 in the second test, which indicates the loss of protection by the adsorbed films.

The joints used in the second tests were rinsed with saline and then the third pendulum tests were conducted lubricated with HA solutions with and without human serum proteins (albumin and γ-globulin, Wako Pure Chemical Industries, Japan) and liposome of phospholipid Lα-dipalmitoyl phosphatidyl-choline (Lα-DPPC), Wako Pure Chemical Industries, Japan. DPPC was selected as it is the predominant phospholipid in synovial fluid and on cartilage surfaces. Sizes of the diameters of liposomes as multilamellar vesicles observed by the negative staining method in TEM were distributed from 0.1 to 0.5 μm. In Fig. 3.33, the extent of effectiveness of addition of synovia

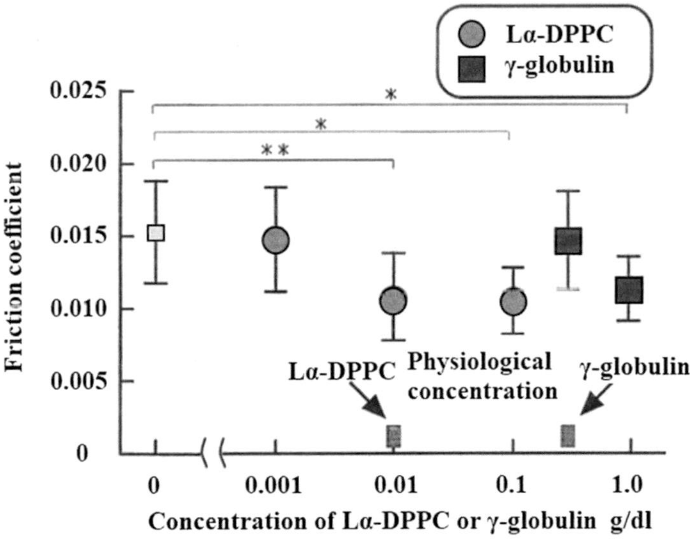

Figure 3.33 Effectiveness of addition of phospholipid (Lα-DPPC) liposomes and serum protein (γ-globulin) on frictional behaviors. Friction of cartilage surfaces treated with detergent were improved by the addition of Lα-DPPC with concentrations of over 0.01 g/dL or γ-globulin over 1.0 g/dL. $N = 6$, error bars indicate SD, $*P < .005$, $**P < .01$.
Modified from Murakami, T., Higaki, H., Sawae, Y., Ohtsuki, N., Moriyama, S., & Nakanishi, Y. (1998). Adaptive multimode lubrication in natural synovial joints and artificial joints. *Proceedings of the Institution of Mechanical Engineers, Part H: Journal of Engineering in Medicine*, *212*(1), 23–35. https://doi.org/10.1243/0954411981533791

constituents on friction is shown. Here, the third test with the same lubricant (0.2 g/dL HA) showed little effective. It is noticed at the third test that the addition of Lα-DPPC at 0.01 g/dL or γ-globulin at 1.0 g/dL was effective in reducing friction of articular cartilage surfaces after removal of adsorbed films. Lα-DPPC of similar concentration to that in the human joints was sufficiently effective, but a higher concentration of γ-globulin than the physiological level was needed. On the contrary, the addition of albumin of even 3.0 g/dL was ineffective, probably due to its less hydrophobic amino acid residues (the ratio of nonpolar residues: 59.3% against high ratio of γ-globulin of 74.1%) (Higaki et al., 1998). It is noted that γ-globulin has high similarity to the ratio of hydrophobic nonpolar residues of the lubricating glycoprotein-1 (LGP-1) which has boundary lubricating ability in synovial fluid (Swann, Slayter, & Silver, 1981).

These results support the boundary lubricating abilities of phospholipids and γ-globulin as amphiphilic molecules. Particularly, phospholipid Lα-DPPC, which showed sufficient effectiveness at physiological concentration for HA solution of low concentration, is expected to play a major role in boundary lubrication. However, other reciprocating tests for porcine articular cartilage against glass plates coated by various Langmuir–Blodgett films lubricated with HA solution (Higaki et al., 1998) exhibited that the mixed bilayer film of Lα-DPPC and γ-globulin showed better frictional properties than the Lα-DPPC bilayer for 30 min running. Here, 5 to 10 bilayers of Lα-DPPC alone exhibited similar low frictional properties to 5 to 10 mixed bilayers. In natural synovial joints under various daily activities, not only phospholipids but proteins or glycoproteins appear to function as effective boundary lubricants depending on the rubbing severity.

In these pendulum tests, the influence of single constituent in HA solution on friction was examined but the combined effect of various constituents should be taken into consideration to elucidate the lubricating ability of synovia constituents. As reported by Nakashima et al. (2005), HA solution containing 1.4 mass% albumin and 0.7 mass% γ-globulin (albumin:globulin = A/G ratio of 2:1), or 0.7 mass% albumin and 1.4 mass% γ-globulin (A/G ratio of 1:2) showed low friction and minimal wear for poly(vinyl alcohol) (PVA) hydrogels as artificial cartilage with the aid of the layered adsorbed film formation. The details of this phenomenon are described in Section 5.4 (in Chapter 5). In these cases, it is suggested that the γ-globulin forms protective adsorbed layer on cartilage surface and albumin plays as low shearing layer. The adsorbed film formation during rubbing was confirmed by in situ observation (Yarimitsu et al., 2007, 2009). Although the effects of layered structure of adsorbed films of this simulated synovial fluid on cartilage surface have not yet been observed, it is expected that similar layered structures might have been formed on rubbing cartilage surfaces.

In the next section, the influence of combination of proteins and phospholipid in saline or HA solution is investigated.

3.4.5.2 Influence of rehydration and roles of adsorbed films on friction for articular cartilage in boundary lubrication

Focusing on the intimate local contact zone of hydrated cartilage surface, the effectiveness of rehydration and repair of adsorbed films formed on gellike layer at the uppermost superficial zone in articular cartilage was examined in repeated reciprocating tests including the restarting processes after interruption and unloading to evaluate the influence of rehydration and roles of adsorbed films (Murakami et al., 2011, 2013, 2017).

The reciprocating test for the sliding pair of the upper stationary ellipsoidal articular cartilage specimen and the lower reciprocating flat glass plate (slide glass) was conducted in the reciprocating tester. To examine clearly the effectiveness of adsorbed films composed of synovia constituents, the severer condition with stationary loading condition without migration of contact zone for articular cartilage was selected because the interstitial pressure subsides for shorter time as indicated by the biphasic FE analysis (Fig. 3.21). An upper intact cartilage specimen with subchondral layer was prepared from a femoral condyle in a porcine knee joint (6–7 months old). The lubricants are saline solutions containing 0.15M NaCl (Otsuka Pharmaceutical Factory Inc., Japan), saline solution of 0.5 mass% HA (molecular weight: 9.2×10^5), HA solutions containing 0.7 mass% or 1.4 mass% bovine serum albumin (Wako Pure Chemical Industries Ltd., Japan) and/or human serum γ-globulin and/or 0.01 mass% Lα-DPPC as an phospholipid liposome. In order to prevent bacterial growth in protein solutions as lubricant, 0.3 mass% sodium azide was added for protein solutions.

The reciprocating test was conducted at a sliding speed of 20 mm/s for rectangular reciprocating mode and at a stroke of 35 mm at a constant load of 9.8 N. The glass plate was cleaned ultrasonically in a solution of 0.5 vol% Triton X-100, distilled water, and ethanol, and then dried. The lubricants were supplied in liquid bath. At room temperature, the reciprocating sliding was started immediately after loading, and interrupted after 514 cycles at sliding distance of 36 m for running time of 30 min, and then the unloading state was maintained for 5 min. Subsequently, the reciprocating test was restarted immediately after reloading and continued for a further sliding distance of 36 m. The restarting processes after unloading were repeated three times. The changes in friction force were continuously monitored to compare the differences in time-dependent frictional behaviors.

Time-dependent frictional behaviors for intact cartilage lubricated with saline and saline solutions of albumin, γ-globulin, HA, and DPPC as single constituent are shown in Fig. 3.34. As expected for intact articular cartilage, the initial friction coefficient is very low as about 0.01 for all lubricants. However, the friction gradually increases with sliding distance until the sliding is interrupted at 36 m. The final values at 36 m sliding are different as the order of γ-globulin > saline > albumin > HA > DPPC. It is noted that the addition of a single constituent into saline usually reduced the friction level at the final stage of 36 m sliding except γ-globulin. For γ-globulin, initial friction is lower than saline but the friction gradually increases to a higher level with thinning of lubricating film. At reloading-restarting after 5 min unloading at 36 m sliding, the restarting friction is remarkably reduced from the previous high level at interruption, but it is slightly higher than the initial friction. This friction reduction was considered

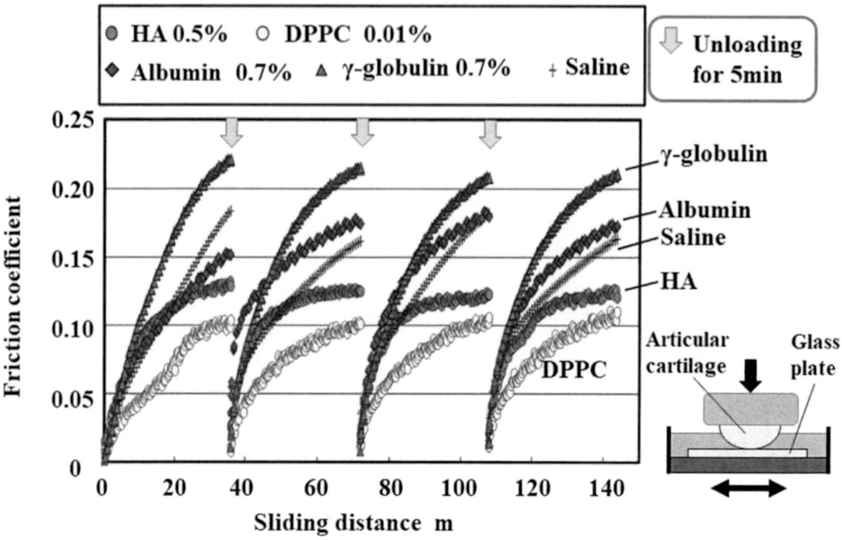

Figure 3.34 Influence of single synovia constituent on changes in friction for intact cartilage during repeated reciprocating test including unloading periods for 5 min. At severe rubbing conditions at higher friction, DPPC and HA reduce friction compared with saline, but γ-globulin and albumin (after second run) increase friction.
Modified from Murakami, T., Yarimitsu, S., Nakashima, K., Sawae, Y., & Sakai, N. (2013). Influence of synovia constituents on tribological behaviors of articular cartilage. *Friction*, *1*(2), 150−162. https://doi.org/10.1007/s40544-013-0010-6

to have been brought by the recovery of both the hydration and some deformation of articular cartilage, in which the hydration lubrication and biphasic lubrication become partly effective accompanied with adsorbed film formation although initial adsorbed film may have been partly removed. In the second reciprocating sliding process, the friction again gradually increases with sliding distance. In the second to fourth processes where the cartilage surface was partly injured, albumin showed higher friction than the saline. It is considered that high friction with γ-globulin or albumin is caused by the bonding effect due to interactions with globular proteins at very thin film conditions, where the surface protection with adsorbed films is little expected.

The γ-globulin possess a strong hydrophobic nature and in aqueous solutions adsorbs easily onto solid surfaces as indicated by fluorescent images (Murakami et al., 2009), thus protecting the bulk material, but at very thin film conditions show strong bonding effect. On the contrary, albumin with superior hydrophilic property constructs the layer with low shearing resistance and appears to desorb easily compared with γ-globulin.

As the addition of single synovia constituent on improvement of steady or final friction at each 36 m sliding gave small, limited effect, the effectiveness of the combination of different synovia constituents was examined (Murakami et al., 2013). In these cases to focus the reduction in final friction at each 36 m sliding, the synergistic effect of γ-globulin and HA as a viscosity improver was confirmed, but the coexistence of albumin and HA showed adverse interaction probably due to the repulsive properties of negatively charged molecules (friction data are not shown). Furthermore, the coexistence of DPPC and albumin or γ-globulin reduced friction compared with DPPC alone in saline.

Therefore, as shown in Fig. 3.35, the effectiveness of DPPC to HA solutions with and without proteins was evaluated. It is noted that even the addition of 0.01 mass% DPPC alone to HA solution exhibited a remarkable reduction in friction. Particularly, it is noteworthy that the addition of 0.01 mass% DPPC accompanied with 1.4 mass% albumin and 0.7 mass% γ-globulin in HA solution (simulated synovial fluid) demonstrated the lowest frictional behaviors as about 0.01 without gradual increase until each 36 m sliding. It is confirmed that the friction does not increase but maintains a very low steady level. On the contrary, the addition of DPPC with albumin alone or γ-globulin alone in HA solution shows higher friction than DPPC alone in HA solution but lower than saline solution.

Next, wear on cartilage surfaces was evaluated by comparison of changes in surface micrographs.

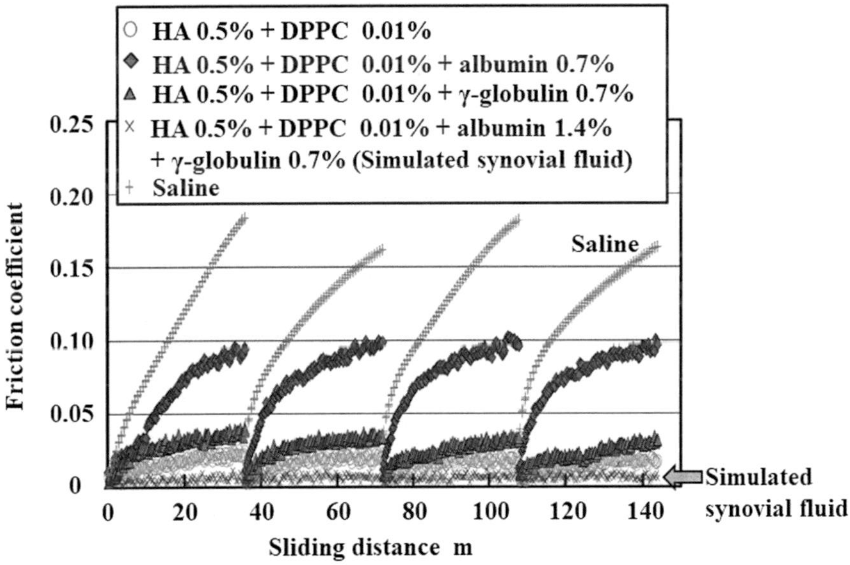

Figure 3.35 Influence of combinations of synovia constituents on changes in friction for intact cartilage during repeated reciprocating test including unloading periods for 5 min.Simulated synovial fluid (HA: 0.5 mass% + DPPC: 0.01 mass% + albumin: 1.4 mass% + γ-globulin: 0.7 mass%) showed superior boundary lubrication ability. Modified from Murakami, T., Yarimitsu, S., Nakashima, K., Sawae, Y., & Sakai, N. (2013). Influence of synovia constituents on tribological behaviors of articular cartilage. *Friction*, *1*(2), 150–162. https://doi.org/10.1007/s40544-013-0010-6

As shown in Fig. 3.36, intact articular cartilage has smooth surface with some irregularity but cartilage specimen after reciprocating test lubricated with HA solution containing 0.01 mass% DPPC exhibited the wear with clear scratches in rubbing direction on surfaces although the friction level was the secondary lower. On the contrary, little wear were confirmed for lubricant as HA solution containing 0.01 mass% DPPC, 1.4 mass% albumin, and 0.7 mass% γ-globulin, where superficial gellike layer may have been slightly removed but has no scratching. This lubricant contains similar important constituents to natural synovial fluid and is called the simulated synovial fluid. Natural synovial fluids have similar composition as hyaluronate solution containing lubricating constituents such as 1.25 mass% albumin and 0.75 mass% globulin (including α-, β-, and γ-globulin) as medium values (Sasada et al., 1988),1.1 mass% albumin and 0.7 mass% globulin (Oates et al., 2006) or 1.9 mass% albumin, 1.1 mass% globulin, and 0.01 mass% DPPC (Higaki et al., 1998).

As described above, it is suggested that the adequate adsorbed films with low friction and surface protective ability appear to be formed on cartilage surfaces lubricated with the simulated synovial fluid which contains four kinds of lubricating synovia constituents at appropriate concentrations. It is suggested that the synergistic lubricating film composed of the layered adsorbed films of proteins in HA solution containing 1.4 mass% albumin and 0.7 mass% γ-globulin (Nakashima et al., 2005) and low shearing phospholipid layer exhibited low friction and minimal wear for articular cartilage surfaces.

3.4.6 Gel film lubrication

In lubrication mechanism of natural synovial joints, the articular cartilage as bearing surfaces appear to play important roles together in close coordination with synovial

(a) Surface before test (b) Surface with HA+DPPC (c) Surface with HA+DPPC +albumin +γ-globulin

Figure 3.36 Influence of synovia constituents on wear of articular cartilage. (a) Intact cartilage surface (b) With HA + DPPC, surface scratching is observed although friction level is secondary lower. (c) With HA + DPPC + albumin+γ-globulin, superficial gellike layer is probably removed but no scratching is observed.
Modified from Murakami, T., Yarimitsu, S., Nakashima, K., Sawae, Y., & Sakai, N. (2013). Influence of synovia constituents on tribological behaviors of articular cartilage. *Friction, 1*(2), 150–162. https://doi.org/10.1007/s40544-013-0010-6

fluid which has superior lubricating ability (Katta et al., 2008). Articular cartilage is not simple, soft elastic layer but has biphasic property with fiber-reinforced hierarchic structure (Fig. 3.1) and lubricious surface such as lamina splendens (Gardner, 1972; MacConaill, 1951; Orford and Gardner, 1985). As shown in transmission electron microscopy (TEM) image of transection of articular cartilage (Fig. 3.2) (Higaki et al., 1998), top surface is constructed by acellular and noncollagenous amorphous layer and covered with continuous adsorbed films composed of mainly phospholipid bilayer and fibrous molecules.

To examine the changes in lubrication mechanism after removal of adsorbed film, the changes in frictional behavior of intact articular cartilage against reciprocating glass plate were observed by severe rubbing in reciprocating motion lubricated with saline at slow speed by Murakami et al. (1999).

The reciprocating tests for intact articular cartilage specimen prepared from porcine femoral chondyle against a flat glass plate were conducted at slow sliding speed of 5 mm/s and stroke length of 35 mm. Stationary load of 4.9 N (average contact stress is 0.13 MPa) was applied. As friction level increased with sliding distance, the tests were interrupted at the specified distance and the rubbing surfaces of cartilage specimen were observed by atomic force microscopy (AFM) in tapping mode in saline.

The changes in frictional behavior and AFM images of articular cartilage surfaces with friction models are shown in Figs. 3.37 and 3.38, respectively. The intact cartilage surface (Fig. 3.38a) has concave-convex surfaces of maximum height as about 1−2 μm with gentle slopes and no fibrous structure is observed, as shown in AFM image. It is noted in Fig. 3.37 that initial friction coefficient is lower than 0.01 but gradually increases with repetition of rubbing. At short sliding distance of 0.35 m, friction

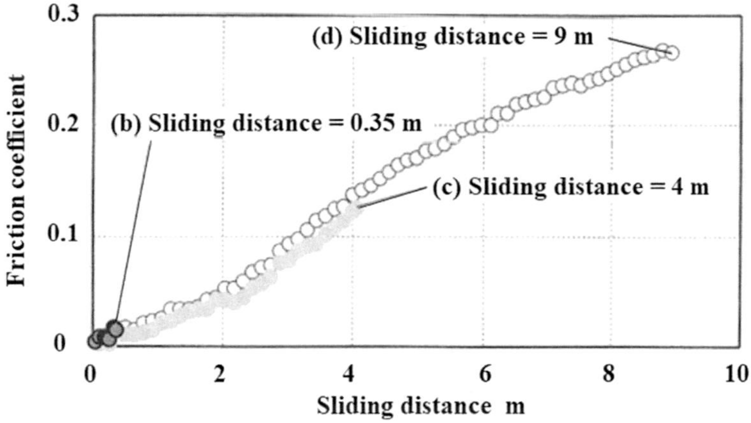

Figure 3.37 Frictional behaviors of articular cartilage against glass plate lubricated with saline. Friction tests were stopped at 0.35 m, 4 m, and 9 m sliding, and then the rubbed surfaces were observed by AFM.
Modified from Murakami, T., Sawae, Y., Horimoto, M., & Noda, M. (1999). Role of surface layers of natural and artificial cartilage in thin film lubrication. *Tribology Series*, *36*, 737−747. https://doi.org/10.1016/s0167-8922(99)80092-3

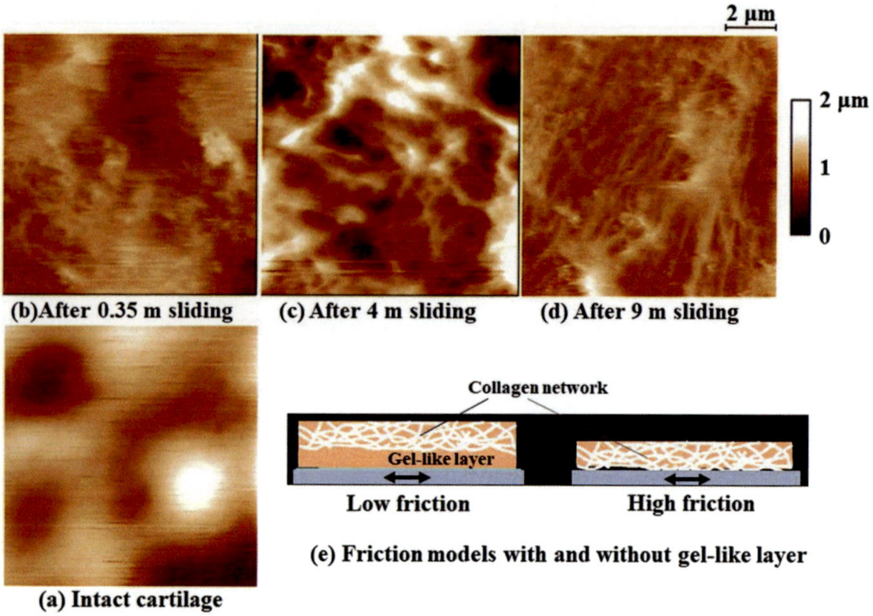

Figure 3.38 Surfaces of articular cartilage imaged with AFM in saline and friction model
(a) Intact cartilage surface (b) After 0.35 m sliding, the adsorbed films and upper part of gellike layer are removed. (c) After 4 m sliding, surface is significantly roughened. (d) After 9 m sliding, the collagen network is clearly observed. (e) Friction models for low friction with gellike layer and high friction without gellike layer are depicted.
Modified from Murakami, T., Sawae, Y., Horimoto, M., & Noda, M. (1999). Role of surface layers of natural and artificial cartilage in thin film lubrication. *Tribology Series*, *36*, 737–747. https://doi.org/10.1016/s0167-8922(99)80092-3

coefficient is still low as about 0.02 even after removal of adsorbed films, where slight wear on cartilage surface was noticed in AFM image (Fig. 3.38b). At 4 m sliding, friction coefficient increased above 0.1 and surface was significantly roughened (Fig. 3.38c). After 9 m sliding, friction further increased and the collagenous fibers became evident on rubbing surface (Fig. 3.38d).

These experimental results indicate that the underlying noncollagenous gellike layer with adequate thickness can maintain low friction and protect the cartilage surface as shown in Fig. 3.38e. The content of gellike layer was identified as mainly proteoglycan by the enzyme treatment with chondroitinase ABC (Sawae et al., 2000). Proteoglycan is the main extracellular matrix formed inside the cartilage. After the digestion of gel layer, the gellike morphology disappeared and collagenous fibrous network became exposed on the surface. If the surface damage of this gel layer is restricted at mild level, the proteoglycan gel layer is expected to be repaired by supply of proteoglycan tissue produced by chondrocytes. Hyaluronic acid and other adsorbed molecules seem to be supplied by synovial fluid.

The same reciprocating test was conducted under unlubricated condition, and similar time-dependent frictional behavior was observed. However, the friction level was about half compared with lubricated test. This fact suggests that the proteoglycan gel layer itself can act as low shearing gel lubricant like solid lubricant. Surface structure covered with adsorbed films on underlying proteoglycan gellike layer is considered as a failsafe system in synovial joints.

This gellike layer is considered to correspond to the surface amorphous layer (SAL) containing glycosaminoglycan (GAG), protein, and lipid in experiment by Graindorge et al. (2006), but the removal of the SAL with sodium dodecyl sulfate (SDS) treatment was found to have no effect on the start-up and dynamic friction in their cartilage-on-cartilage rubbing test. In contrast, in their biphasic FE analyses (Graindorge et al., 2005; Graindorge, Ferrandez, Jin, et al., 2006), the effectiveness of fluid load support in model with SAL had been indicated for shock load and dynamic loading. On the reason why the removal of SAL did not increase friction in their experiment, it was noticed that SAL containing lipids and sulfated GAGs may have been replenished during loading and friction testing process under mild condition. At their test the removal of SAL without damaging the underlying cartilage with SDS treatment may have enabled to replenish the surface layer with similar biphasic behavior to the original SAL with the aid of rehydration (Graindorge, Ferrandez, Jin, et al., 2006).

At severe condition in reciprocating test of cartilage against glass plate at continuous loading (Murakami et al., 1999), the removal of proteoglycan gellike layer showed clear increase in friction. Under both mild and severe conditions, the noncollagenous gellike superficial thin layer is likely to play important roles as protective barrier with biphasic and low shear properties, and/or as surface material desirable for adsorption of lubricating synovia constituents.

3.4.7 Super-lubricity based on adaptive multimode lubrication

Natural synovial joints should sustain healthy states with extremely low friction and minimal damage/wear to actualize various daily activities under different operating conditions, where it is expected that the adaptive multimode lubrication mechanism becomes effective according to the severity of operating conditions. In this section, the important features of biphasic, boundary and gel-film lubrication mechanisms are mainly described considering the implication of hydration lubrication in thin film conditions for slow movement activities at which the hydrodynamic lubrication mode has small effect.

As shown in biphasic FE analysis, the high interstitial fluid pressure can be maintained during reciprocating motions under migrating (on-off) contact to articular cartilage (Fig. 3.19), which enables the unloaded region in cartilage to rehydrate. In such case, the biphasic lubrication mode plays a main role. Furthermore, the effect of fiber-reinforcement in STZ layer of cartilage on frictional behavior was investigated and it was shown that STZ model maintained lower friction level than the standard model under migrating contact condition (Fig. 3.23). In STZ model, high interstitial fluid pressure zone at 100 s running is wider and the negative pressure is noticed in front of leading zone which contributes to rehydration of cartilage surface (Fig. 3.24).

As a consequence, the fluid phase load partitioning of STZ model at 100 s was 86.2% with low friction coefficient of about 0.03 (Fig. 3.23) while that for standard model was 82.5% with an increase in friction coefficient as 0.04. Besides, it is confirmed that lower level in von Mises stress in superficial surface zone of conjunction for STZ model was observed. It is considered that the reinforcement of superficial tangential zone with anisotropic collagen fiber network restricted the horizontal stretch of the tissue of the superficial layer which enhanced the interstitial fluid pressurization.

Next, we discuss the effectiveness of adsorbed film formation of synovia constituents on friction and wear of intact ellipsoidal articular cartilage sliding against a flat glass plate in boundary lubrication. In reciprocating experiment, the effect of adsorbed film formation on friction was examined under stationary contact condition where the interstitial fluid pressurization had subsided. In the repeated reciprocating tests of 36 m sliding with unloading for 5 min and reloading, time-dependent frictional behaviors were observed as shown in Figs. 3.34 and 3.35. The reduction of restarting friction indicates the extent of recovery of biphasic lubrication and hydration lubrication accompanied with friction reduction by adsorbed films after rehydration of cartilage and partial recovery of cartilage deformation. As shown in Fig. 3.34, the addition of single protein such as albumin or γ-globulin improved the restarting friction but did show higher friction than saline at 36 m sliding, while DPPC exhibited friction reduction for full 36 m sliding. In contrast, a combination of HA as a viscous constituent and phospholipid and/or protein exhibited lower friction at restart and at 36 m sliding (Fig. 3.35). Particularly, it is noteworthy that the simulated synovial fluid as HA solution of 1.4 mass% albumin, 0.7 mass% γ-globulin (A/G ratio = 2:1), and 0.01 mass% DPPC maintained extremely low friction, which contains main lubricating synovia constituents except lubricin. Furthermore, this simulated synovial fluid was the most effective in friction reduction for partially damaged cartilage (Murakami et al., 2013) and in wear reduction of PVA hydrogel as artificial cartilage (Murakami et al., 2014; Murakami, Yarimitsu, et al., 2015). It is expected that some form of layered structure with low shearing zone might have been constructed in adsorbed films of this simulated synovial fluid. This was suggested on the basis of the fact that the layered structure of albumin and γ-globulin was observed by fluorescent staining method for HA solution of 0.7 mass% albumin/1.4 mass% γ-globulin or 1.4 mass% albumin/0.7 mass% γ-globulin which showed low friction and minimal wear for PVA hydrogel against itself (Nakashima et al., 2005). Although in situ observation of adsorbed film formation for HA solution with proteins indicated the appropriate adsorbed film formation on cartilage surfaces enhanced by rubbing action (Murakami et al., 2009), the experimental validation for a layered structure with simulated synovial fluid has not yet been conducted and is planned as a future study.

On the lubricating ability of lubricin (Swann et al., 1985), the effects to reduce friction have been reported, but lubricin seems to mainly contribute to connect the proteoglycan molecules and/or hyaluronan to collagen fiber in surface layer as indicated by Jahn et al. (2016) (Fig. 3.39). Mirea et al. (2013) suggested that it anchors lipid layers on the cartilage. As shown in TEM image of intact cartilage (Fig. 3.2), the adsorbed films on the proteoglycan gel layer appear to be composed of bilayer (or including mono/multilayer) of phospholipids accompanied with proteins, lubricin, and fibrous

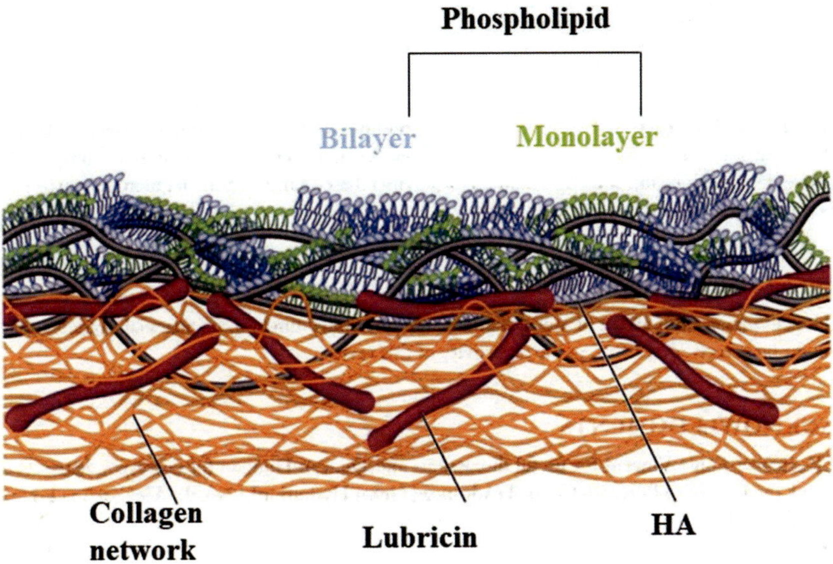

Figure 3.39 Structure of cartilage boundary lubricant layer. This structure was proposed by Seror et al. (2015). Lubricin in the outer superficial zone and at the cartilage surface interacts with and immobilizes hyaluronic acid (HA) at the surface; the surface-attached HA in turn complexes with phosphatidylcholine (PC) lipids to form monolayer or bilayer assemblies exposing phosphocholine groups. These exposed groups slide past similar groups from the opposing surface with low friction (μ values down to 10^{-3}) up to high pressures (100 atm or more) via the hydration lubrication mechanism.
Modified with permission from Jahn, S., Seror, J., & Klein, J. (2016). Lubrication of articular cartilage. *Annual Review of Biomedical Engineering*, *18*, 235–258. https://doi.org/10.1146/annurev-bioeng-081514-123305

molecules. Clarification of detailed structure of adsorbed films on cartilage surface is expected to be actualized in future research.

After removal of adsorbed film at local direct contacts under severe condition, the underlying proteoglycan gellike layer should act as low shearing layer and protective barrier as gel film lubrication described in Section 3.4.6, or play a lubricating role in hydration lubrication. Under severe rubbing some parts of proteoglycan gellike layer may be damaged, in which the restoration by appropriate supply of proteoglycan molecules or related constituents via chondrocytes and/or synovial fluid becomes important. The simulated synovial fluid was most effective in friction reduction for damaged gellike layer (Murakami et al., 2013).

Hydration lubrication mechanism is expected to contribute to load carrying with low friction for hydrophilic gellike layer of cartilage in aqueous environment under wider operating conditions as described below. Ikeuchi (2007) defined the hydration lubrication as "water lubrication of surface layer with hydrophilic polymers fixed to the surface," with load support by water and shearing of water molecules with low stress for rubbing of both cartilage-on-cartilage and cartilage-on-hydrated surface.

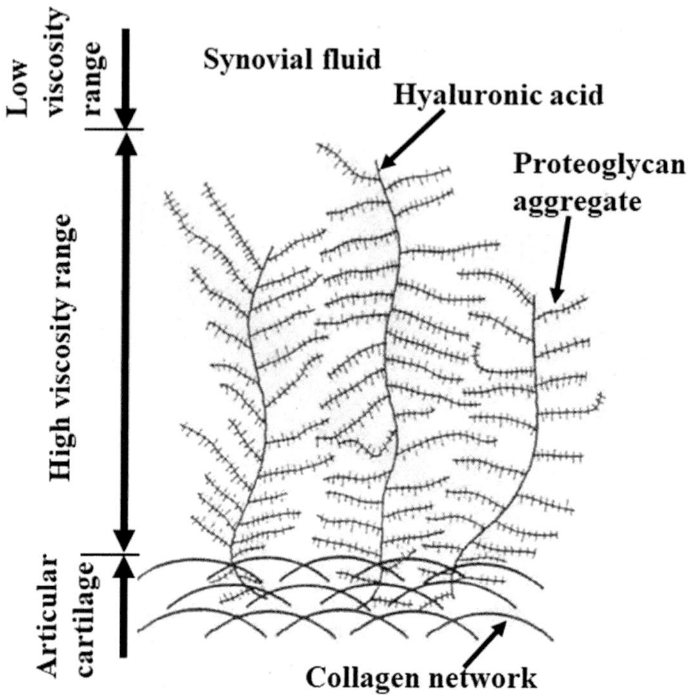

Figure 3.40 Schematic representation of surface gel hydration lubrication. It is expected that the high viscous region around hydrated proteoglycan surface gel layer of cartilage extends the valid region of fluid film lubrication.
Modified from Sasada, T. (2000). Lubrication of human joints — Nature of joint friction and "surface gel hydration lubrication". *Japanese Journal of Clinical Biomechanics*, *21*, 17–22.

For the effective hydration lubrication to maintain very low friction even at high contact pressure, it is pointed out by Klein et al. (2013) (Klein, 2006; Lin & Klein, 2022; Jahn et al., 2016; Raviv & Klein, 2002) that hydration layers strongly held by the charges they surround in aqueous media can support large pressures without being squeeze out and may be very rapidly relaxing to show a fluid response to shear.

As additional feature, Sasada (1996) proposed "Surface gel hydration lubrication" to focus the important role of high viscous region around hydrated proteoglycan surface gel layer to extend the valid region of fluid film lubrication (Fig. 3.40) (Ishikawa et al., 2001; Sasada, 2000). Surface gel hydration lubrication plays an important role in reducing friction and protecting articular cartilage particularly under thin film severe condition, but more research is required to elucidate the detailed mechanism including the interaction between adsorbed films and viscous hydrated layers.

As mentioned above, in natural synovial joints, various lubrication modes are expected to become effective depending on the severity of operating conditions, as adaptive multimode lubrication mechanism (Fig. 3.41). The effectiveness of each lubrication mode appears to be comprehensively controlled by the lubrication

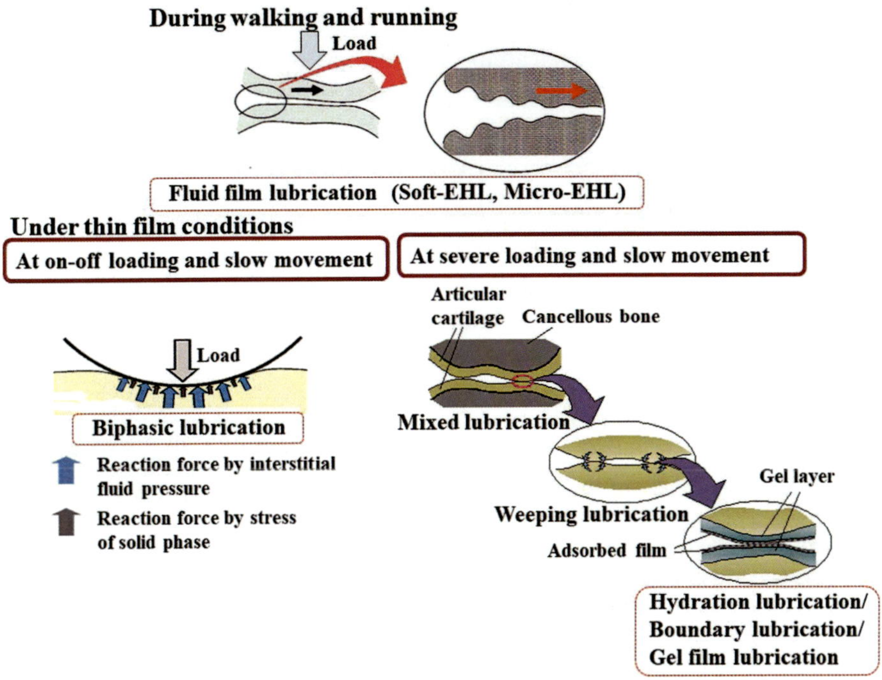

Figure 3.41 Schema for adaptive multimode lubrication mechanism in synovial joints. Natural synovial joints function as low friction and minimum wear bearings with synergistic combination of different lubrication mechanisms under various daily activities.
Modified from Murakami, T., Yarimitsu, S., Sakai, N., Nakashima, K., Yamaguchi, T., & Sawae, Y. (2017). Importance of adaptive multimode lubrication mechanism in natural synovial joints. *Tribology International*, *113*, 306–315. https://doi.org/10.1016/j.triboint.2016.12.052

parameter such as [(Lubricant viscosity) × (Sliding speed)/Load], as shown in Stribeck curve in Fig. 3.42. In this graph, it is indicated that the hydration lubrication mechanism can play a lubricating role in wider operating conditions, although other lubrication modes also may function with an overlap. It is important that various lubrication mechanisms can act synergistically in natural synovial joints as failsafe and robust systems corresponding to changes in operating variables.

In Sections 3.4.5 and 3.4.6, most of results for rubbing of articular cartilage against smooth rigid body were evaluated as models under severer rubbing conditions than that for cartilage-on-cartilage in natural synovial joints such as pendulum tests for porcine shoulder joints in Section 3.4.5; therefore, it is expected that most of lubricating mechanisms effective for cartilage-on-rigid body can be commonly applied for the rubbing pair of cartilage and cartilage in natural whole synovial joints. In rubbing system of healthy articular cartilage against itself, some beneficial lubricating mechanisms can be added. In combination of compliant biphasic materials, the contact stress can be reduced accompanied with improvement of congruity and the rehydration

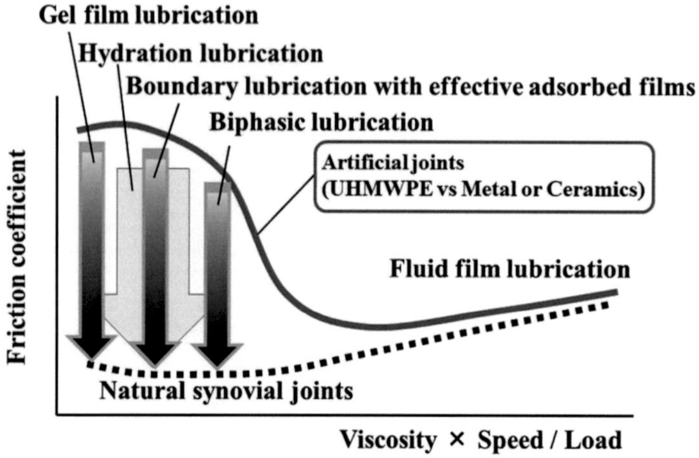

Figure 3.42 Stribeck curve for adaptive multimode lubrication. In healthy natural synovial joints, low friction is sustained with the synergistic combinations of various lubrication modes in wider operating conditions.
Modified from Murakami, T., Yarimitsu, S., Sakai, N., Nakashima, K., Yamaguchi, T., & Sawae, Y. (2017). Importance of adaptive multimode lubrication mechanism in natural synovial joints. *Tribology International*, *113*, 306–315. https://doi.org/10.1016/j.triboint.2016.12.052

of cartilage can be made easier. The interaction between rubbing surfaces covered with hydrophilic adsorbed films on underlying proteoglycan layers becomes favorable. In the pendulum tests of porcine shoulder joints (Higaki et al., 1998; Murakami et al., 1998), the effectiveness of adsorbed film formation by HA solutions with 0.01 mass% DPPC or 1.0 mass% γ-globulin was confirmed as some common features. The micro-movement of cartilage with appropriate surface texture will be beneficial for rehydration and recovery of adsorbed film formation at inevitable unloading phase for cartilage-on-cartilage.

In the related field, the concept of adaptive multimode lubrication described above has been successfully applied to the biomimetic design of artificial hydrogel cartilage materials with superior lubricity (Murakami et al., 1993; Murakami, Sakai, et al., 2015). The elucidation of adaptive multimode lubrication mechanism of the tribological systems in other biological tissues and organs will be required in near future.

3.5 Wear and degradation mechanism in natural joints

Healthy synovial joints sustain smooth movement with low friction and minimum wear under various daily activities for a whole human life. However, under excessively severe loading conditions, at accidental failures or at diseased situations including inflammation of synovium (synovial membrane), articular cartilage will be degraded and worn or damaged. In cases with loss of adequate lubricating mechanisms and lowering of lubricating ability with lubricants, the rubbing process is likely to enhance the wear of articular cartilage at direct contact areas. Fig. 3.35 indicates the insufficiency of protection by adsorbed films, and Fig. 3.37 shows the wear process of the uppermost superficial surface zone of articular cartilage. Articular cartilage is avascular tissue and has extremely low ability to repair. If the worn or damaged zones of cartilage have not been repaired against further rubbing actions, friction can be risen and the severe shearing and deformation will occur at direct contact zones, which will result in progress of wear and surface failure.

As shown in Fig. 3.43, the extent of friction reduction effect with the same lubricating constituents is depressed for damaged cartilage surfaces with partially removed

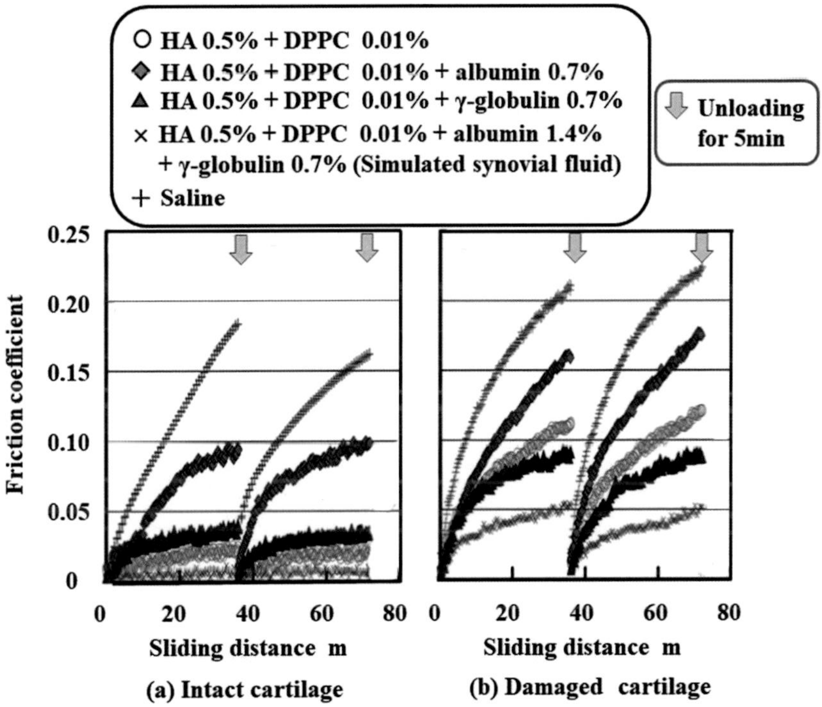

Figure 3.43 Influence of combinations of synovia constituents on changes in friction for intact and damaged cartilage specimens during repeated reciprocating test including unloading periods for 5 min. (a) Intact cartilage: Simulated synovial fluid shows the lowest frictional behaviors as about 0.01 without gradual increase. Simulated synovial fluid showed little wear. (b) Damaged cartilage: Compared with intact cartilage, the friction levels are generally increased and the order of friction level is partly changed. Simulated synovial fluid shows the lowest friction.

Modified from Murakami, T., Yarimitsu, S., Nakashima, K., Sawae, Y., & Sakai, N. (2013). Influence of synovia constituents on tribological behaviors of articular cartilage. *Friction*, *1*(2), 150–162. https://doi.org/10.1007/s40544-013-0010-6

proteoglycan brushlike layer compared with intact cartilage surfaces. For damaged cartilage, the friction levels are generally increased and the order of friction level is partly changed, compared with intact cartilage. It is suggested for damaged cartilage that the protective role of γ-globulin with strong adsorption ability becomes superior to that of DPPC in HA solution.

The degradation and damage of articular cartilage tissue and the deterioration of lubricating ability in synovial fluid with denatured synovia constituents can raise the friction level. It should be taken into consideration that an increase in friction for damaged cartilage enhances the rubbing severity in a vicious cycle. Therefore, it is expected to sustain low friction and minimal wear in situation to take advantage of the adaptive multimode lubrication mechanism including the restoration systems.

References

Alberts, B., Johnson, A., Lewis, J., Morgan, D., Raff, M., Roberts, K., & Walter, P. (2021). *Molecular biology of the cell*. W.W. Norton & Company.

Ateshian, G. A. (1997). A theoretical formulation for boundary friction in articular cartilage. *Journal of Biomechanical Engineering, 119*(1), 81–86. https://doi.org/10.1115/1.2796069

Ateshian, G. A. (2009). The role of interstitial fluid pressurization in articular cartilage lubrication. *Journal of Biomechanics, 42*(9), 1163–1176. https://doi.org/10.1016/j.jbiomech.2009.04.040

Ateshian, G. A., Wang, H., & Lai, W. M. (1998). The role of interstitial fluid pressurization and surface porosities on the boundary friction of articular cartilage. *Journal of Tribology, 120*(2), 241–248. https://doi.org/10.1115/1.2834416

Blewis, M. E., Nugent-Derfus, G. E., Schmidt, T. A., Schumacher, B. L., & Sah, R. L. (2007). A model of synovial fluid lubricant composition in normal and injured joints. *European Cells and Materials, 13*, 26–38. https://doi.org/10.22203/ecm.v013a03

Charnley, J. (1959). The lubrication of animal joints. In *Proc. Symposium on Biomechanics* (pp. 12–19). Institution of Mechanical Engineers.

Charnley, J. (1960). The lubrication of animal joints in relation to surgical reconstruction by arthroplasty. *Annals of the Rheumatic Diseases, 19*, 10–19. https://doi.org/10.1136/ard.19.1.10

Chen, M., Briscoe, W. H., Armes, S. P., & Klein, J. (2009). Lubrication at physiological pressures by polyzwitterionic brushes. *Science, 323*(5922), 1698–1701. https://doi.org/10.1126/science.1169399

Cook, A. (1974). *A study of the rheological characteristics of some fluids and tissues in synovial joint lubrication*. Ph.D. Thesis. The University of Leeds.

Crocket, R. (2014). Tribology of natural articular joints. In N. Spencer, D. (Ed.), *Aqueous lubrication Natural and biomimetic approaches* (pp. 1–32). World Scientific Publishing.

Dintenfass, L. (1963). Lubrication in synovial joints: A theoretical analysis. *The journal of bone and joint surgery American, 45*, 1241–1256. https://doi.org/10.2106/00004623-196345060-00011

Dowson, D., & Jin, Z.-M. (1986). Micro-elastohydrodynamic lubrication of synovial joints. *Engineering in Medicine, 15*(2), 63–65. https://doi.org/10.1243/EMED_JOUR_1986_015_019_02

Dowson, D., Unsworth, A., & Wright, V. (1970). Analysis of 'boosted lubrication' in human joints. *Journal of Mechanical Engineering Science, 12*(5), 364–369. https://doi.org/10.1243/jmes_jour_1970_012_060_02

Dowson, D., & Wright, V. (1973). Bio-Tribology. In T. Davenport, C. (Ed.), *The rheology of lubricants* (pp. 81–88). Applied Science Publishers Ltd.

Dowson, D. (1966-67). Modes of lubrication in human joints. *Proceedings of the Institution of Mechanical Engineers, 181*(Pt3J), 45–54.

Eyre, D. (2002). Collagen of articular cartilage. *Arthritis Research, 4*(1), 30–35. https://doi.org/10.1186/ar380

Fein, R., S. (1966-67). Are synovial joints squeeze film lubricated? *Proceedings of the Institution of Mechanical Engineers, 181*(Pt3J), 125–128.

Forster, H., & Fisher, J. (1996). The influence of loading time and lubricant on the friction of articular cartilage. *Proceedings of the Institution of Mechanical Engineers, Part H: Journal of Engineering in Medicine, 210*(2), 109–119. https://doi.org/10.1243/pime_proc_1996_210_399_02

Forster, H., & Fisher, J. (1999). The influence of continuous sliding and subsequent surface wear on the friction of articular cartilage. *Proceedings of the Institution of Mechanical Engineers, Part H: Journal of Engineering in Medicine, 213*(4), 329–345. https://doi.org/10.1243/0954411991535167

Fujie, H., & Imade, K. (2015). Effect of low tangential permeability in the superficial layer on the frictional property of articular cartilage. *Biosurface and Biotribology, 1*, 125–129.

Fujie, H., Morishita, S., & Yarimitsu, S. (2018). Effect of collagen-induced residual stress on the frictional property of articular cartilage. *Biosurface and Biotribology, 4*(2), 68–71. https://doi.org/10.1049/bsbt.2018.0008

Gardner, D. L. (1972). The influence of microscopic technology on knowledge of cartilage surface structure. *Annals of the Rheumatic Diseases, 31*(4), 235–258. https://doi.org/10.1136/ard.31.4.235

Gladstone, J. R., & Medley, J. B. (1990). Paper X (ii) Comparison of theoretical and experimental values for friction of lubricated elastomeric surface layers under transient conditions. *Tribology Series, 17*(C), 241–250. https://doi.org/10.1016/S0167-8922(08)70262-1

Graindorge, S., Ferrandez, W., Ingham, E., Jin, Z., Twigg, P., & Fisher, J. (2006). The role of the surface amorphous layer of articular cartilage in joint lubrication. *Proceedings of the Institution of Mechanical Engineers, Part H: Journal of Engineering in Medicine, 220*(5), 597–607. https://doi.org/10.1243/09544119JEIM122

Graindorge, S., Ferrandez, W., Jin, Z. M., Ingham, E., & Fisher, J. (2006). The natural synovial joint: A finite element investigation of biphasic surface amorphous layer lubrication under dynamic loading conditions. *Proceedings of the Institution of Mechanical Engineers, Part J: Journal of Engineering Tribology, 220*(8), 671–681. https://doi.org/10.1243/13506501JET95

Graindorge, S., Ferrandez, W., Jin, Z., Ingham, E., Grant, C., Twigg, P., & Fisher, J. (2005). Biphasic surface amorphous layer lubrication of articular cartilage. *Medical Engineering and Physics, 27*(10), 836–844. https://doi.org/10.1016/j.medengphy.2005.05.001

Higaki, H., & Murakami, T. (1996). Role of constituents in synovial fluid and surface layer of articular cartilage in joint lubrication (part 2): The boundary lubricating ability of proteins. *Japanese Journal of Tribology, 40*(7), 691–700.

Higaki, H., Murakami, T., Nakanishi, Y., Miura, H., Mawatari, T., & Iwamoto, Y. (1998). The lubricating ability of biomembrane models with dipalmitoyl phosphatidylcholine and γ-globulin. *Proceedings of the Institution of Mechanical Engineers, Part H: Journal of Engineering in Medicine, 212*(5), 337–346. https://doi.org/10.1243/0954411981534114

Hills, B. A. (1989). Oligolamellar lubrication of joints by surface active phospholipid. *Journal of Rheumatology, 16*(1), 82–91.

Hills, B. A. (2000). Boundary lubrication in vivo. *Proceedings of the Institution of Mechanical Engineers, Part H: Journal of Engineering in Medicine, 214*(1), 83–94. https://doi.org/10.1243/0954411001535264

Hills, B. A., & Butler, B. D. (1984). Surfactants identified in synovial fluid and their ability to act as boundary lubricants. *Annals of the Rheumatic Diseases, 43*(4), 641–648. https://doi.org/10.1136/ard.43.4.641

Hosoda, N., Sakai, N., Sawae, Y., & Murakami, T. (2008). Depth-dependence and time-dependence in mechanical behaviors of articular cartilage in unconfined compression test under constant total deformation. *Journal of Biomechanical Science and Engineering, 3*(2), 209–220. https://doi.org/10.1299/jbse.3.209

Ikeuchi, K. (1995). The role of synovial fluid in joint lubrication. *Tribology Series, 30*(C), 65–69. https://doi.org/10.1016/S0167-8922(08)70617-5

Ikeuchi, K. (2007). Origin and future of hydration lubrication. *Proceedings of the Institution of Mechanical Engineers, Part J: Journal of Engineering Tribology, 221*(3), 301–305. https://doi.org/10.1243/13506501JET214

Ikeuchi, K., Oka, M., & Gi, K. (1989). An experimental study of deformation and squeeze film effect in a synovial joint. *Transactions of the Japan Society of Mechanical Engineers Series C (In Japanese), 55*(516), 2123–2130. https://doi.org/10.1299/kikaic.55.2123

Ikeuchi, K., Oka, M., & Mori, H. (1989). A simulation of the squeeze film effect in a hip joint. *Transactions of the Japan Society of Mechanical Engineers Series C (In Japanese), 55*(510), 508–515. https://doi.org/10.1299/kikaic.55.508

Ishikawa, Y., Sasada, T., & Ikeuchi, K. (2001). Surface gel hydration lubrication – Proposal of a new idea for joint lubrication. In JAST (Ed.), *Proc. ITC Nagasaki 2000* (Vol. 2, pp. 1491–1496). JAST.

Jahn, S., Seror, J., & Klein, J. (2016). Lubrication of articular cartilage. *Annual Review of Biomedical Engineering, 18*, 235–258. https://doi.org/10.1146/annurev-bioeng-081514-123305

Johnson, K. L., Greenwood, J. A., & Poon, S. Y. (1972). A simple theory of asperity contact in elastohydro-dynamic lubrication. *Wear, 19*(1), 91–108. https://doi.org/10.1016/0043-1648(72)90445-0

Jones, E. S. (1936). Joint lubrication. *Lancet*, 1043–1044.

Jurvelin, J. S., Buschmann, M. D., & Hunziker, E. B. (2003). Mechanical anisotropy of the human knee articular cartilage in compression. *Proceedings of the Institution of Mechanical Engineers, Part H: Journal of Engineering in Medicine, 217*(3), 215–219. https://doi.org/10.1243/095441103765212712

Katta, J., Jin, Z., Ingham, E., & Fisher, J. (2008). Biotribology of articular cartilage – A review of the recent advances. *Medical Engineering & Physics, 30*, 1340–1363.

Klein, J. (2006). Molecular mechanisms of synovial joint lubrication. *Proceedings of the Institution of Mechanical Engineers, Part J: Journal of Engineering Tribology, 220*(8), 691–710. https://doi.org/10.1243/13506501JET143

Klein, J. (2013). Hydration lubrication. *Friction, 1*(1), 1–23. https://doi.org/10.1007/s40544-013-0001-7

Kobayashi, S., Yonekubo, S., & Kurogouchi, Y. (1995). Cryoscanning electron microscopic study of the surface amorphous layer of articular cartilage. *Journal of Anatomy, 187*(2), 429–444.

Krishnan, R., Kopacz, M., & Ateshian, G. A. (2004). Experimental verification of the role of interstitial fluid pressurization in cartilage lubrication. *Journal of Orthopaedic Research, 22*(3), 565–570. https://doi.org/10.1016/j.orthres.2003.07.002

Kumar, P., Oka, M., Toguchida, J., Kobayashi, M., Uchida, E., Nakamura, T., & Tanaka, K. (2001). Role of uppermost superficial surface layer of articular cartilage in the lubrication mechanism of joints. *Journal of Anatomy, 199*(3), 241–250. https://doi.org/10.1017/S0021878201008032

Lewis, P. R., & McCutchen, C. W. (1959). Experimental evidence for weeping lubrication in mammalian joints. *Nature, 184.*

Lin, W., & Klein, J. (2022). Hydration lubrication in Biomedical applications: From cartilage to hydrogels. *Accounts of Materials Research, 3*(2), 213–223. https://doi.org/10.1021/accountsmr.1c00219

Li, L. P., Soulhat, J., Buschmann, M. D., & Shirazi-Adl, A. (1999). Nonlinear analysis of cartilage in unconfined ramp compression using a fibril reinforced poroelastic model. *Clinical Biomechanics, 14*(9), 673–682. https://doi.org/10.1016/S0268-0033(99)00013-3

MacConaill, M. A. (1951). The movements of bones and joints; the mechanical structure of articulating cartilage. *The Journal of Bone and Joint Surgery, 33*(2), 251–257. https://doi.org/10.1302/0301-620x.33b2.251

Mansour, J. M., & Mow, V. C. (1977). On the natural lubrication of synovial joints: Normal and Degenerate. *Transactions of the ASME Journal of Lubrication Technology, 99*, 163–173.

Maroudas, A. (1966-67). Hyaluronic acid films. *Proceedings of the Institution of Mechanical Engineers, 181*(Pt3J) 122–124.

McConaill, M. A. (1932). The function of intra-articular fibrocartilages with special references to the knee and inferior radio-ulnar joints. *Journal of Anatomy, 66*, 210–227.

McCutchen, C. W. (1959). Mechanism of animal joints: Sponge-hydrostatic and weeping bearings. *Nature, 184*(4695), 1284–1285. https://doi.org/10.1038/1841284a0

McCutchen, C. W. (1962). The frictional properties of animal joints. *Wear, 5*(1), 1–17. https://doi.org/10.1016/0043-1648(62)90176-X

McCutchen, C. W. (1966–67.). Physiological lubrication. *Proceedings of the Institution of Mechanical Engineers, 181*(Pt3J) 55-62.

Medley, J. B., & Dowson, D. (1984). Lubrication of elastic-isoviscous line contacts subject to cyclic time-varying loads and entrainment velocities. *A S L E Transactions, 27*(3), 243–251. https://doi.org/10.1080/05698198408981567

Medley, J. B., Dowson, D., & Wright, V. (1984). Transient elastohydrodynamic lubrication models for the human ankle joint. *Engineering in Medicine, 13*(3), 137–151. https://doi.org/10.1243/EMED_JOUR_1984_013_035_02

Mirea, D. A., Trunfio-Sfarghiu, A. M., Matei, C. I., Munteanu, B., Piednoir, A., Rieu, J. P., Blanchin, M. G., & Berthier, Y. (2013). Role of the biomolecular interactions in the structure and tribological properties of synovial fluid. *Tribology International, 59*, 302–311. https://doi.org/10.1016/j.triboint.2012.06.015

Mow, V. C., Hou, J. S., Owens, J. M., & Ratcliffe, A. (1990). Biphasic and quasilinear viscoelastic theories for hydrated tissues. In A. Ratcliffe, S. L.-Y. Woo, & V. C. Mow (Eds.), *Biomechanics of diarthroidal joint* (Vol. 1, pp. 215–260). Springer, New York, NY.

Mow, V. C., Kuei, S. C., Lai, W. M., & Armstrong, C. G. (1980). Biphasic creep and stress relaxation of articular cartilage in compression: Theory and experiments. *Journal of Biomechanical Engineering, 102*(1), 73–84. https://doi.org/10.1115/1.3138202

Murakami, T. (1990). The lubrication in natural synovial joints and joint prostheses. *JSME International Journal. Series 3, Vibration, Control Engineering, Engineering for Industry, 33*(4), 465–474. https://doi.org/10.1299/jsmec1988.33.465

Murakami, T. (2007). Musculoskeletal system. In JSME (Ed.), *JSME Mechanical Engineers' Handbook, Design β8:Bioengineering (In Japanese)* (pp. 48–57). Tokyo: JSME.

Murakami, T. (2012). Importance of adaptive multimode lubrication mechanism in natural and artificial joints. *Proceedings of the Institution of Mechanical Engineers, Part J: Journal of Engineering Tribology, 226*(10), 827–837. https://doi.org/10.1177/1350650112451377

Murakami, T., Hayakawa, Y., Higaki, H., & Sawae, Y. (1997). Tribological behavior of sliding pairs of articular cartilage and Bioceramics. In *Proceedings of International Conference on New Frontiers in Biomechanical Engineering* (pp. 233–236).

Murakami, T., Higaki, H., Ando, H., & Nakanishi, Y. (1997). Viscous property of synovial fluids and related solutions and its role in joint lubrication. *Transactions of the Japan Society of Mechanical Engineers, Part C (In Japanese), 63*(607), 750–756. https://doi.org/10.1299/kikaic.63.750

Murakami, T., Higaki, H., Sawae, Y., Ohtsuki, N., Moriyama, S., & Nakanishi, Y. (1998). Adaptive multimode lubrication in natural synovial joints and artificial joints. *Proceedings of the Institution of Mechanical Engineers, Part H: Journal of Engineering in Medicine, 212*(1), 23–35. https://doi.org/10.1243/0954411981533791

Murakami, T., Nakashima, K., Sawae, Y., Sakai, N., & Hosoda, N. (2009). Roles of adsorbed film and gel layer in hydration lubrication for articular cartilage. *Proceedings of the Institution of Mechanical Engineers, Part J: Journal of Engineering Tribology, 223*(3), 287–295. https://doi.org/10.1243/13506501JET536

Murakami, T., Nakashima, K., Yarimitsu, S., Sawae, Y., & Sakai, N. (2011). Effectiveness of adsorbed film and gel layer in hydration lubrication as adaptive multimode lubrication mechanism for articular cartilage. *Proceedings of the Institution of Mechanical Engineers, Part J: Journal of Engineering Tribology, 225*(12), 1174–1185. https://doi.org/10.1177/1350650111415756

Murakami, T., Ohtsuki, N., & Higaki, H. (1993). The adaptive multimode lubrication in knee prostheses with compliant layer during walking motion. *Tribology Series, 25*(C), 673–682. https://doi.org/10.1016/S0167-8922(08)70420-6

Murakami, T., Sakai, N., Sawae, Y., Tanaka, K., & Ihara, M. (2004). Influence of proteoglycan on time-dependent mechanical behaviors of articular cartilage under constant total compressive deformation. *JSME International Journal, Series C: Mechanical Systems, Machine Elements and Manufacturing, 47*(4), 1049–1055. https://doi.org/10.1299/jsmec.47.1049

Murakami, T., Sakai, N., Yamaguchi, T., Yarimitsu, S., Nakashima, K., Sawae, Y., & Suzuki, A. (2015). Evaluation of a superior lubrication mechanism with biphasic hydrogels for artificial cartilage. *Tribology International, 89*, 19–26. https://doi.org/10.1016/j.triboint.2014.12.013

Murakami, T., Sawae, Y., Horimoto, M., & Noda, M. (1999). Role of surface layers of natural and artificial cartilage in thin film lubrication. *Tribology Series, 36*, 737–747. https://doi.org/10.1016/s0167-8922(99)80092-3

Murakami, T., Yarimitsu, S., Nakashima, K., Sakai, N., Yamaguchi, T., Sawae, Y., & Suzuki, A. (2015). Biphasic and boundary lubrication mechanisms in artificial hydrogel cartilage: A review. *Proceedings of the Institution of Mechanical Engineers, Part H: Journal of Engineering in Medicine, 229*(12), 864–878. https://doi.org/10.1177/0954411915611160

Murakami, T., Yarimitsu, S., Nakashima, K., Sawae, Y., & Sakai, N. (2013). Influence of synovia constituents on tribological behaviors of articular cartilage. *Friction, 1*(2), 150–162. https://doi.org/10.1007/s40544-013-0010-6

Murakami, T., Yarimitsu, S., Nakashima, K., Yamaguchi, T., Sawae, Y., Sakai, N., & Suzuki, A. (2014). Superior lubricity in articular cartilage and artificial hydrogel cartilage.

Proceedings of the Institution of Mechanical Engineers, Part J: Journal of Engineering Tribology, 228(10), 1099–1111. https://doi.org/10.1177/1350650114530273

Murakami, T., Yarimitsu, S., Sakai, N., Nakashima, K., Yamaguchi, T., & Sawae, Y. (2017). Importance of adaptive multimode lubrication mechanism in natural synovial joints. *Tribology International, 113*, 306–315. https://doi.org/10.1016/j.triboint.2016.12.052

Nakashima, K., Sawae, Y., & Murakami, T. (2005). Study on wear reduction mechanisms of artificial cartilage by synergistic protein boundary film formation. *JSME International Journal Series C, 48*(4), 555–561. https://doi.org/10.1299/jsmec.48.555

Oates, K. M. N., Krause, W. E., Jones, R. L., & Colby, R. H. (2006). Rheopexy of synovial fluid and protein aggregation. *Journal of the Royal Society Interface, 3*(6), 167–174. https://doi.org/10.1098/rsif.2005.0086

Ogston, A. G., & Stanier, J. E. (1953). The physiological function of hyaluronic acid in synovial fluid; viscous, elastic and lubricant properties. *The Journal of Physiology, 119*(2–3), 244–252. https://doi.org/10.1113/jphysiol.1953.sp004842

Orford, C. R., & Gardner, D. L. (1985). Ultrastructural histochemistry of the surface lamina of normal articular cartilage. *The Histochemical Journal, 17*(2), 223–233. https://doi.org/10.1007/BF01003221

Pawaskar, S. S., Jin, Z. M., & Fisher, J. (2007). Modelling of fluid support inside articular cartilage during sliding. *Proceedings of the Institution of Mechanical Engineers, Part J: Journal of Engineering Tribology, 221*(3), 165–174. https://doi.org/10.1243/13506501JET241

Radin, E. L., Swann, D. A., & Weisser, P. A. (1970). Separation of a hyaluronate-free lubricating fraction from synovial fluid. *Nature, 228*(5269), 377–378. https://doi.org/10.1038/228377a0

Raviv, U., & Klein, J. (2002). Fluidity of bound hydration layers. *Science, 297*(5586), 1540–1543. https://doi.org/10.1126/science.1074481

Reynolds, O. (1886). On the theory of lubrication and its applications to Mr. Beauchamp Tower's experiments including an experimental determination of the viscosity of olive oil. *Philosophical Transactions of the Royal Society, 177*, 157–234.

Roberts, B. J., Unsworth, A., & Mian, N. (1982). Modes of lubrication in human hip joints. *Annals of the Rheumatic Diseases, 41*(3), 217–224. https://doi.org/10.1136/ard.41.3.217

Sakai, N., Hagihara, Y., Furusawa, T., Hosoda, N., Sawae, Y., & Murakami, T. (2012). Analysis of biphasic lubrication of articular cartilage loaded by cylindrical indenter. *Tribology International, 46*(1), 225–236. https://doi.org/10.1016/j.triboint.2011.03.016

Sakai, N., Hagihara, Y., Hashimoto, C., Komori, M., Sawae, Y., & Murakami, T. (2015). An estimation of mechanical propertes of articular cartilage for biphasic finite element analyses. *Journal of Biomechanical Science and Engineering, 10*(4). https://doi.org/10.1299/jbse.15-00228, 15-00228-15–00228.

Sakai, N., Hashimoto, C., Yarimitsu, S., Sawae, Y., Komori, M., & Murakami, T. (2016). A functional effect of the superficial mechanical properties of articular cartilage as a load bearing system in a sliding condition. *Biosurface and Biotribology, 2*(1), 26–39. https://doi.org/10.1016/j.bsbt.2016.02.004

Sakai, N., Sawae, Y., Murakami, T., Kurohara, Y., & Okamoto, M. (2005). The development of micro testing machine equipped with microscope and visualization method for investigation of cartilaginous tissue (in Japanese). *Japanese Journal of Clinical Biomechanics, 26*, 59–63.

Sasada, T. (1996). Biotribology. *Toraibarojisuto/Journal of Japanese Society of Tribologists (In Japanese), 41*(1), 50–52.

Sasada, T. (2000). Lubrication of human joints — Nature of joint friction and "surface gel hydration lubrication" (In Japanese). *Japanese Journal of Clinical Biomechanics, 21*, 17–22.

Sasada, T., Tsukamoto, Y., & Mabuchi, K. (1988). *Biotribology — Friction and lubrication of synovial joints*. Sangyo Tosyo.

Sawae, Y., Murakami, T., Higaki, H., & Moriyama, S. (1996). Lubrication property of total knee prostheses with PV A hydrogel layer as artificial cartilage. *JSME International Journal, Series C: Dynamics, Control, Robotics, Design and Manufacturing, 39*(2), 356–364. https://doi.org/10.1299/jsmec1993.39.356

Sawae, Y., Murakami, T., Matsumoto, K., & Horimoto, M. (2000). Study on morphology and lubrication of articular cartilage surface with atomic force microscopy. *Japanese Journal of Tribology, 45*(1), 51–62.

Schwarz, I. M., & Hills, B. A. (1998). Surface-active phospholipid as the lubricating component of lubricin. *British Journal of Rheumatology, 37*(1), 21–26. https://doi.org/10.1093/rheumatology/37.1.21

Seror, J., Zhu, L., Goldberg, R., Day, A., J., & Klein, J. (2015). Supramolecular synergy in the boundary lubrication of synovial joints. *nature communications, 6*. https://doi.org/10.1038/ncomms7497

Swann, D. A., Hendren, R. B., Radin, E. L., & Sotman, S. L. (1981). The lubricating activity of synovial fluid glycoproteins. *Arthritis & Rheumatism, 24*(1), 22–30. https://doi.org/10.1002/art.1780240104

Swann, D. A., Silver, F. H., Slayter, H. S., Stafford, W., & Shore, E. (1985). The molecular structure and lubricating activity of lubricin isolated from bovine and human synovial fluids. *Biochemical Journal, 225*(1), 195–201. https://doi.org/10.1042/bj2250195

Swann, D. A., Slayter, H. S., & Silver, F. H. (1981). The molecular structure of lubricating glycoprotein-I, the boundary lubricant for articular cartilage. *Journal of Biological Chemistry, 256*(11), 5921–5925.

Tallian, T. E. (1967). On competing failure modes in rolling contact. *ASLE Transactions, 10*(4), 418–439. https://doi.org/10.1080/05698196708972201

Tanner, R. I. (1966). An alternative lubrication mechanism for the lubrication of synovial joints. *Physics in Medicine and Biology, 11*(1), 119–127.

Unsworth, A., Dowson, D., & Wright, V. (1975). Some new evidence on human joint lubrication. *Annals of the Rheumatic Diseases, 34*(4), 277–285. https://doi.org/10.1136/ard.34.4.277

Walker, P. S., Dowson, D., Longfield, M. D., & Wright, V. (1968). "Boosted lubrication" in synovial joints by fluid entrapment and enrichment. *Annals of the Rheumatic Diseases, 27*(6), 512–520. https://doi.org/10.1136/ard.27.6.512

Wilson, W., Van Donkelaar, C. C., Van Rietbergen, B., Ito, K., & Huiskes, R. (2004). Stresses in the local collagen network of articular cartilage: A poroviscoelastic fibril-reinforced finite element study. *Journal of Biomechanics, 37*(3), 357–366. https://doi.org/10.1016/S0021-9290(03)00267-7

Wu, J. Z., Herzog, W., & Epstein, M. (1997). Evaluation of the finite element software ABAQUS for biomechanical modelling of biphasic tissues. *Journal of Biomechanics, 31*(2), 165–169. https://doi.org/10.1016/S0021-9290(97)00117-6

Yarimitsu, S., Nakashima, K., Sawae, Y., & Murakami, T. (2007). Study on the mechanisms of wear reduction of artificial cartilage through in situ observation on forming protein boundary film. *Tribology Online, 2*(4), 114–119. https://doi.org/10.2474/trol.2.114

Yarimitsu, S., Nakashima, K., Sawae, Y., & Murakami, T. (2009). Influences of lubricant composition on forming boundary film composed of synovia constituents. *Tribology International, 42*(11–12), 1615–1623. https://doi.org/10.1016/j.triboint.2008.11.005

Zappone, B., Ruths, M., Greene, G. W., Jay, G. D., & Israelachvili, J. N. (2007). Adsorption, lubrication, and wear of lubricin on model surfaces: Polymer brush-like behavior of a glycoprotein. *Biophysical Journal, 92*(5), 1693–1708. https://doi.org/10.1529/biophysj.106.088799

Biotribology of artificial joints

4

4.1 Importance of tribology in artificial joints

4.1.1 Development of various artificial joints

Various artificial joints have been applied as effective medical treatments for joint diseases with restriction in motion, deficiency in load support ability and severe pain in osteoarthritis, rheumatoid arthritis, avascular necrosis of the femoral head, femoral neck fracture, bone tumor, and so on. Furthermore, it should be kept in mind that the recovery of daily activities after joint arthroplasty brings about the expansion of the self-motivating life fields, the improvement of the systemic metabolism, and the promotion of the spiritual ambition. Therefore, artificial joints with appropriate function are very valuable medical devices.

In historical trials of clinical application of artificial joints, the availability of appropriate biomaterials which are biocompatible with the environment in human body has been an absolute requisite condition for success with recovery of joint function. In clinical applications, various biomaterials which have bioinert, corrosion-resistant, and strong properties were used, as shown in Table 4.1. In this table the main materials and devices developed until 1971 are mentioned, but thereafter new materials and devices with improved properties have been developed as described in this chapter. Inventions of various new materials such as corrosion-resistant metals, polymers, ceramics, and composites which are applicable in human body brought about new medical treatments using artificial joints.

4.1.2 Roles of tribology in artificial joints

As mentioned in Chapter 1, tribology includes friction, lubrication, wear, bearing design, and so on. In development of artificial joints, it was required that new artificial joints have low friction, good lubrication, and minimal wear with wide range of motion under various daily activities. However, it was difficult to attain good lubrication for rubbing of artificial materials with different properties from natural articular cartilage. Furthermore, it was by no means easy to achieve an ideal balance between low friction and low wear.

For example, in metal-on-metal total hip replacement as developed by McKee and Farrer, the mechanical durability was improved by using Co-Cr-Mo alloy (vitallium) in replace of stainless steel, but for considerable cases the loosening occurred due to large frictional torque (caused by high friction of metal/metal at direct contacts and large femoral head) and insufficient fixation property.

In contrast, in 1958, Charnley (Charnley, 1979) could actualize the low friction hip arthroplasty by applying artificial hip joint composed of polytetrafluoroethylene

Biotribology of Natural and Artificial Joints. https://doi.org/10.1016/B978-0-12-823669-7.00001-6

Table 4.1 Early history of materials and devices for artificial joints.

Date	Inventor	Material and device	Invention of material
1890	T. Glück	Ivory: Hinged knee joint, ball-and-socket hip joint	
1912			Stainless steel
1929			Co-Cr-Mo alloy
1936			PMMA by mixing
1938			PTFE
1938	M. N. Smith-Petersen	Co-Cr-Mo alloy: Acetabular cup (mold arthroplasty)	
1938	P. Wiles	Stainless steel: Metal-on-metal hip joint	
1943	A. T. Moore	Co-Cr-Mo alloy: Artificial femoral head	
1946	J. Judet and R. Judet	Acrylic resin (PMMA): Artificial femoral head	
1951	G. K. McKee and J. Watson-Farrar	Stainless steel: Metal-on-metal hip joint	
1951	Kaier and J.A. Jansen	PMMA: Bone cement for fixing acrylic cup to subchondral bone of femoral head	
1955	(Ruhrchemie AG)		UHMWPE
1956	G. K. McKee and J. Watson-Farrar	Co-Cr-Mo alloy: Metal-on-metal hip joint (Thompson stem)	
1957	Leventhal	Titanium: Artificial femoral head	
1958	J. Charnley	PTFE/Stainless steel: Artificial hip joint, PMMA: Bone cement	
1962	J. Charnley	UHMWPE/Stainless steel: Artificial hip joint	
1970	P. Boutin	Alumina/alumina: Artificial hip joint	
1971	H. Oonishi	Cross-linked UHMWPE: Artificial hip joint	

PMMA, Polymethylmethacrylate; *PTFE*, Polytetrafluoroethylene; *UHMWPE*, Ultra-high molecular weight polyethylene.

(PTFE) acetabular cup and stainless-steel femoral head, because of lower friction property of PTFE as a self-lubricating material and lower frictional torque for smaller femoral head (7/8 inch ($\fallingdotseq$ 22 mm) diameter). He investigated frictional behavior of natural synovial joints and indicated that the boundary lubrication predominated in natural synovial joints. At that time, it was suggested that hydrodynamic action is ineffective under high loading and slow movement. Therefore, he selected at first the combination of PTFE as low friction material against metal in artificial hip joint for low friction hip arthroplasty. In clinical usage, however, significant wear of PTFE at wear depth of 1−3 mm/year accompanied with severe tissue reactions due to excessive wear debris occurred, and the clinical uses of this low friction arthroplasty using PTFE cup were interrupted. Then, in 1962 fortunately he could obtain new material and change the rubbing material from PTFE acetabular cup to new material: ultra-high

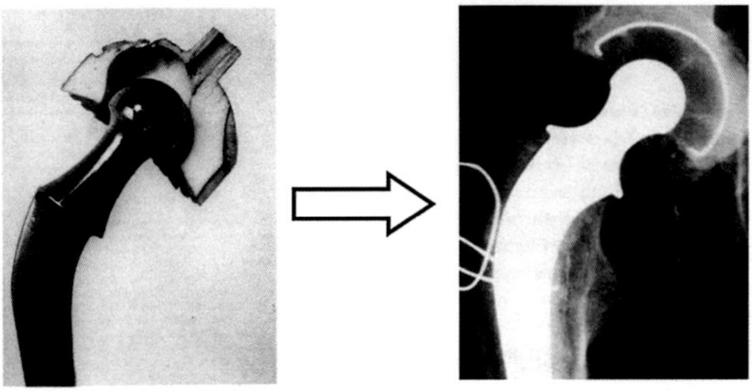

(a) 3 years after implanting
[PTFE/Stainless steel]

(b) 11 years after implanting
[UHMWPE/Stainless steel]

Figure 4.1 Improvement of wear resistance by application of UHMWPE for PTFE in total hip arthroplasty. (a) Excessive wear of PTFE at wear depth of 1–3 mm/year occurred in combination of PTFE acetabular cup and stainless-steel femoral head. (b) Reduction of wear of UHMWPE as wear depth of 0.05–0.2 mm/year was attained in combination of UHMWPE acetabular cup and stainless-steel femoral head. Modified with permission from Charnley, J. (1979) *Low friction arthroplasty of the hip. Springer-Verlag.*

molecular weight polyethylene (UHMWPE) (RCH 1000) cup with superior wear resistance against stainless-steel femoral head and could reduce wear of artificial hip joint (Fig. 4.1). Friction of UHMWPE against stainless steel was a little higher than PTFE, but the annual wear depth of UHMWPE cup was improved to 0.05–0.2 mm/year. Since successful clinical application of Charnley's hip joint, the clinical applications of total joint arthroplasty have been expanded all over the world.

Tribological properties of both superior wear resistance and low friction are funda mentally requisite design factors for artificial joints.

The clinical performances of artificial joints have been improved with further application of new technologies for implant designs, materials, and operation systems. At the present stage, the performance of artificial joints appears to be approaching to the ability of natural synovial joints but have several limitations to attain the same performance as natural joints.

The comparison between artificial joints (as hard-on-UHMWPE) and natural joints is shown in Table 4.2. Fundamental limitation is the difference in rubbing materials as anticorrosive metal/UHMWPE or bioceramic/UHMWPE versus cartilage/cartilage. Material properties of artificial joints are very different (harder) and geometrical congruities are usually lower, and therefore maximum contact pressure of artificial joints become higher. Synovial fluid after joint replacement has to some extent similar constituents to intact synovial fluid but seems to be deficient in some lubricating constituents and less viscous. Anticorrosive metals such as Co-Cr-Mo alloy, stainless steel, and titanium alloy have Young's modulus of 100–210 GPa, bioceramics such as alumina and zirconia have Young's modulus of 200–340 GPa. UHMWPE has Young's modulus of 0.5–1.5 GPa, which is stiffer at two or three order than that of articular cartilage. Therefore, the minimum fluid film thickness for hard-on-UHMWPE

Table 4.2 Comparison between artificial joints as hard-on-soft and natural joints.

	Artificial joints	Natural joints
Rubbing materials	Anticorrosive metal/UHMWPE, bioceramic/UHMWPE	Cartilage/cartilage
Lubricant	Synovial fluid, body fluid	Synovial fluid
Maximum contact pressure	5–50 MPa	1–5 MPa
Friction coefficient	0.03–0.1	0.003–0.02
Lubrication mode	Mixed lubrication, boundary lubrication	Adaptive multimode lubrication
Durability	15–30 years	70–80 years

Modified from Murakami, T. (2013). *Biotribology of natural and artificial joints. In Frontiers of biomechanics (In Japanese).* (JSME, Ed.; pp. 167–226). Kyoritsu Shuppan Co., Ltd.

joints is estimated to be thinner than 0.1 μm for total hip and knee prostheses during stance phase of walking (Auger et al., 1995; Jalali-Vahid et al., 2001). The fluid film lubrication mode is mixed or boundary lubrication mode and thus local direct contact occurs, resulting in medium friction and considerable wear. Thus, most of lubrication modes are mixed or boundary lubrication with higher friction in contrast to superior adaptive multimode lubrication in natural joints which is described in Chapter 3. In longer operation in mixed or boundary lubrication, considerable wear occurred and wear particles were produced due to repeated asperity contacts, as various forms such as adhesive wear, abrasive wear, corrosive wear, and fatigue wear including delamination. Usually wear of UHMWPE surfaces was considerable but metal and ceramic rubbing surfaces were also worn depending on the rubbing severity. It was indicated that the adverse tissue reactions to wear debris causes osteolysis which results in joint loosening and implant failure (Ingham & Fisher, 2005; Willert & Semlitsch, 1977). Therefore, the improvement of lubrication modes including the promotion of fluid film lubrication, the reduction of friction, and the minimizing wear with higher wear resistance and improved biocompatibility have been performed from the tribological viewpoints.

Durability of many artificial joints is estimated as 15–30 years depending on the situations such as types (materials and design) of joint prostheses and operating conditions for patients.

4.2 Lubrication modes in artificial joints

The indication of lubrication mode in lubricated rubbing surfaces is predicted by the film parameter (lambda ratio): Λ, defined by Eq. (4.1), representing the ratio of theoretical minimum fluid film thickness: $h_{\min}$ to the composite surface roughness of both rubbing surfaces: σ (Johnson et al., 1972; Tallian, 1967).

$$\Lambda = \frac{h_{\min}}{\sigma} \tag{4.1}$$

$$\text{where } \sigma = \sqrt{Sq_1^2 + Sq_2^2} \tag{4.2}$$

Sq_1, Sq_2: Root mean square roughness.
The lubrication modes are discriminated as shown below:
$\Lambda \geq 3$; fluid film lubrication.
$1 \leq \Lambda < 3$; mixed lubrication.
$\Lambda < 1$; boundary lubrication or mixed lubrication with significant contacts.

4.2.1 Hip joint replacements as hard-on-hard and hard-on-UHMWPE

Since successful clinical application of total hip replacements composed of metallic femoral head and UHMWPE acetabular cup by Charnley, hard-on-UHMWPE hip prostheses had been clinically applied to patients with osteoarthritis, rheumatoid arthritis and other joint diseases. Then, most of patients were able to go through their daily activities and were relieved from severe pains. In the long run, however, in certain cases, the revisions of joint replacements were performed due to joint loosening, mechanical failure, dislocation, infection, and others. Various improvements to overcome these problems had been conducted. On the aseptic loosening, it was pointed out that adverse biological reaction to wear debris causes loosening and failure (Willert & Semlitsch, 1977), and the improvement of tribological performance of joint prostheses with higher wear resistance, low friction, appropriate surface properties, and adequate mechanical strength were required. Furthermore, with an increasing need to implant hip prostheses in younger, more active patients, the applications of alternative materials including hard-on-hard combinations such as metal-on-metal and ceramic-on-ceramic were performed. In this section, the lubrication modes in both hard-on-hard and hard-on-UHMWPE are discussed.

4.2.1.1 Experimental evaluation of lubrication modes in hip joint replacements as hard-on-hard and hard-on-UHMWPE

4.2.1.1.1 Evaluation of lubrication modes on the basis of friction characteristics

Lubrication modes for metal-on-metal, ceramic-on-ceramic, and metal-on-UHMWPE hip joints were investigated by measuring friction at the bearing surfaces using different lubricants in hip joint simulator. Experimental results on friction properties for metal-on-metal (Co-Cr-Mo/Co-Cr-Mo), ceramic-on-ceramic (Al_2O_3/Al_2O_3), and metal-on-plastic (CoCrMo/UHMWPE) lubricated with carboxymethyl cellulose (CMC) fluids ($0.001 <$ viscosity: $\eta < 0.154$ Pa s) are shown as Stribeck plots in Fig. 4.3 (Scholes & Unsworth, 2000). Here, friction factor f is defined as

$$f = \frac{T}{rW} \tag{4.3}$$

where T is the frictional torque between the bearing surfaces, r is the femoral head radius, and W is the load applied. The friction factor is similar to the friction coefficient but varies slightly depending on the elastic contact of the hip joint. The friction was measured throughout the loading cycle (maximum: 2 kN, minimum: 100 N). In Fig. 4.3, five sampling points were taken from the high-load (2 kN) high-speed (entraining velocity: $u = 0.015$ m/s) stage of the loading cycle (equivalent to the stance phase of walking). For Stribeck analysis, the friction factor was plotted against the Sommerfeld number z defined as

$$z = \frac{\eta u r}{W} \tag{4.4}$$

In Stribeck curve as shown in Figs. 1.8 and 3.11, a decrease in the friction factor with increase in Sommerfeld number is indicative of a mixed lubrication regime in which the load is carried in part by the contact between the asperities of the rubbing surfaces and also by the fluid film pressure generated within the lubricant. In contrast, a rising trend of friction factor is indicative of a full fluid film lubrication regime where the rubbing surfaces are completely separated by the fluid film. Therefore, the frictional resistance is generated solely by the shear within the fluid and usually friction coefficient can be low.

The levels of friction factors for different material combinations in CMC are as follows:

metal-on-metal > metal-on-UHMWPE > ceramic-on-ceramic joints.

In metal-on-metal joints (Fig. 4.2a), high friction factors (about 0.3) and some trend of falling friction with increasing Sommerfeld number suggest mixed lubrication or some boundary lubrication.

In ceramic-on-ceramic joints (Fig. 4.2b), consistently low friction factors (about 0.002) with increasing trend with rising Sommerfeld number are indicative of a full fluid film lubrication regime.

In metal-on-UHMWPE joints (Fig. 4.2c), a falling friction factor with increasing Sommerfeld number indicates the feature of mixed lubrication regime. In their related study on metal-on-UHMWPE joints (Unsworth et al., 1987), similar trend was observed.

In the comparison mentioned above, the lubrication modes were evaluated on the basis of the friction properties for CMC lubricant viscosity range from 0.001 to 0.154 Pa s. From the viewpoints of the film factor (lambda ratio) for lubricant viscosity of 0.01 Pa s, the lubrication modes are predicted as follows:

$$\text{Metal} - \text{on} - \text{metal}: \quad \Lambda = \frac{0.05}{0.080} = 0.62 < 1 \text{ (boundary or mixed lubrication)}$$

$$\text{Ceramic} - \text{on} - \text{ceramic}: \quad \Lambda = \frac{0.06}{0.010} = 6.0 > 3 \text{ (full fluid film lubrication)}$$

$$\text{Metal} - \text{on} - \text{UHMWPE}: \quad \Lambda = \frac{0.09}{1.29} = 0.070 < 1 \text{ (boundary or mixed lubrication)}$$

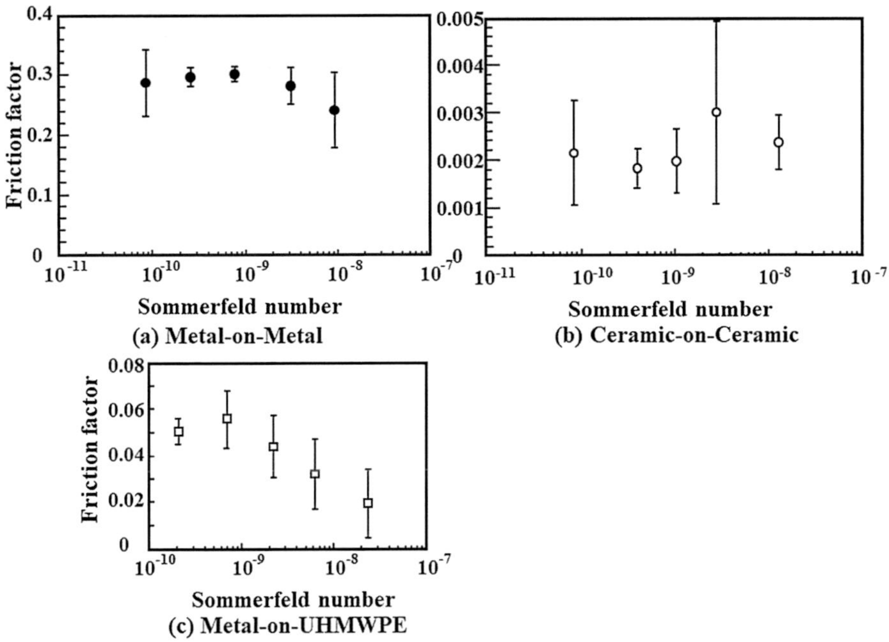

Figure 4.2 Stribeck plots for metal-on-metal, ceramic-on-ceramic, and metal-on-UHMWPE joints with CMC fluids. (a) Metal-on-metal joint: friction level is high and the trend of falling friction with increasing Sommerfeld number is observed. (b) Ceramic-on-ceramic joint: friction level is very low and the trend of rising friction with increasing Sommerfeld number is observed. (c) Metal-on-UHMWPE joint: friction level is medium and the trend of falling friction with increasing Sommerfeld number is observed. Error bars indicate standard deviation.

Modified from Scholes, S. C., & Unsworth, A. (2000). Comparison of friction and lubrication of different hip prostheses. *Proceedings of the Institution of Mechanical Engineers, Part H: Journal of Engineering in Medicine*, *214*(1), 49–57. https://doi.org/10.1243/0954411001535237

Here, CMC fluid has similar viscosity level to synovial fluid but different properties. Therefore, friction properties of these hip joints lubricated with synovial fluid and serum solution were examined.

In Fig. 4.3, friction properties lubricated with synovial fluids are shown. Pooled human synovial fluid taken from patients with arthritis (including rheumatoid arthritis and inflammatory arthritis) was used either filtered (1 μm filter) or unfiltered (the viscosities of both were 0.006 Pa s). Friction level for metal-on-metal joint was lower than in CMC fluid but still higher than friction for ceramic-on-ceramic and metal-on-UHMWPE joints. It is noticed that friction for ceramic-on-ceramic was significantly higher than that in CMC fluid. Ceramic-on-ceramic joint lubricated with serum (CMC solutions of bovine serum from 8.3% to 100%) showed an increase of one order in friction factor compared with CMC fluid. These differences indicate the important roles of proteinaceous constituents with boundary lubricating ability in synovial fluid.

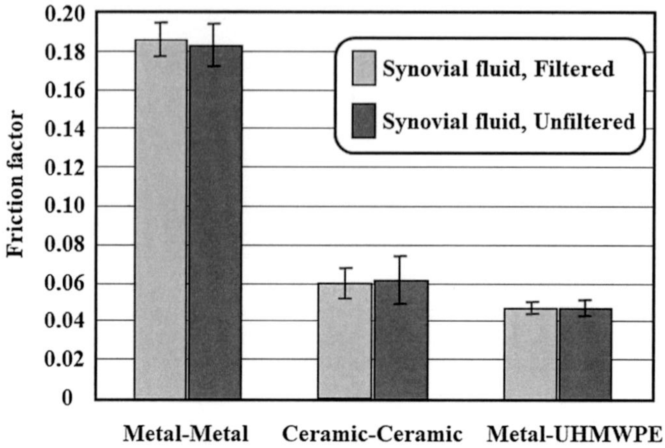

Figure 4.3 Friction properties for metal-on-metal, ceramic-on-ceramic, and metal-on-UHMWPE joints lubricated with filtered and unfiltered synovial fluid. Friction factor for metal-on-metal joint was a little reduced but friction level for ceramic-on-ceramic joint increased compared with friction levels in CMC fluids. Friction factor for metal-on-UHMWPE maintained similar level. Error bars indicate standard deviation.
Modified from Scholes, S. C., & Unsworth, A. (2000). Comparison of friction and lubrication of different hip prostheses. *Proceedings of the Institution of Mechanical Engineers, Part H: Journal of Engineering in Medicine, 214*(1), 49–57. https://doi.org/10.1243/0954411001535237

Protein adsorption on these rubbing surfaces was confirmed by protein gel technique. More protein was adsorbed on to UHMWPE and ceramic surfaces than on to the metal surface. In these cases, main adsorbed molecule was serum albumin. It is considered that protein-protein interaction appeared to generate higher resistance than shearing resistance of fluid film for ceramic-on-ceramic joint and lower resistance than the frictional resistance of metal-metal interaction for metal-on-metal joints. In lubrication mechanism of total hip prostheses with small radial clearance, it is required to evaluate it by considering the influences of both fluid film formation and adsorbed film formation in human body environment. Here, it should be considered that the proteinaceous adsorbed films have benefits to protect rubbing surfaces, resulting in reduced wear as another aspect.

4.2.1.1.2 Evaluation of lubrication modes on the basis of lubricating film measurement

To evaluate lubrication mode in metal-on-metal total replacement hip joint during walking motion, Dowson et al. (2000) investigated the conditions of lubrication and contact between a metallic femoral head and metallic acetabular cup in a hip joint simulator with an electric resistivity technique. Total hip prosthesis of 36 mm diameter and 40 μm radial clearance made of high carbon Co-Cr alloy was used under walking condition of 0.93 Hz cycle frequency with the maximum load of 2.87 kN (about four times body weight) in proteinaceous and nonproteinaceous lubricants.

In simulator tests, proteinaceous fluids with boundary lubricating ability were much more effective in promoting separation of the metallic surfaces with running time. For 25% serum solution whose Λ values during swing phase were between 0.6 and 1.5, at initial stage the voltage trace was steady zero (direct contact) even during light loading swing phase, but with running time some higher voltages for considerable separation were observed in the vicinity of the toe-off position. For the 100% serum lubricant, the predicted Λ values during swing phase were between 1 and 2.85 suggesting the mixed lubrication applied but showed higher separation in the swing phase at 20 min traces, and exhibited significant separation except drops at peak loads during stance phase ($\Lambda < 1.1$). It is suggested that the prevalent mixed lubrication regime arises from a combination of protein-based boundary lubrication and a super-imposed fluid-film action with elastohydrodynamic lubrication. In contrast, no separation was detected when silicone oil (viscosity: 0.01 Pa s) was used, and this fact means that elastohydrodynamic action alone is unable to generate adequate protective films.

As another method to measure fluid film thickness, an ultrasonic reflection method (Dwyer-Joyce et al., 2003, 2004) was used. It was shown in friction simulator test for large diameter metal-on-metal hip joint lubricated with 25% and 100% serum that the fluid film thickness in large clearance bearing was reduced and friction was increased with incidence of squeaking (Brockett et al., 2008). A negative trend of increasing friction with reducing film thickness, as the swing phase load increased, was observed. This feature suggests that these metal-on-metal hip joints operate in mixed lubrication regime.

4.2.1.2 Lubrication analyses for evaluation of lubrication modes in hip joint replacements as hard-on-hard and hard-on-UHMWPE

4.2.1.2.1 Elastohydrodynamic lubrication analysis for equivalent ball-on-plane models

As a simplified method to evaluate the lubrication modes in hip joint replacements from the viewpoint of film parameter (lambda ratio: Λ), analysis of fluid film lubrication and contact mechanics for equivalent ball-on-plane models was performed by Jin et al. (1997).

In an equivalent ball-on-plane model (Fig. 4.5), the equivalent radius of the ball R was determined from the following equation:

$$R = \frac{R_1 R_2}{R_2 - R_1} \tag{4.5}$$

where R_1 and R_2 are the radii of the femoral head and socket (acetabular cup), respectively. In their study, changes in minimum elastohydrodynamic film thickness h_{min} for different radial clearances ($R_2 - R_1 = 2.5{-}500\ \mu m$) were examined for femoral head of $R_1 = 14$ mm. The equivalent radius ranged from 0.41 to 78.4 m, some of which should be rejected by limitation with excessive contact width.

The lubricant viscosity was selected as 0.005 Pa s, because synovial fluid in artificial joints during walking should be assumed as isoviscous and Newtonian fluid at higher shear rate than $10^5\ s^{-1}$ and lower pressure than 100 MPa.

The load and speed in hip joints vary considerably during walking motion, but the previous lubrication studies of natural hip and ankle joints (Medley et al., 1984; Jin et al., 1993) indicated that the predicted lubricating film thickness stays relatively constant caused by combined effect of entraining (in swing phase) and squeeze film (in stance phase) actions. In addition, the predicted cyclic film thickness is similar to the predictions based upon entraining action alone using the average load and average speed. Therefore, the load and speed were assumed to be constant in their lubrication analysis. The sliding speed is about 30 mm/s, and then the entraining speed was selected 15 mm/s as half of sliding speed. The average load was assumed as 2500 N (three or four times body weight).

In this analysis for hard-on-hard and UHMWPE-on-hard hip joints, the bearing surfaces were assumed to be semi-infinite and corresponding elastohydrodynamic film thickness was estimated using the following formula by Hamrock and Dowson (1978):

$$\frac{h_{\min}}{R} = 2.80\left(\frac{\eta u}{E'R}\right)^{0.65}\left(\frac{w}{E'R^2}\right)^{-0.21} \tag{4.6}$$

where $h_{\min}$: minimum lubricating film thickness.

η: viscosity of lubricant,

u: entraining velocity ($= u_1/2$).

E': equivalent elastic modulus;

$$\frac{1}{E'} = \frac{1}{2}\left(\frac{1-\nu_1^2}{E_1} + \frac{1-\nu_2^2}{E_2}\right) \tag{4.7}$$

w: load.

ν_1: Poisson's ratio of femoral head,

ν_2: Poisson's ratio of acetabular cup.

The contact half width can be predicted from Hertzian theory as follows:

$$a = \sqrt[3]{\frac{3wR}{2E'}} \tag{4.8}$$

where a: radius of contact area.

The predictions of the minimum film thickness are plotted against different radial clearances for metal-on-metal, ceramic-on-ceramic, and UHMWPE-on-metal in Fig. 4.4. It is observed that the minimum film thickness increases as the radial clearance decreases for all material combinations. For a given radial clearance, a decrease of the elastic modulus results in an increase of the minimum film thickness.

As estimated by Eq. (4.8), the predicted contact half width for all the material combinations increases as the radial clearance decreases. As shown in Fig. 4.5 the

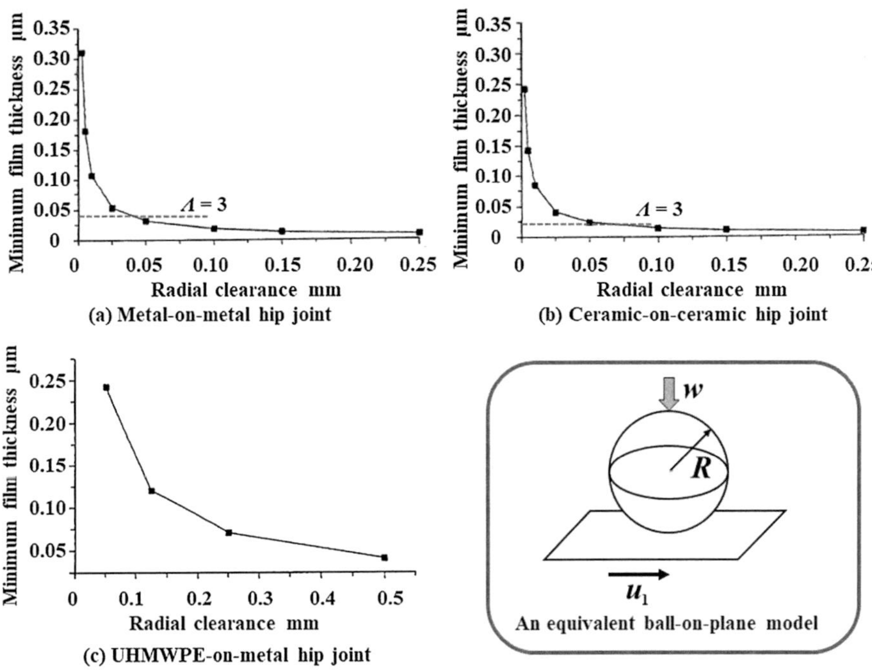

Figure 4.4 Prediction of minimum film thickness for total hip replacements for different radial clearances. (a) Metal-on-metal hip joints: $w = 2.5$ kN, $R_1 = 14$ mm, $E_1 = 230$ GPa, $E_2 = 230$ GPa, $\nu_1 = 0.3$, $\nu_2 = 0.3$, $u = 0.015$ m/s, $\eta = 0.005$ Pa s. Minimum film thickness for $\Lambda = 3$ is indicated. (b) Ceramic-on-ceramic hip joints: $w = 2.5$ kN, $R_1 = 14$ mm, $E_1 = 400$ GPa, $E_2 = 400$ GPa, $\nu_1 = 0.3$, $\nu_2 = 0.3$, $u = 0.015$ m/s, $\eta = 0.005$ Pa s. Minimum film thickness for $\Lambda = 3$ is indicated. (c) UHMWPE-on-stainless steel hip joints: $w = 2.5$ kN, $R_1 = 14$ mm, $E_1 = 200$ GPa, $E_2 = 1$ GPa, $\nu_1 = 0.3$, $\nu_2 = 0.4$, $u = 0.015$ m/s, $\eta = 0.005$ Pa s. Modified from Jin, Z. M., Dowson, D., & Fisher, J. (1997). Analysis of fluid film lubrication in artificial hip joint replacements with surfaces of high elastic modulus. *Proceedings of the Institution of Mechanical Engineers, Part H: Journal of Engineering in Medicine, 211*(3), 247–256. https://doi.org/10.1243/0954411971534359

minimum film thickness is plotted against the contact half width, and an increase of the contact half width results in an increase of minimum film thickness. Here, it should be taken into consideration to prevent the adverse influence of starvation at inlet zone in cases where the contact half width becomes comparable to the inlet zone distance.

The critical conditions for full fluid film lubrication (as $\Lambda = 3$) is indicated for metal-on-metal and ceramic-on-ceramic hip joints in Fig. 4.4a and b, respectively. For UHMWPE-on-metal hip joint in Fig. 4.4c, the critical level for $\Lambda = 3$ is located out of the range, and at $\Lambda \ll 1$, it operates in boundary lubrication or mixed lubrication regime with significant solid-solid contacts. From these results, it is suggested for metal-on-metal hip joints with radial clearance of 50 or 100 μm that it is difficult to attain full fluid film lubrication and operate in mixed lubrication regime. In contrast, for ceramic-on-ceramic hip joints with very smooth surface finish and small radial clearance of 10 μm or so, it is expected to operate in full fluid film lubrication regime.

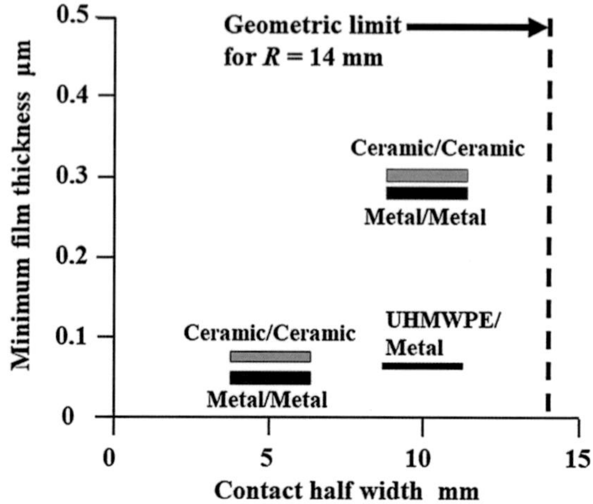

Figure 4.5 Comparison of the predicted minimum film thickness for metal-on-metal, ceramic-on-ceramic, and UHMWPE-on-metal hip joints as a function of the predicted contact half width. Conditions at $w = 2.5$ kN, $R_1 = 14$ mm, $u = 0.015$ m/s, $\eta = 0.005$ Pa s. Modified from Jin, Z. M., Dowson, D., & Fisher, J. (1997). Analysis of fluid film lubrication in artificial hip joint replacements with surfaces of high elastic modulus. *Proceedings of the Institution of Mechanical Engineers, Part H: Journal of Engineering in Medicine, 211*(3), 247–256. https://doi.org/10.1243/0954411971534359

However, the present simple analysis for ball-on-plane model has various limitations on the inlet boundary condition, the assumption of the semi-infinite solid for the bearing materials, the effect of the finite thickness of bearing materials, lubricant rheology, and so on. Consequently, at the next step, the elastohydrodynamic analysis for ball-on-cup model is mentioned below.

4.2.1.2.2 Elastohydrodynamic lubrication analysis for ball-in-socket models

The elastohydrodynamic lubrication analysis of metal-on-metal hip prostheses under steady state entraining motion was conducted for ball-in-socket model by Jagatia and Jin (2001), in which the Reynolds equation in a spherical coordinate was adopted for fluid film lubrication analysis. The corresponding elastic deformation was calculated by the finite element method for metal-on-metal ball-in-socket joint and backing materials such as cement layer, cancellous and cortical bones underneath the acetabular cup. Both the Reynolds and the elasticity equations were solved simultaneously using the Newton–Raphson finite difference method. The analysis for Co-Cr alloy-on-Co-Cr alloy hip prosthesis of 14 mm radius femoral head with 30 µm radial clearance was performed at the loads from 100 to 2500 N, at the angular velocity of 2 rad/s for walking, with viscosity of 0.01 Pa s. The acetabular cup of 9.5 mm thickness was assumed to be rigidly fixed to pelvic bone by a cement layer of 2 mm thickness.

The predicted film profiles under various loads from 100 to 2500 N are shown in Fig. 4.6, where minimum film thickness can be determined. The corresponding pressure distributions are shown in Fig. 4.7. It is noticed that the predicted film

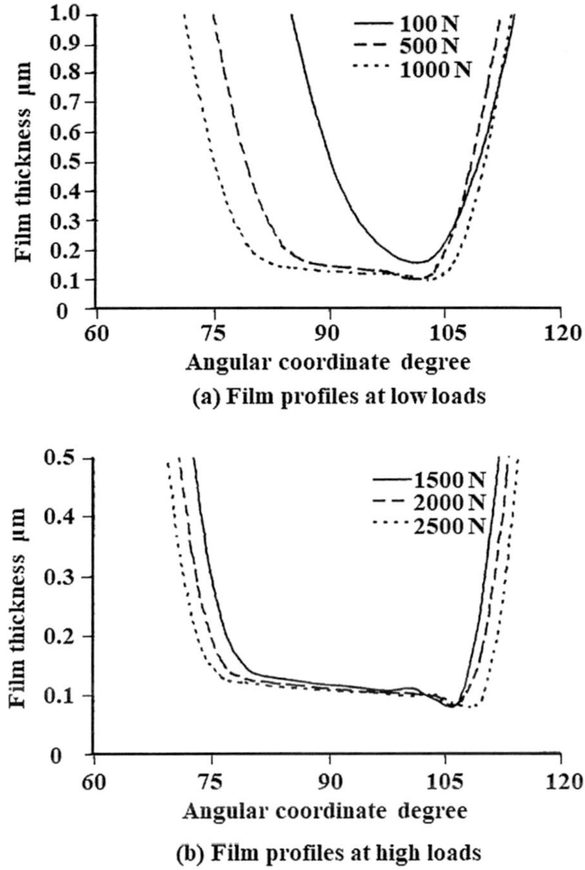

Figure 4.6 Prediction of the film profile through the center of the contact along the entraining direction under various loads in metal-on-metal hip joint replacement. Thinning of minimum film thickness and the movement of corresponding location with the load can be observed.
Modified from Jagatia, M., & Jin, Z. M. (2001). Elastohydrodynamic lubrication analysis of metal-on-metal hip prostheses under steady state entraining motion. *Proceedings of the Institution of Mechanical Engineers, Part H: Journal of Engineering in Medicine*, *215*(6), 531–541. https://doi.org/10.1243/0954411011536136

profiles and pressure distributions follow the typical characteristics of soft elastohydrodynamic lubrication conditions. The minimum film thickness along the entraining direction moves toward the exit and decreases as the load increases. The contact pressure increases with the load and approaches to that under the condition of dry contact.

In Fig. 4.8, the predicted minimum film thickness in ball-in-socket model by Jagatia and Jin (2001) for metal-on-metal hip joint replacements with 9.5 mm thick acetabular cup is plotted against load, and compared with Hamrock and Dowson formulae based upon an equivalent ball-on-plane model (Hamrock & Dowson, 1978). It is noteworthy

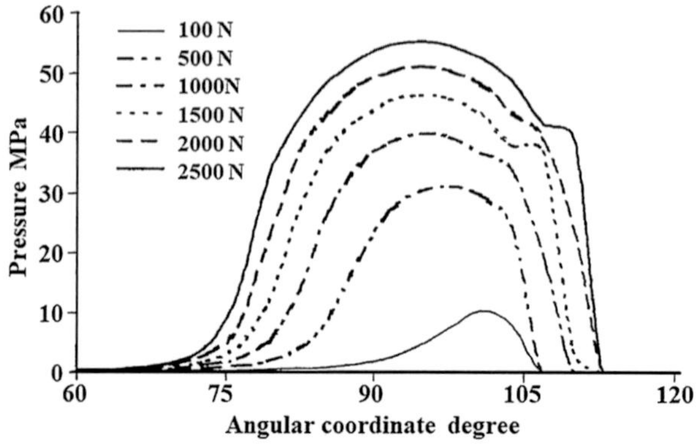

Figure 4.7 Prediction of the pressure distribution through the center of the contact along the entraining direction under various loads in metal-on-metal hip joint replacement. The changes in contact pressure distribution with the load can be observed.
Modified from Jagatia, M., & Jin, Z. M. (2001). Elastohydrodynamic lubrication analysis of metal-on-metal hip prostheses under steady state entraining motion. *Proceedings of the Institution of Mechanical Engineers, Part H: Journal of Engineering in Medicine*, *215*(6), 531–541. https://doi.org/10.1243/0954411011536136

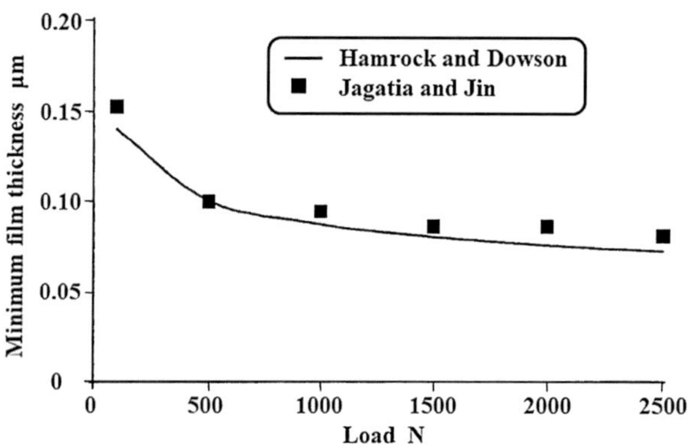

Figure 4.8 Comparison of the minimum film thickness in metal-on-metal hip joint replacement predicted in the study by Jagatia and Jin and using the Hamrock and Dowson formulae. An excellent agreement between the predicted minimum film thickness and Hamrock and Dowson formulae is observed.
Modified from Jagatia, M., & Jin, Z. M. (2001). Elastohydrodynamic lubrication analysis of metal-on-metal hip prostheses under steady state entraining motion. *Proceedings of the Institution of Mechanical Engineers, Part H: Journal of Engineering in Medicine*, *215*(6), 531–541. https://doi.org/10.1243/0954411011536136

that the predicted minimum film thickness exhibits an excellent agreement with Hamrock and Dowson formulae, and the backing materials underneath the acetabular cup have a small effect on the predicted contact pressure and the elastic deformation at the bearing surfaces in this metal-on-metal hip prosthesis.

In ceramic-on-ceramic hip joint as hard-on-hard bearing, lubricating fluid film with thinner thickness than metal-on-metal hip joint of similar geometrical design can be formed (Mak & Jin, 2002). However, the surface roughness of ceramic components is smaller than metallic component. Therefore, it is expected in cases with $\Lambda > 3$ in which ceramic-on-ceramic hip prosthesis operates in full fluid film lubrication regime.

To evaluate the lubrication mode in hip joint replacement consisting of an UHMWPE acetabular cup against a metallic or ceramic femoral head, the analysis of elastohydrodynamic lubrication was conducted using ball-in-socket model by Jalali-Vahid et al. (2001). The cup was assumed to be stationary while the ball was assumed to rotate at a steady angular velocity and under a constant load. It was found in this analysis that the predicted lubricating film thickness increases with a decrease in the radial clearance: c, an increase in the femoral head radius: R_1, an increase in UHMWPE thickness: d, and a decrease in UHMWPE elastic modulus E. Effects of femoral head radius and radial clearance on minimum film thickness are shown in Fig. 4.9. For metal-on-UHMWPE with composite surface roughness of 1.0 μm or so, minimum film thickness of 0.1−0.2 μm suggests that the corresponding lubrication mode is boundary lubrication or mixed lubrication with significant contacts because of $\Lambda = 0.1$ to $0.2 < 1.0$. Although it may be expected that UHMWPE surface becomes smoother after appropriate running-in with mild wear in use for longer years in certain cases, it will still remain in mixed lubrication regime at $\Lambda < 3.0$.

The elastohydrodynamic analysis described above was conducted under steady state entraining motion for ball-in-socket model. As it is well known, however, the load and speed in hip joints during normal walking are not constant but change in cyclic form. Jalali-Vahid and Jin (2002) performed the transient elastohydrodynamic lubrication analysis of UHMWPE hip joint replacements under cyclic load and speed conditions. They found that the predicted minimum film thickness maintains around the constant value, despite a large change in either angular velocity or load. It is considered that this is attributed to the combined effect of entraining and squeeze film actions. Furthermore, it was indicated that the predicted film thickness under the quasistatic condition based on the average load and speed is very close to the transient film thickness predicted under cyclic variation in either the load or the speed.

4.2.1.2.3 Elastohydrodynamic lubrication analysis for 3D physiological loading and motion conditions

To evaluate appropriately the actual tribological performance of total hip replacements in clinical use, the elastohydrodynamic lubrication analyses for not only under vertical load and flexion-extension motion condition but under 3D loading and motions are required. Gao et al. (2009) investigated the effect of 3D physiological loading and motion of gait cycles in all three directions on the elastohydrodynamic lubrication of metal-on-metal total hip replacement with a femoral head diameter of 28 mm, cup wall thickness of 9.5 mm and the radial clearance of 30 μm. The hip implant and associated 3D loading and motions are shown in Fig. 4.10. The metallic acetabular cup was assumed to be firmly fixed to the pelvic bone through an equivalent layer and anatomically inclined from the horizontal plane (x axis) with an angle of $-\pi/4$. The rotations

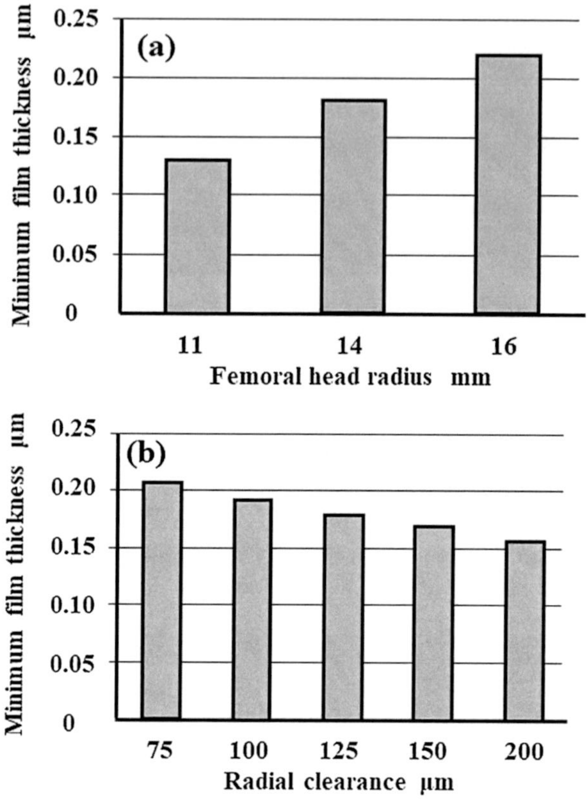

Figure 4.9 Predicted minimum film thickness in UHMWPE hip joint replacements. (a) Effect of femoral head radius R_1 on the predicted minimum film thickness: $\omega = 2$ rad/s, $w = 2.5$ kN, $d = 10$ mm, $\eta = 0.005$ Pa s, $E = 1.0$ GPa, $c = R_2 - R_1 = 120$ μm. (b) Effect of radial clearance c on the predicted minimum film thickness: $\omega = 2$ rad/s, $w = 2.5$ kN, $d = 10$ mm, $\eta = 0.005$ Pa s, $E = 1.0$ GPa, $R_1 = 14$ mm.
Modified from Mak, M. M., & Jin, Z. M. (2002). Analysis of contact mechanics in ceramic-on-ceramic hip joint replacements. *Proceedings of the Institution of Mechanical Engineers, Part H: Journal of Engineering in Medicine*, *216*(4), 231–236. https://doi.org/10.1243/09544110260138718

of the femoral head around the *x*, *y*, and *z* axes represented flexion-extension (FE), internal-external rotation (IER), and abduction-adduction (AA), respectively. Adopted 3D forces and angular velocities of 3D rotation for gait cycle based on measured data by Bergman et al. (2001) is shown in Fig. 4.11. The synovial fluid was assumed to be Newtonian, isoviscous (0.001 Pa s), and incompressible.

In the EHL analysis under 3D loading and motion, the introduction of multigrid method into a spherical coordinate system and combining with the spherical fast Fourier transform (SFFT) technique for the elastic deformation has significantly enhanced the capability of numerical methods. Variation of the minimum, center, and average

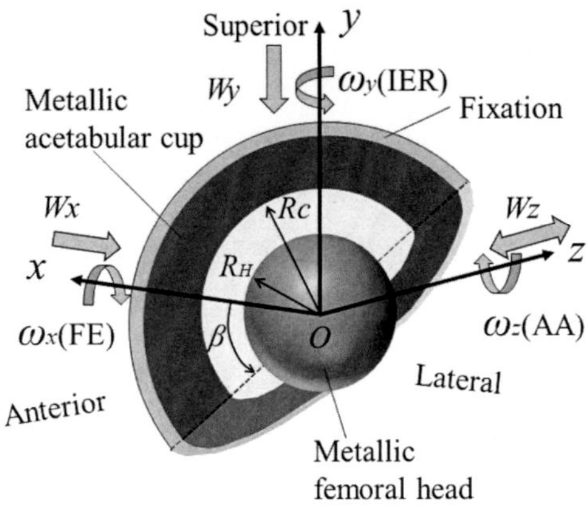

Left hip

Figure 4.10 An anatomic configuration of a MOM hip joint replacement under 3D load and rotation. 3D load (medial-lateral, w_x; superior-inferior, w_y; anterior-posterior, w_z) 3D rotation (flexion-extension, ω_x; internal-external rotation, ω_y; adduction-abduction, ω_z).
Modified from Gao, L., Wang, F., Yang, P., & Jin, Z. (2009). Effect of 3D physiological loading and motion on elastohydrodynamic lubrication of metal-on-metal total hip replacements. *Medical Engineering and Physics*, *31*(6), 720–729. https://doi.org/10.1016/j.medengphy.2009.02.002

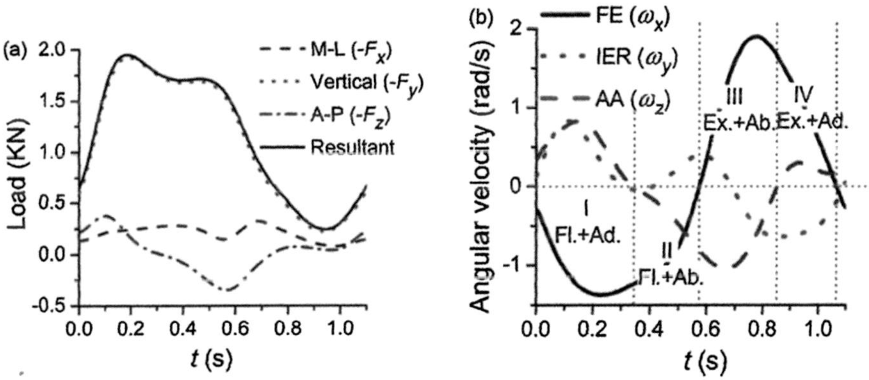

Figure 4.11 Adopted 3D forces and angular velocities of 3D rotation for gait cycle. (a) 3D forces (b) angular velocities of 3D rotation.
Modified from Gao, L., Wang, F., Yang, P., & Jin, Z. (2009). Effect of 3D physiological loading and motion on elastohydrodynamic lubrication of metal-on-metal total hip replacements. *Medical Engineering and Physics*, *31*(6), 720–729. https://doi.org/10.1016/j.medengphy.2009.02.002

film thicknesses during convergent walking cycle is shown in Fig. 4.12 for under (a) the vertical load and FE motion and (b) 3D load and 3D motion. Under (a) the vertical load and FE motion, two main troughs in the minimum film thickness were observed around 0.2 s at peak load and around 0.6 s at peak load and the low FE

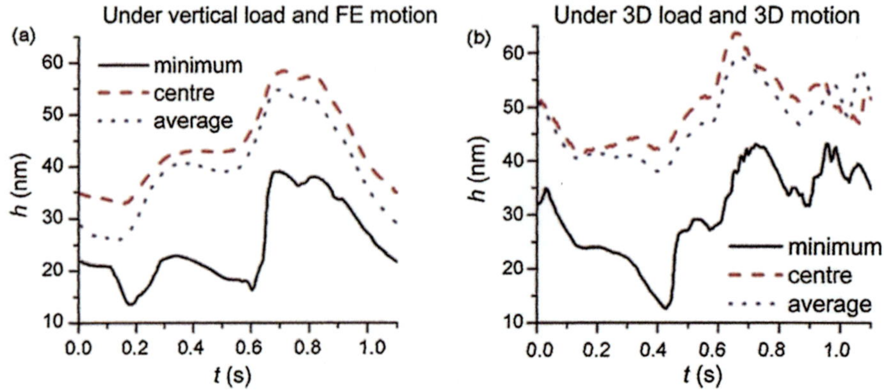

Figure 4.12 Variation of the minimum, center, and average film thickness during convergent walking cycle. (a) Under the vertical load and FE motion (b) Under 3D load and 3D motion.

Modified from Gao, L., Wang, F., Yang, P., & Jin, Z. (2009). Effect of 3D physiological loading and motion on elastohydrodynamic lubrication of metal-on-metal total hip replacements. *Medical Engineering and Physics*, *31*(6), 720–729. https://doi.org/10.1016/j.medengphy.2009.02.002

angular velocity. In contrast, under (b) 3D loading and motion, these troughs became almost flat and the corresponding film thickness increased from about 14–17 nm to 24–28 nm due to horizontal translation of contact driven by the horizontal A-P and M-L load components. This increased the center and average film thickness (Fig. 4.12). However, it should be noticed that the horizontal translation also caused an adverse effect on lubrication when its direction changed, as rapid reduction of film thickness at 0.32 and 0.74 s (Fig. 4.12b) after the A-P load reversal (Fig. 4.11a). During the whole walking cycle both the average and the center film thickness were significantly increased compared with the range under 1D loading and motion.

As mentioned, based on the experimental and analytical studies, most of total hip prostheses appear to operate in the corresponding lubrication regime depending on the material combination and design of prostheses.

Ceramic-on-ceramic (CoC): fluid film lubrication or mixed lubrication.

Metal-on-metal (MoM): mixed lubrication.

Hard (metal or ceramic)-on-UHMWPE: boundary or mixed lubrication.

As improved methods for lubrication mode in metal-on-metal, the superiority of ceramic-on-metal (CoM) was indicated by Meng et al. (2013). The film thickness of CoM hip joint is thicker than CoC hip but thinner than MoM hip. However, CoC and CoM are more likely to benefit full fluid film lubrication due to higher Λ ratio than MoM hip joint.

In hip prostheses in clinical use, the proteinaceous adsorbed films are likely to control the actual lubrication mode, as discussed in 4.3. In model test for metallic femoral head rubbed in serum solution with optical interferometry, a thin adsorbed (boundary) protein film under cyclic loading and a thicker protein-containing lubricant film than predicted film at slow sliding were observed (Myant & Cann, 2013). Furthermore, the changes in surface roughness and properties of rubbing surfaces with running-in or creep deformation and wear have considerable influence. In addition, the changes in alignment may enhance the mechanical severity. The appropriate evaluation from the tribological viewpoint will be effective to extend the durability of joint prostheses.

4.2.2 Knee joint replacements as hard-on-UHMWPE

Most of knee prostheses have simplified anatomical geometry without meniscus component with or without patella component replacement. Therefore, contact areas in femoral-tibial joint and patellar-femoral joint are limited as Hertzian contact within smaller areas and the maximum contact pressure becomes higher than total hip prostheses as Hard-on-UHMWPE spherical joints. Therefore, it is expected that the rubbing severity at lubrication modes in knee joint replacements will be higher than that in hip joint replacements. The evaluations of lubrication modes are described below on the basis of experimental measurements and numerical simulations.

4.2.2.1 Experimental evaluation of lubrication modes in knee joint replacements as hard-on-UHMWPE

To evaluate the actual fluid film formation in knee prostheses with UHMWPE tibial components in knee joint simulator under walking condition, the electric resistance method for conductive femoral and tibial components was applied by using the electric circuit shown in Fig. 4.13 (Murakami, 1990; Murakami & Ohtsuki, 1987; Murakami et al., 1993). Nonconductive UHMWPE was coated with metallic thin layer via sputter coating method. As lubricant, the nonconductive lubricant was used. To evaluate the

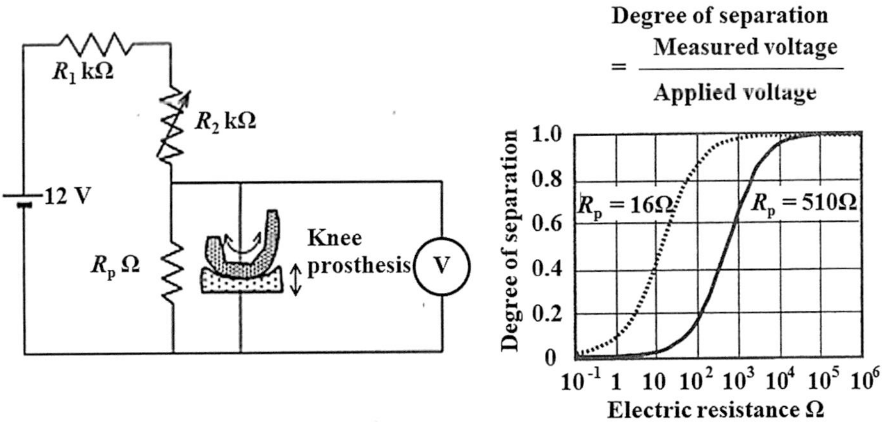

Figure 4.13 Electric circuit for measurement of degree of separation between rubbing surfaces. Electric circuit for evaluating fluid film formation between femoral and tibial components in knee prostheses is shown. The degree of separation indicates the extent of fluid film formation.

Modified from Murakami, T., Ohtsuki, N., & Higaki, H. (1993). The adaptive multimode lubrication in knee prostheses with compliant layer during walking motion. *Tribology Series, 25*(C), 673–682. https://doi.org/10.1016/S0167-8922(08)70420-6

effectiveness of fluid film formation, the degree of separation is defined as the ratio of measured voltage to the applied electric voltage. The degree of separation = 1 means full separation, and 0 means intimate local contact. In this electric circuit, the electric resistance between rubbing surfaces at 0.5 as the degree of separation is equal to parallel resistance R_p. Therefore, it is possible to evaluate the fluid film formation corresponding to the level of insulating intervening substance between rubbing surfaces by selecting the value of R_p.

Here, the changes in fluid film formation in anatomical knee and cylindrical knee (Radius of metallic femoral component = 30.0 mm, Radius of concave UHMWPE tibial component = 30.3 mm) as fixed bearings were evaluated lubricated with silicone oils in knee joints simulator using electric resistance method. S-100 and S-10000 mean silicone oils with kinetic viscosity of 100 and 10,000 mm^2/s, at 25°C, respectively. Experimental results are shown in Fig. 4.14.

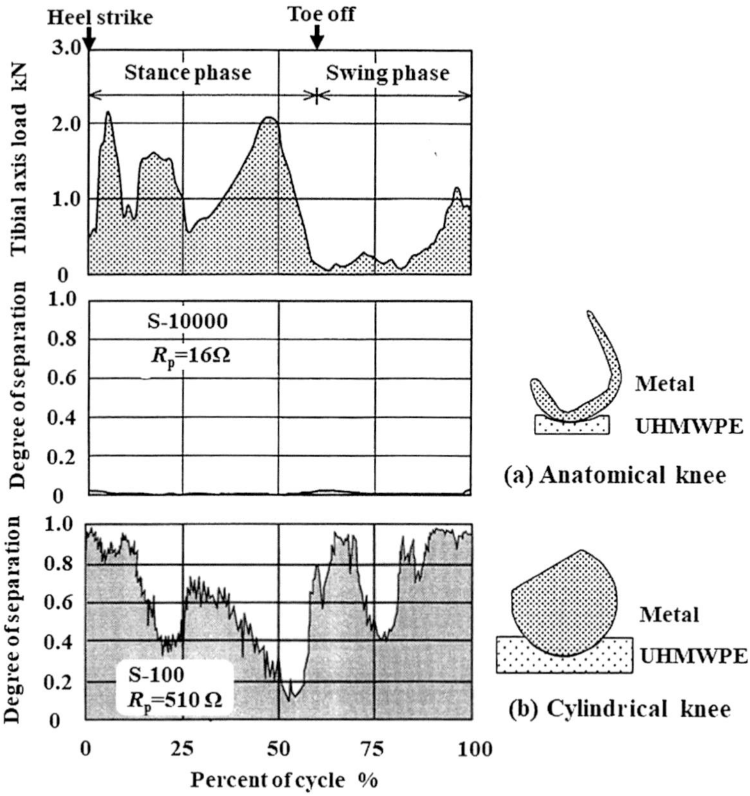

Figure 4.14 Fluid film formation in knee prostheses during walking in knee joint simulator. (a) Anatomical knee: Fluid film formation is little observed during walking even with high viscosity lubricant (S-10000). Some signals with very low degree of separation are observed around heel strike and after toe-off. (b) Cylindrical knee: Significant fluid film formation is confirmed. Degree of separation decreases during stance phase and recovers during swing phase with some drops at stroke ends where sliding speed becomes zero.
Modified from Murakami, T., Ohtsuki, N., & Higaki, H. (1993). The adaptive multimode lubrication in knee prostheses with compliant layer during walking motion. *Tribology Series, 25*(C), 673–682. https://doi.org/10.1016/S0167-8922(08)70420-6

In anatomical knee prosthesis, signals indicating very thin film formation with very low degree of separation in mixed lubrication were observed only around heel strike and after toe-off even with very viscous lubricant S-10000 (Fig. 4.14a). It indicates that significant direct contacts occur particularly at the second and third peak loads during stance phase. These low level signals disappeared with decreasing viscosity.

In contrast, perfect separation was almost established with high viscosity oil S-10000 in cylindrical knee with better geometrical congruity. With medium viscosity S-100 of about 0.1 Pa s, considerable elastohydrodynamic film appears to be formed even in loading stance phase (Fig. 4.14b). At the first peak load, considerable film is preserved, which is attributed to squeeze film action after film recovery during preceding swing phase. However, with decreasing viscosity, significant direct contacts occurred particular during loading stance phase. Therefore, most of knee prostheses appear to operate in severe mixed lubrication or boundary lubrication regime.

The apparent contact areas in anatomical and cylindrical knee joint models which were measured by the pressure sensitive paper ($P > 0.5$ MPa, Fuji Film) under extended condition at 2 kN are shown in Fig. 4.15. It is noticed that higher contact pressure occurs in anatomical knee. In cylindrical knee with high conformity at lower contact pressure, it is expected that some fluid film formation becomes effective but the range of motions may be limited and may produce high torque at the bone-implant interface by high geometrical congruity. In fixed bearing knee prostheses, the optimum geometric design has been explored to improve the satisfaction of patients. In contrast, in mobile-bearing knee prostheses (Fig. 2. 29) (Huang et al., 2007), the combination of congruity and mobility allows both low contact stress and low constraint, and considerable fluid film formation is expected under tibiofemoral malalignment conditions.

4.2.2.2 Numerical analyses for evaluation of lubrication modes in knee joint replacements

The knee requires a large range of flexion (150−160°) for normal activities and it is required that artificial knee joints have low conformity between femoral and tibial components. Therefore, artificial knee joints composed of hard (metal or ceramic)-

	(a) Anatomical knee	**(b) Cylindrical knee**
Apparent contact area: A	$A = 144\ \mathrm{mm}^2$	$A = 1530\ \mathrm{mm}^2$
Average contact pressure: p	$p = 13.9$ **MPa**	$p = 1.3$ **MPa**

Figure 4.15 Apparent contact area and average contact pressure on UHMWPE tibial components at extended position at 2 kN. Cylindrical knee prosthesis with high geometrical congruity exhibits wider contact area and lower contact pressure than anatomical knee prosthesis with low congruity.
Modified from Murakami, T., Ohtsuki, N., & Higaki, H. (1993). The adaptive multimode lubrication in knee prostheses with compliant layer during walking motion. *Tribology Series*, 25(C), 673−682. https://doi.org/10.1016/S0167-8922(08)70420-6

on-UHMWPE operate at higher contact pressure for smaller contact areas and appear to operate in the mixed or boundary lubrication as described above. For these knee prostheses, the analytical studies on fluid film lubrication are limited in few cases. Auger et al. estimated minimum fluid film thickness h_{min} by Hamrock and Dowson formula (1978) in its line contact form (Dowson et al., 1981).

$$\frac{h_{min}}{R} = 7.43\left(\frac{\eta u}{E'R}\right)^{0.65}\left(\frac{w}{E'RL_1}\right)^{-0.21}\left(\frac{R}{L_1}\right)^{0.21} \tag{4.9}$$

where L_1 is bearing length (width of femoral component), and other variables are the same in Eq. (4.6). In this analysis, a cylindrical geometry was selected to model the knee prosthesis with a femoral metallic cylinder of radius $R_1 = 47.88$ mm articulating against a UHMWPE semi-infinite tibial concave UHMWPE component of radius $R_2 = 53.34$ mm. Here, $L_1 = 58.17$ mm. The minimum film thickness for $w = 2.0$ kN, $u = 0.0382$ m/s, and $\eta = 0.001$ Pa s was predicted as 0.073 μm, which is smaller than the composite surface roughness (>1.0 μm). Thus, $\Lambda < 0.073$. The lubrication mode is suggested to be boundary or severe mixed lubrication regime.

Mongkolwongrojn et al. (2010) performed the transient elastohydrodynamic lubrication analysis in artificial knee joint with non-Newtonian fluids. In numerical simulation, a perturbation method, Newton–Raphson method and multigrid method with full approximation technique were employed to solve simultaneously both the time-dependent Reynolds equation with non-Newtonian fluid based on a Carreau model (η_0: low shear strain rate viscosity), and the elasticity equation. Numerical simulations were carried out for artificial knee joint design by Bartel et al. The changes in minimum film thickness during slow walking in artificial knee joint are shown for Newtonian and non-Newtonian fluids with different viscosities in Fig. 4.16. The minimum film thickness reaches a very low value in the transient state at the beginning of the first walking cycle. Then, the fluid film develops in the second walking cycle as almost steady state condition. It is noticed that the minimum film thickness is lowest around toe-off at highest load and zero velocity, whereby in squeeze film action the film thickness never descends to zero. The film thickness decreases with lowering of viscosity and for non-Newtonian fluid with the shear thinning effect. In steady walking motion as $h_{min} < 0.1$ μm, lubrication mode is suggested to be severe mixed or boundary lubrication regime. Therefore, wear resistance of rubbing materials and boundary lubrication ability of lubricants become important.

4.3 Boundary lubrication in artificial joints

As mentioned previously on lubrication modes, most of the artificial joints operate in mixed or boundary lubrication mode in various daily activities and direct contacts between rubbing surfaces occur. Therefore, the protective properties of boundary lubricants or alternate surface layers become important for minimizing wear and friction to extend the durability of artificial joints.

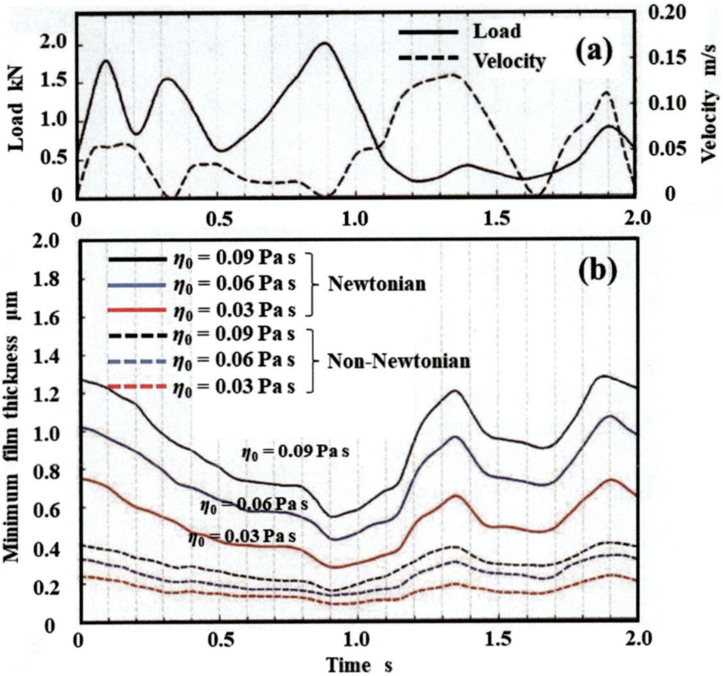

Figure 4.16 Numerical results for transient minimum film thickness in artificial knee joint during walking for Newtonian and non-Newtonian fluids with different viscosities. (a) Load and velocity during walking cycle at period of 2.0 s. (b) The changes in minimum film thickness in the second walking cycle.
Modified from Mongkolwongrojn, M., Wongseedakaew, K., & Kennedy, F. E. (2010). Transient elastohydrodynamic lubrication in artificial knee joint with non-Newtonian fluids. *Tribology International*, *43*(5–6), 1017–1026. https://doi.org/10.1016/j.triboint.2009.12.041

Lubricants as periprosthetic synovial fluid after joint replacement contain synovia constituents such as hyaluronic acid, proteins, glycoproteins and phospholipids, which form adsorbed films on metallic, ceramic, and polymeric surfaces of joint prostheses and have considerable influence on friction and wear in boundary lubrication (Sawae et al., 1998). Appropriate adsorbed film formation on rubbing surfaces reduces friction and protects the rubbing surfaces in mild mixed lubrication. However, in cases the adsorbed serum protein molecules such as albumin and γ-globulin enhance the adhesion in direct contact, the adsorbed films raise friction and may increase or decrease wear. Protein films suppress the formation of transfer layer of UHMWPE. It is shown in wear tests for UHMWPE in hip joint simulator (Wang et al., 1998) and multidirectional pin-on-disk tester (Saikko, 2003) that wear rate of UHMWPE increased with increasing protein concentration in the low protein concentration range.

Sawae et al. (2008) indicated the combined effect of proteins and phospholipids on wear of UHMWPE in multidirectional pin-on-plate test. In this wear test, phospholipids acted as a boundary lubricant and reduced polyethylene wear in the low protein lubricants. In lubricants with protein within physiological level, however, UHMWPE wear increased with increasing lipid concentrations. Increased interaction between protein and lipid molecules and lipid diffusion to polyethylene surface might be responsible for the increased wear. Greenbaum et al. (2004) examined the effect of diffused lipid molecules on the mechanical properties and wear characteristics of UHMWPE. Their experimental results indicated that the lipid molecules that diffused into the bulk of material had a significant plasticizing effect on both conventional and cross-linked UHMWPE.

Protein boundary film formation on rubbing surfaces can be confirmed by in situ observation with fluorescence microscopy (Yarimitsu et al., 2007), and the differences in boundary film formation of synovia constituents can be evaluated. Nečas et al. (2021) observed the protein lubricating film in contact zone between metallic femoral component and transparent polymer tibial component using fluorescent microscopy with intensity measurement in knee joint simulator. In this study, agglomerations of proteins near the scratches and edge of contact were observed and the difference in film thickness at extended and flexed positions was confirmed. The stabilizing phenomenon by HA and phospholipids on the interaction between albumin and γ-globulin was indicated.

The actual phenomena of boundary lubrication in artificial joints have not been sufficiently clarified, and it is expected to be elucidated by various approaches (Stevenson & Cann, 2021).

4.4 Wear in artificial joints

The occurrence of local direct contact between rubbing materials in insufficient fluid film lubrication can cause wear problems. In artificial joints, wear debris-induced osteolysis (Ingham & Fisher, 2005; Willert & Semlitsch, 1977) brought severe clinical problems such as joint loosening. In rubbing surfaces of the artificial joints, various types of wear occur as single or combined type. To solve wear problems it is important to elucidate the real wear mechanism and explore the appropriate methods to reduce wear. To extend the durability of rubbing elements or devices, the minimization of wear is required on the basis of understanding of actual wear phenomena. The quantitative evaluation of wear is performed based on the wear measurements by (a) weight loss and (b) dimensional change (Dumbleton, 1981). Here, weight loss should be estimated from the weight change with subtraction of the weight increase due to liquid absorption. Dimensional change should be estimated from total change in dimension with subtraction of the decrease due to creep and elastic/plastic deformation and with correction for changes due to liquid absorption.

4.4.1 Wear mechanisms

Wear mechanisms in artificial joints are divided broadly into the following four categories. In actual rubbing surfaces, various wear mechanisms have mutual interaction.

4.4.1.1 Fundamental wear mechanisms

4.4.1.1.1 Adhesive wear

Adhesive wear is defined as "wear by transference of material from one surface to another during relative motion, due to a process of solid-phase welding." (OECD, 1969), as mentioned in Section 1.4.1. In general, the same materials tend to adhere at intimate contact. To suppress the adhesive wear, the combination of different materials as rubbing pair is recommended as in UHMWPE against metal or ceramics. However, when the rubbing force acts after UHMWPE adhered to the mating material, the shearing occurs not in interface itself but within the softer UHMWPE if the junction is stronger than one of the materials. Thus, some part of UHMWPE moves to mating material as transfer film (Fig. 4.17a and b). In repeated rubbing process, mixing of small transfer pieces and forward-and-backward transfer occur and transferred pieces pile up to larger transfer film. If the transferred film integrated on surface cannot withstand to the shearing force during rubbing, it is released as wear debris. If the stable transfer films are formed, wear can be decreased. After joint arthroplasty, the joint prosthesis is lubricated with body fluid or synovial fluid which contains proteins and phospholipids. If the rubbing surfaces of the prosthesis are covered with adsorbed films, the adhesion is prevented and transfer is prevented (Fig. 4.17c and d). But, if the adsorbed films act as bonding molecules, some transfer can occur (Fig. 4.17e and f). In this case, high wear of polyethylene is observed in lubricant with proteins at high concentration than in saline. In actual wear process, more complicated behaviors accompanied with transfer phenomena and deterioration of rubbing materials are likely to produce wear debris.

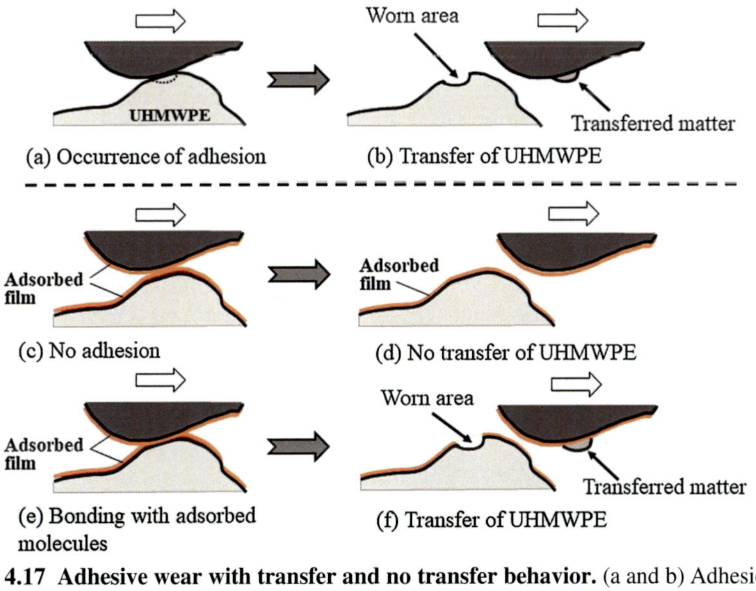

Figure 4.17 Adhesive wear with transfer and no transfer behavior. (a and b) Adhesion and transfer of UHMWPE, (c and d) No transfer of UHMWPE, (e and f) Bonding with adsorbed molecules and transfer of UHMWPE.

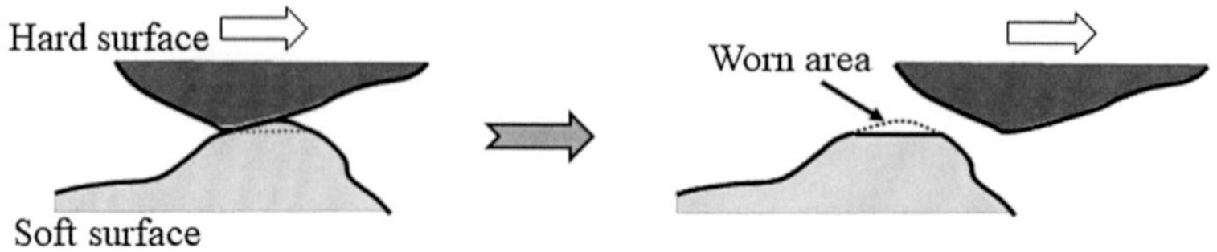

Figure 4.18 Two-body abrasive wear. Hard protuberances penetrate into soft material and plow the soft surface, or act as microcutting tool.

4.4.1.1.2 Abrasive wear

Abrasive wear occurs when a hard rough surface or hard particles run against soft surface. In the wear mode as two-body abrasive wear, hard protuberances penetrate into soft material and plow the soft surface, or act as microcutting tool (Fig. 4.18). For hard material of joint prosthesis, smoother finish of rubbing surface is required. In hard-on-UHMWPE, metallic component is likely to generate abrasive wear because usually metallic surface has higher surface roughness than ceramic material.

In three-body abrasive wear, hard particles indent, groove, and cut material out of the soft rubbing surfaces. However, if the hard particles including wear debris or cement debris were embedded in the soft rubbing surface, the extruded particles can cause heavy damage to the hard material.

4.4.1.1.3 Fatigue wear/delamination

Fatigue wear is defined as "removal of particles detached by fatigue arising from cyclic stress variations" (OECD, 1969). On the rubbing surfaces at high stress under cyclic and/or fluctuating heavy load, the delamination or pitting occurs as a result of subsurface or surface cracking and its propagation. The repeated loading and unloading cycles at rolling-sliding contact induce the formation of subsurface or surface cracks. In artificial joints, the surface failure with large laminated fragment is called delamination and the small pit is called pitting. The delaminations are observed in UHMWPE tibial insert in total knee prostheses, as mentioned in Section 4.4.2.1. Fatigue wear at asperity appears to produce pitting, as described in Section 4.4.2.2.

4.4.1.1.4 Corrosive wear

Corrosive wear is defined as "a wear process in which chemical or electrochemical reaction with the environment predominates" (OECD, 1969). In many cases, the mechanochemical mechanism combined of chemical reaction and mechanical action enhances the corrosive wear. The inside of the human body is a corrosive environment, and it is required for artificial joints to have adequate corrosion resistance. Anticorrosive metals such as stainless steel, Co-Cr-Mo alloy, Ti alloy, and oxidized zirconium, and bioinert ceramics such as alumina, zirconia, zirconia-toughened alumina, and UHMWPE are generally used as artificial joint materials.

Here, tribocorrosion is defined as the chemical, electrochemical, and mechanical processes leading to a degradation of materials in tribological contacts operating in a corrosive environment (Yan et al., 2006). It should be taken into consideration that tribological contacts can accelerate the corrosion rate, and released wear debris can also be corroded in the human body.

It should be noticed in not only metal-on-metal THA but metal-on-UHMWPE THA that the taper corrosion at metal femoral head-stem neck junction which releases a metal ion and/or debris can give adverse local tissue reaction. The appropriate strategy to suppress corrosion in head-neck junction is required on the basis of identification of various corrosion mechanisms including galvanic corrosion, crevice corrosion, fretting, intergranular corrosion, and pitting corrosion.

4.4.1.1.5 Influence of surface roughness on wear mechanisms

The lubrication mode in lubricated rubbing surfaces is predicted by the film parameter (lambda ratio): Λ, defined by Eq. (4.1), representing the ratio of theoretical minimum fluid film thickness to composite surface roughness. At mixed or boundary lubrication mode at $\Lambda < 3$, considerable direct contacts between rubbing surfaces occur according to level of Λ. Therefore, the surface roughness is prepared to be minimized for rubbing surfaces.

Wear occurs usually at direct contact zones with other materials, and therefore surface topographies of rubbing materials have influence on wear behaviors. Particularly, the surface roughness of hard material affects the wear characteristics of UHMWPE. The changes in wear factor (wear rate) against the counterface arithmetic average roughness R_a under wet (in water) and dry conditions are shown in Fig. 4.19 (Dowson et al., 1985). It is noticed that wear factor has minimum R_a values in the range of

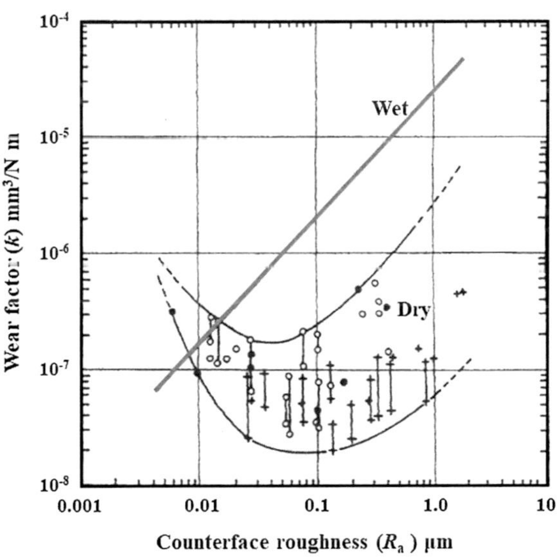

Figure 4.19 Wear factor against the counterface arithmetic average roughness R_a under wet (in water) and dry conditions. Optimum surface roughness minimizing wear is recognized under dry condition, but simple increasing wear is confirmed under wet condition.
Reprinted with permission from Dowson, D., Diab, M. M. E.-H., Gillis, B. J., & Atkinson, J. R. (1985). Influence of counterface topography on the wear of ultra high molecular weight polyethylene under wet and dry conditions. In *Polymer wear and its control*. American Chemical Society (ACS) (pp. 171–187). Copyright 1985 ACS. https://doi.org/10.1021/bk-1985-0287.ch012.

0.05−0.1 μm under dry condition. These phenomena appear to be caused by wear increase due to adhesive action with lowering surface roughness and wear increase due to abrasive and/or fatigue actions with larger surface roughness. In contrast, under wet (nonprotein) conditions the wear rate of polyethylene increased steadily with the effective counterface roughness. Furthermore, it is noted that the wear rate under wet conditions becomes higher than under dry conditions in the larger surface roughness. The results in Fig. 4.19 demonstrates the fundamental feature for the dependency of wear rate of UHMWPE on counterface roughness, but the possibility of the transfer film formation from polyethylene to counterface can affect the wear phenomena. Under lubricated conditions with protein containing lubricants, the adsorbed protein films have the ability to prevent the transfer film formation and may reduce or increase wear. The influence of lubricants on wear is complex and the appropriate evaluation is required for actual wear behaviors. On extremely smooth counterface, the adhesion of polyethylene might be enhanced, which may result in increase in wear. In multidirectional sliding test for UHMWPE pin on Co-Cr alloy plate lubricated with diluted bovine serum solution, minimum wear was observed at medium surface roughness (R_a: around 0.033 μm) for both conventional and highly cross-linked (irradiated with 1000 kGy) polyethylene (Sawae, 2009).

In the design process of artificial joints, it is required to explore the optimum surface roughness with minimum wear.

4.4.1.2 Wear of UHMWPE

Since successful clinical application of total hip replacements composed of ultra-high molecular weight polyethylene (UHMWPE) acetabular cup and metallic femoral head by Charnley in 1962 (Charnley, 1979), UHMWPE have been used as liners in hip, knee, ankle, shoulder, elbow, and other joint prostheses. UHMWPE with extraordinarily high molecular weight possesses superior mechanical toughness and wear resistance as approximately 0.1 mm/year for total hip prosthesis, which has a 30-fold reduction in wear rate compared with PTFE as described in Section 4.1. However, in longer clinical usage of total hip replacements composed of metal or ceramic femoral head and conventional UHMWPE acetabular cup, UHMWPE wear debris was widely recognized to play a central role in initiating osteolysis, which resulted in loosening of the prostheses. Therefore, various strategies to improve wear property of conventional UHMWPE have been developed, as described in 4.5.

The detailed information on UHMWPE biomaterials is included in *UHMWPE Biomaterials Handbook, Third Edition* (Kurtz, 2016). In this section, main features on UHMWPE wear are mentioned. In clinical evaluation of wear in total hip prostheses, the penetration of the femoral head into UHMWPE cup in X-ray images is measured. Here, it should be kept in mind that the penetration includes both wear depth and creep deformation. In clinical study with radiostereometric analysis (RSA) (Glyn-Jones et al., 2008), it was indicated that creep mainly occured in the first year with an exponential pattern. It was suggested that the penetration after one year is predominantly due to wear.

On wear mechanism in UHMWPE for total joint replacements, the following important factors should be noticed.

4.4.1.2.1 Multidirectional motion

In hip and knee joints during walking, flexion/extension motion is accompanied by internal/external rotation and abduction/adduction. Therefore, the rubbing area of articular surfaces moves in multidirectional motion. In multidirectional motion, the UHMWPE rubbing surface experiences both shear and tensile stresses in multiple directions. In this case, strengthening in one particular direction results in weakening in a perpendicular direction. Therefore, wear increases as sliding proceeds in multidirectional motion. In contrast, in linear-tracking motion, molecular orientation in UHMWPE leads to strain-hardening of the rubbing surfaces. In this case, wear resistance increases as sliding proceeds (Wang, 2001).

In Fig. 4.20, AFM images of UHMWPE worn surfaces in unidirectional and multidirectional pin-on-disk tests lubricated with diluted serum solution are shown (Sawae & Murakami, 1999). In unidirectional test, the worn pattern in perpendicular to sliding direction is observed, but in multidirectional test, fine fibrils in various directions are observed. In unidirectional rubbing, molecular orientation in UHMWPE appears to lead to strain-hardening of the rubbing surfaces, and the wear rate was more than one order lower than multidirectional test.

Considering these wear behaviors, it is expected that the application of cross-linking to UHMWPE becomes an effective means to improve the wear resistance of UHMWPE in multidirectional sliding. Cross-linking retards chain mobility and hence reduces the degree of molecular orientation during sliding. In addition, cross-linking increases the density of the C—C chemical bonds between adjacent molecular chains and hence makes it more difficult to split one molecule from another.

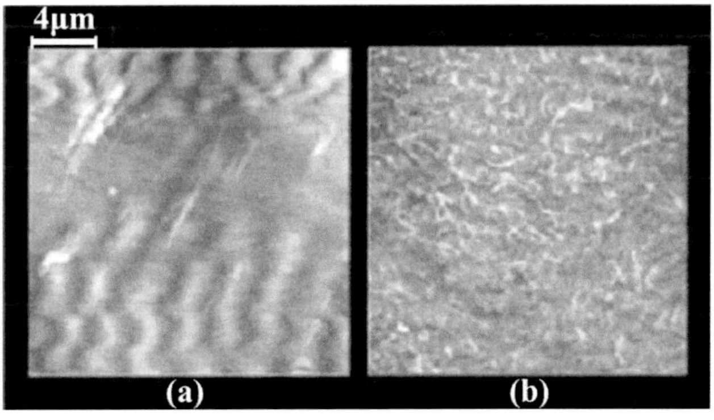

Figure 4.20 AFM images of worn UHMWPE surface in unidirectional and multidirectional tests. (a) In unidirectional test, the worn pattern in perpendicular to sliding direction is observed. (b) In multidirectional test, fine fibrils in various directions are observed.
Modified from Sawae, Y., & Murakami, T. (1999). Wear evaluation of UHMWPE wear in multi-directional tests. In *Preprint of JAST annual meeting* (pp. 91—94).

4.4.1.2.2 Oxidation in UHMWPE

The previous challenge to reduce oxidation in UHMWPE had been associated with short-term clinical failures after gamma sterilization in air and long-term shelf aging in air. The standard gamma radiation sterilization (25−40 kGy) leads to the generation of free radicals through the cleavage of C−C and C−H bonds within amorphous and crystalline phases in UHMWPE (Costa et al., 1998). The generated free radicals may act with each other to form the cross-links or may react with oxygen to cause oxidation degradation. The elevated oxidation resulting from the use of gamma sterilization in air reduced static strength, elongation and fatigue resistance of UHMWPE, and brought severe wear damages of knee prostheses (Collier et al., 1996). (Zhang et al. 2015) investigated the effect of radiation dose (50−1000 kGy) on the depth-dependent oxidation and wear of shelf-aged gamma-irradiated UHMWPE. The gamma irradiated UHMWPEs were annealed after irradiation but they oxidized after long-term (7 years) shelf-aging. The experimental results of multidirectional sliding pin-on-plate wear tests showed that when the oxidation index (OI: the ratio of the area of carbonyl absorption peak centered at 1720 cm^{-1} (>C=O absorption) to that centered at 1370 cm^{-1} (C−H absorption)) > 1, the wear rate significantly increased. When $1 < OI < 3$, the wear rate is mainly controlled by the OI level.

At present, gamma irradiation has been performed in an inert environment and the thermal treatment of radiation-cross-linked UHMWPE reduced the concentration of free radicals generated during the ionizing radiation process. Furthermore, the packaging process has been improved.

Since introduction of highly cross-linked and thermally treated polyethylene (HXLPE), in case of annealing as thermal treatment below the crystalline melt transition, the residual free radicals have the potential to oxidize in vivo (Kurtz et al., 2011). In the second generation of HXLPE, various improvements such as the addition of vitamin E have been applied.

4.4.1.2.3 Influence of lubricants

As mentioned in Section 4.3 Boundary lubrication in artificial joints, periprosthetic lubricants after joint replacement contain synovia constituents such as hyaluronic acid, proteins, and phospholipids, which form adsorbed films on metallic, ceramic, and polymeric surfaces of joint prostheses and have considerable influence on wear in contact surfaces in boundary or mixed lubrication. Appropriate adsorbed film formation on rubbing surfaces reduces friction and protects the rubbing surfaces in mild mixed lubrication. However, in cases the adsorbed serum protein molecules such as albumin and γ-globulin enhance the adhesion in direct contact, the adsorbed films raise friction and may increase or decrease wear. Protein films suppress the formation of transfer layer of UHMWPE and affect friction and wear behaviors. In many cases, wear rate of UHMWPE increases with increasing protein concentration, but wear properties should be evaluated on the basis of actual observation of worn surfaces and wear debris with consideration on mutual interaction between lubricant constituents.

4.4.2 Wear phenomena

4.4.2.1 Wear phenomena in artificial hip joints

Total hip replacements are divided as hard-on-soft system such as metal-on-UHMWPE and ceramic-on-UHMWPE hip replacements, and hard-on-hard systems such as ceramic-on-ceramic (CoC) and metal-on-metal (MoM) hip replacements.

As described in this chapter, the hard-on-UHMWPE hip replacements appear to operate in mixed or boundary lubrication mode in most of daily activities. Therefore, some wear can occur at severely rubbing surfaces of UHMWPE component, sometimes of hard materials. In longer clinical usage of hard-on-UHMWPE hip replacements, serious problems about wear debris–induced osteolysis and consequent loosening caused the revision procedure, but the improvement of material properties of UHMWPE with cross-linking and related technologies appear to have given desirable solutions. Improvements by the second-generation highly cross-linked polyethylene are described in Section 4.5.

In comparison of ceramic-on-UHMWPE and metal-on-UHMWPE, wear rate for alumina-on-UHMWPE decreased to approximately 50% of that of metal-on-UHMWPE in simulator study using of serum proteins in the physiological concentration (Clarke & Gustafson, 2000). Superiority of ceramic-on-UHMWPE was confirmed in clinical comparison of linear wear rates of polyethylene cups. Here, it was pointed out that the other studies of serum-based simulators generally have shown greater wear with alumina-on-polyethylene than with metal-on-polyethylene when nonphysiological, high protein concentration lubricant was used. To use physiological protein concentration becomes important for adequate evaluation.

For hard-on-hard hip prostheses such as metal-on-metal (MoM), ceramic-on-ceramic (CoC), and ceramic-on-metal (CoM) with high precision and high wear resistance, the possibility of fluid film is expected to be actualized under normal walking condition. However, some problems such as stripe wear, streaking, breakage of ceramic component, taper failure with modular stem designs, adverse biological reaction to metallic ions, and others occurred particularly at severe operating conditions under edge loading and/or at malpositioning.

In clinical application of second generation MoM hip replacements and hip resurfacing replacement with expectation of considerable wear reduction, decreased risk of dislocation and increased range of motion, wide applications had serious problems of pseudotumor formation as a result of the production of very fine wear debris and the associated production of metallic ions of Co and Cr (Yan et al., 2013). In most cases of MoM hip joints, the mixed lubrication with direct contact appears to be predominant, and the fundamental improvement including surface modification is required.

As other hard-on-hard systems, ceramic-on-ceramic hip replacements have been applied in considerable cases, with advantage of improvement in material properties in zirconia-toughened alumina. In ceramic-on-ceramic hip joints, there are problems on stripe wear, squeaking, and impingement of the femoral neck on the rim of the ceramic liner leading to chipping, especially in younger and physically active patients.

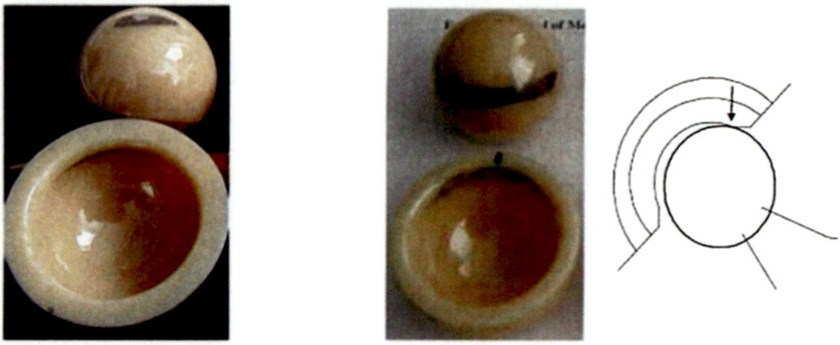

(a) Retrieved CoC bearing after 12 months in vivo **(b) In vitro tested CoC bearing and schematic of microseparation**

Figure 4.21 Stripe wear indicated by pencil rubbing in alumina-on-alumina hip prostheses. In vivo stripe wear in (a) is reproduced by in vitro stripe wear in (b) in hip joint simulator under microseparation condition.
Modified from Mak, M., Jin, Z., Fisher, J., & Stewart, T. D. (2011). Influence of acetabular cup rim design on the contact stress during edge loading in ceramic-on-ceramic hip prostheses. *Journal of Arthroplasty*, *26*(1), 131−136. https://doi.org/10.1016/j.arth.2009.10.019

When the femoral head translates a sufficient distance laterally to contact the rim of the acetabular cup under microseparation condition, "stripe wear" occurs in ceramic-on-ceramic hip joint. The retrieved alumina ceramic bearing with stripe wear is shown in Fig. 4.21a, and an in vitro tested alumina ceramic bearing in hip joint simulator is shown in Fig. 4.21b (Mak et al., 2011). The edge contact between the femoral head and the rim of acetabular cup raises the contact stress, which results in intragranular fracture, showing stripe pattern. In vivo stripe wear was reproduced in hip joint simulator test under microseparation condition (Stewart et al., 2001). The appropriate alignment in arthroplasty to avoid the edge loading, the improvement of composite ceramic material properties, and optimum geometrical design for cup edge appear to resist the stripe wear.

4.4.2.2 Wear phenomena in artificial knee joints

4.4.2.2.1 Elastoplastic contact analysis of UHMWPE tibial components

The wear phenomena of UHMWPE components in knee and hip prostheses are the major restriction factors on the longevity of these implants. Particularly in knee prostheses with low conformity which are operated at high contact pressure, they appear to be used under elastoplastic contact condition and the predominant types of wear on UHMWPE tibial components are delamination and pitting. These fatigue wear patterns of UHMWPE are believed to result from repeated plastic deformation owing to high contact stresses, and the elucidation of these wear mechanisms in clinical application have been required.

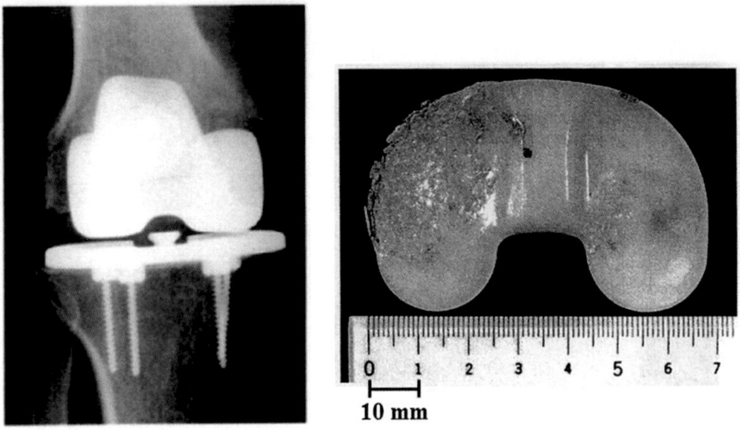

(a) Preoperative radiograph of retrieved knee prosthesis

(b) Retrieved tibial component

Figure 4.22 Preoperative radiograph of retrieved knee prosthesis and retrieved tibial insert. (a) In the radiograph, the clearance gap between femoral component and tibial tray is narrow at medial side. (b) In medial side of tibial insert, severe wear is observed.
Modified from Cho, C. H., Murakami, T., Sawae, Y., Sakai, N., Miura, H., Kawano, T., & Iwamoto, Y. (2004). Elasto-plastic contact analysis of an ultra-high molecular weight polyethylene tibial component based on geometrical measurement from a retrieved knee prosthesis. *Proceedings of the Institution of Mechanical Engineers, Part H: Journal of Engineering in Medicine, 218*(4), 251–259. https://doi.org/10.1243/0954411041561018

Cho et al. (2004) performed the elastoplastic finite element contact analysis of UHMWPE tibial insert based on geometrical measurement for retrieved knee prostheses composed of metallic femoral and UHMWPE tibial components. One of the retrieved knee prostheses, a preoperative radiograph of the retrieved knee prosthesis as Miller Galante I (MG I) porous total knee system (right side, Zimmer Inc., USA, standing position) is shown in Fig. 4.22a. The knee prosthesis was retrieved from a 68-year-old man (162 cm, 93 kg) and the in vivo duration was 133 months. Estimating from the clearance gap between the metallic femoral component and tibial tray, the medial side has a narrower clearance gap than the lateral side. This fact corresponds to the excessive wear including delamination and pitting and high plastic deformation observed in the medial side (Fig. 4.22b). The estimation of three-dimensional contact position on the basis of only a sheet of radiograph is difficult; therefore, the estimation of contact position was conducted by finite element analysis for three-dimensional model based on measured geometrical data.

Their elastoplastic finite element contact analysis was conducted for three-dimensional models of metallic femoral components and UHMWPE tibial inserts of retrieved knee prostheses at extension position in which the rational contact position was explored to maximize the contact area in the loading experiment with changes in translation and rotation. The contact region by FE analysis at 3.0 kN loading corresponded to the measured area by tactile sensor. As material properties of UHMWPE, the linear elastic modulus of 498 MPa and the yield stress of 13.3 MPa were used. The contour plot of von Mises stress on the surface of the UHMWPE tibial insert in elastoplastic FE contact analysis is shown in Fig. 4.23. Contour of von Mises stress and plastic strain for representative element-sets in cross-section of medial and lateral sides is shown in Fig. 4.24. The retrieved medial side demonstrated more excessive wear and higher plastic deformation than the lateral side (Fig. 4.22). For this reason, the medial side had a higher surface conformity for the femoral surface than the lateral side and showed wider contact area and lower contact stress which was lower than the yield stress (Figs. 4.23 and 4.24). Therefore, the worn medial side showed the elastic contact for even 3 kN loading in extension, and additional plastic deformation did not occur. This suggests the decrease of delamination in knee prosthesis with high geometrical congruity. On the contrary, at the lateral side with lower surface conformity, the high contact stress that exceeds the yield stress and the considerable plastic strain were generated below the surface (Fig. 4.24b and c). The maximum plastic strain occurred at about 1.5 mm depth under surface in this lateral side. The subsurface crack at about 1 mm depth was observed in X-ray CT image of the retrieved Press-Fit Condylar (PFC) type UHMWPE tibial component (Fig. 4.25), in which the correspondence for the depth of maximum plastic strain to the crack depth was confirmed in similar FE analysis.

The complex mechanism of delamination as fatigue wear is suggested as below. In the early stage immediately after the total knee arthroplasty, which has low surface conformity for the femoral component, the plastic deformation due to excessive contact stress will occur at the subsurface of UHMWPE tibial insert. Under cyclic loading

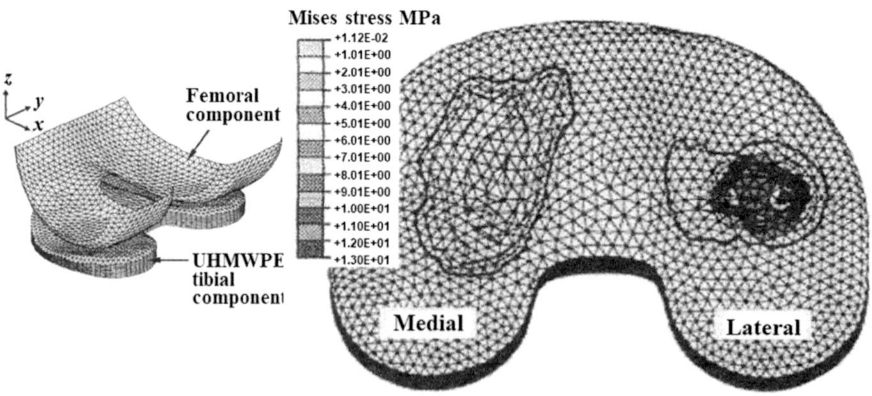

Figure 4.23 Contour plot of von Mises stress on the surface of the retrieved UHMWPE tibial insert in FE analysis. At severely worn medial side Mises stress is low, but Mises stress is high at slightly worn lateral side.

Modified from Cho, C. H., Murakami, T., Sawae, Y., Sakai, N., Miura, H., Kawano, T., & Iwamoto, Y. (2004). Elasto-plastic contact analysis of an ultra-high molecular weight polyethylene tibial component based on geometrical measurement from a retrieved knee prosthesis. *Proceedings of the Institution of Mechanical Engineers, Part H: Journal of Engineering in Medicine*, *218*(4), 251–259. https://doi.org/10.1243/0954411041561018

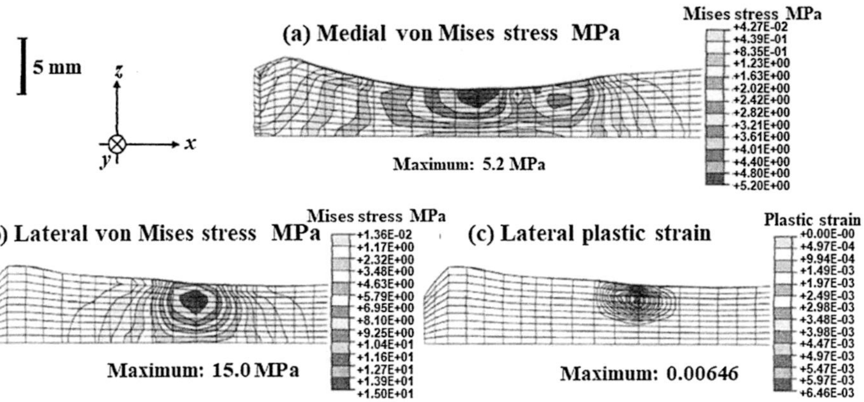

Figure 4.24 Contour plots of von Mises stress and plastic strain in the sagittal cross-section of UHMWPE tibial insert. Mises stress is low at severely worn medial side, but Mises stress is high at slightly worn lateral side. Plastic strain at lateral side is shown.
Modified from Cho, C. H., Murakami, T., Sawae, Y., Sakai, N., Miura, H., Kawano, T., & Iwamoto, Y. (2004). Elasto-plastic contact analysis of an ultra-high molecular weight polyethylene tibial component based on geometrical measurement from a retrieved knee prosthesis. *Proceedings of the Institution of Mechanical Engineers, Part H: Journal of Engineering in Medicine, 218*(4), 251–259. https://doi.org/10.1243/0954411041561018

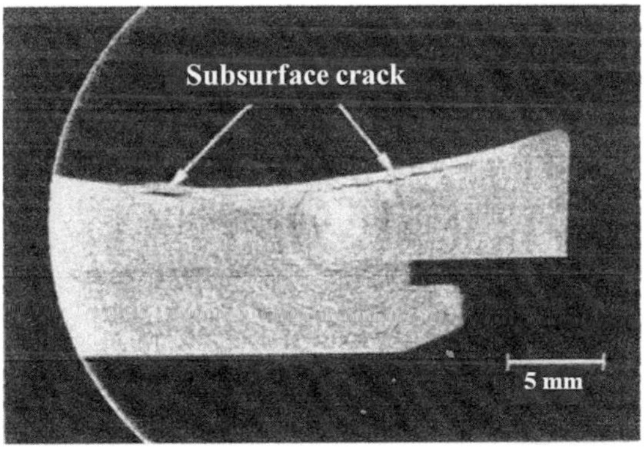

Figure 4.25 X-ray CT image of the retrieved PFC-type UHMWPE tibial insert. At subsurface zone about 1 mm deep, cracks parallel to surface are noticed.
Modified from Cho, C. H., Murakami, T., Sawae, Y., Sakai, N., Miura, H., Kawano, T., & Iwamoto, Y. (2004). Elasto-plastic contact analysis of an ultra-high molecular weight polyethylene tibial component based on geometrical measurement from a retrieved knee prosthesis. *Proceedings of the Institution of Mechanical Engineers, Part H: Journal of Engineering in Medicine, 218*(4), 251–259. https://doi.org/10.1243/0954411041561018

and/or motion, the accumulated residual plastic strain can reach a certain value as ductile limit of the UHMWPE, and then microcracks will be initiated in the subsurface. These microcracks can then propagate at a high rate in a tangential path relative to the surface by the maximum shear stress or shear stress with maximum amplitude that occurs in the subsurface and eventually produces the delamination. The wear particles of UHMWPE will be finally generated by this delamination. The occurrence of delamination appears to be reduced by improvement of UHMWPE material properties, but further researches including the evaluation method with targeted delamination tests (Tomita et al., 1999; Sakoda et al., 2019) are expected.

In actual rubbing surfaces, various types of wear processes occur as adhesive and abrasive wear, microscopic pitting, and others. To minimize wear, detailed observation and analysis for surface failure and deformation is required.

4.4.2.2.2 Influence of microscopic asperities on the wear of UHMWPE in knee prosthesis

On the metallic femoral components of retrieved knee prostheses, a number of microscopic scratches were observed. These microscopic surface asperities appeared to have enhanced microscopic wear of the UHMWPE tibial insert. Cho et al. (2010) performed 2D and 3D finite element elastoplastic contact analyses to investigate the microscopic wear of UHMWPE caused by a metallic microscopic surface asperity. Finite element model for 3D rigid asperity (metallic femoral surface) and elastoplastic UHMWPE plate (tibial insert) are shown in Fig. 4.26a. In Fig. 4.26b and c, von Mises stress and plastic strain for the indentation depth of 0.015 μm are exhibited, respectively. The length of major axis and minor axis of the asperity bottom surface perpendicular

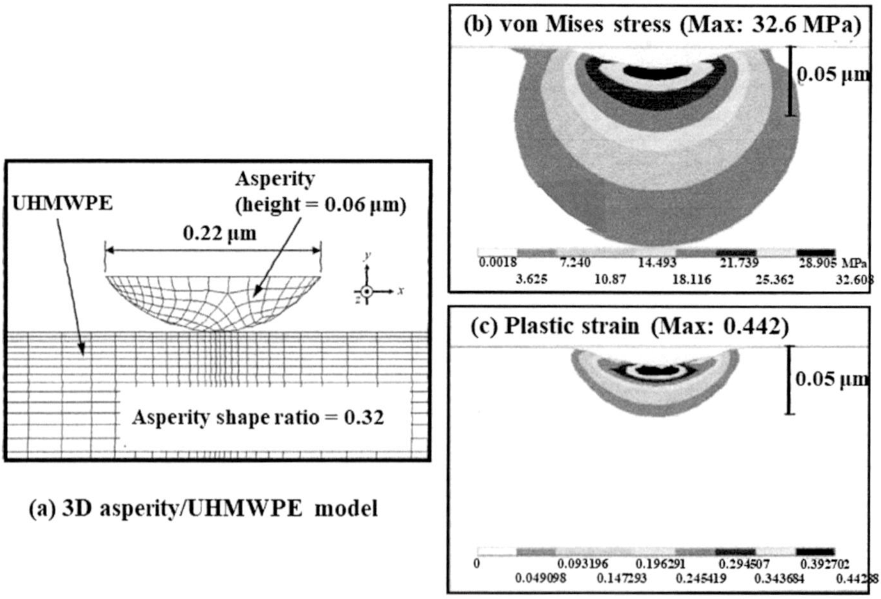

Figure 4.26 3D asperity/UHMWPE FE model with contours for contact stress and plastic strain. In (a), shape ratio is defined as ratio of minor axis to major axis of asperity bottom cross-section perpendicular to indentation direction. Contours for von Mises stress and plastic strain for the indentation depth of 0.15 μm are shown in (b) and (c).

Modified from Cho, C.-H., Murakami, T., & Sawae, Y. (2010). Influence of microscopic surface asperities on the wear of ultra-high molecular weight polyethylene in a knee prosthesis. *Proceedings of the Institution of Mechanical Engineers, Part H: Journal of Engineering in Medicine*, *224*(4), 515–529. https://doi.org/10.1243/09544119jeim690

to the indentation direction were 0.22 and 0.07 μm, respectively. The ratio of minor axis to major axis is defined as shape ratio. It is noticed from FE contact analysis that high contact stress which exceed the yield stress and significant plastic strains occurred in the surface region of the UHMWPE tibial insert for even a small indentation depth of 0.015 μm equivalent to one quarter of the asperity height (0.06 μm). The locations of occurrence of maximum contact stress and maximum plastic strain moved from just below the surface to on the surface as the indentation depth increased. This phenomenon is thought to have caused the generation of the microscopic surface scratches and cracks observed in the retrieved UHMWPE tibial insert. These scratches and cracks may have propagated into the UHMWPE tibial insert and generated pitting as fatigue wear.

It was suggested in the FE analyses that the aspect ratio (asperity height divided by half-width of asperity), shape ratio, and indentation depth of the microscopic surface asperity of metallic femoral component have significant influence on increasing and/or accelerating the wear of UHMWPE. Higher aspect ratios, shape ratios, and indentation depths cause higher contact stresses and plastic strains in UHMWPE. In contrast, in knee prothesis with UHMWPE tibial component with machining surface asperities, it was indicated that the elastic recovery of the machining surface asperities has a significant influence on contact stress and plastic strain in surface of UHMWPE tibial insert under repeated loading-unloading process (Cho et al., 2016).

4.4.3 Wear debris

To establish the longer durability of total joint replacements composed of UHMWPE and metal or ceramic component in clinical application, it is required to prevent the aseptic loosening due to wear debris-induced osteolysis as the active resorption of bone matrix by osteoclasts associated with the wear particles (Amstutz et al., 1992; Catelas et al., 2011; Gallo et al., 2013, 2014; Ingham & Fisher, 2005; Willert & Semlitsch, 1977). The osteolysis is induced by synthesizing various proinflammatory cytokines such as interleukin-1β (IL-1β), tumor necrosis factor-α (TNF-α), and prostaglandin E2, from macrophages and multinucleated giant cells. The synthesized cytokines from macrophages stimulate the surrounding osteoclasts and activate bone resorption (Fig. 4.27) (Masaoka, 2017). Thus, unbalance of a bone remodeling occurs and induces osteolysis and aseptic loosening of the prostheses.

Important point to note is that it is not the wear volume that determines the biological response to the debris, but the concentration of the wear volume that is within the critical size range (0.2–0.8 μm for UHMWPE) for macrophage activation (Ingham & Fisher, 2005).

Gallo et al. (2013) clearly demonstrated the process for periprosthetic osteolysis from the activation by wear debris detached from TKA for several cell lines of innate and adaptive immunity and indicated that mechanical and biological mechanisms do not act separately in the process of osteolysis and aseptic loosening. For example, wear particles from the polyethylene liner trigger a complex host response with increased secretion of joint fluid produced by fibroblastlike type B synoviocytes of the surface layer of pseudosynovial tissue around the TKA joint cavity. High joint fluid

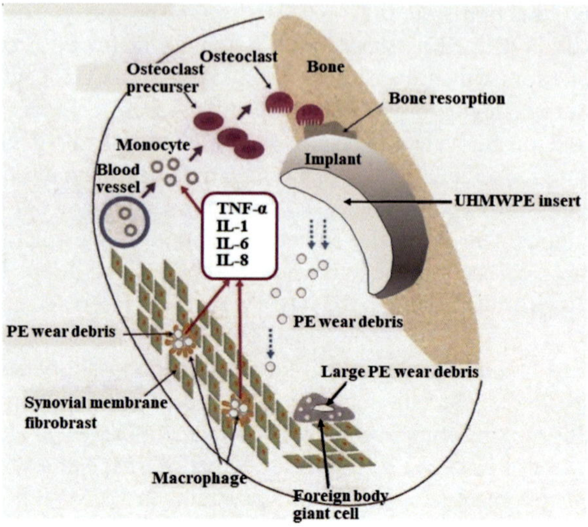

Figure 4.27 Wear debris-induced osteolysis. UHMWPE wear debris plays main role to induce osteolysis. Cytokines released by macrophage activates osteoclasts.
Modified with permission from Masaoka, T. (2017). Wear, oxidation and osteolysis. In Yamamoto K. (Ed.), *Biomaterials for artificial hip joints*. (pp.242–249). Medical View.

pressure can induce tissue ischemia and even direct bone necrosis. Additionally, increased joint fluid pressure could influence the osteoclast differentiation, their cytokine expression and may also lead to a lack of anabolic effect in bone remodeling.

4.4.4 Evaluation of wear properties

Prior to clinical application of new materials and new design for artificial joints, the wear properties should be evaluated by appropriate methods such as screening tests for evaluation of new materials, simulator tests to simulate the walking motion or other human activities, and computer simulation.

4.4.4.1 Simplified wear tests

The screening test in laboratory to evaluate wear properties of new rubbing materials is required to reproduce the wear properties in vivo after arthroplasty. In 1982, J. Charnley indicated at the One Day course on "Laboratory and Clinical Evaluation of Prostheses" held in The University of Leeds that large difference by two orders of magnitude existed between laboratory and clinical wear data on UHMWPE (Medley et al., 2021). The clinical wear was very much larger than laboratory wear. It was suggested that laboratory testing using simplified pin-on-disc sliding motion (rather than the actual three-dimensional motion in human joints) and using a nonphysiological lubricant (water) might have caused these large differences. At that time in 1982, in ASTM (American Society for Testing and Materials) standard F732-82 on reciprocating pin-on-flat evaluation, not multidirectional motion but simple reciprocating motion, was specified as only the initial stage in the full wear characterization of a candidate material. As a lubricant, bovine blood serum was recommended. Furthermore, it was suggested that all candidate materials should be tested in a joint simulator apparatus using prototype prostheses before being used in clinical trials in patients.

As indicated by many studies, the following points should be taken into consideration to simulate the in vivo wear behaviors.

1. Multidirectional rubbing
2. Lubricants containing synovia constituents in artificial joints

In the current ASTM standard F732-17 (2017): Standard test method for wear testing of polymeric materials used in total joint prostheses, specific wear test methods are described for each distinct wear application (for example, linear reciprocating motion, ball-cup ("hip-type") wear, delamination wear, and so forth). In the evaluation of friction and wear in fixed-bearing ball/cup devices for total hip replacement, test method for fixed-bearing ball-cup wear application is adopted. In this method, relative motion between the pin specimen and the flat counterface must be multidirectional. On lubricant, it is mentioned that the specimen shall be lubricated with bovine blood serum unless an alternative medium can be justified, and dilution with deionized water of up to 75% (volume fraction) may be appropriate. Appropriate adjustment of protein concentration is recommended.

4.4.4.2 Simulator tests for wear evaluation

The evaluation of wear properties of total joint arthroplasty articulating surfaces by joint simulator corresponding to International Standards by the International Organization for Standardization (ISO) is recommended before the clinical application of new total joint prostheses.

Case example: ISO 14242-3 (2009) and AMENDMENT 1(2019):

Implants for surgery – Wear of total hip-joint prostheses –

Part 3: Loading and displacement parameters for orbital bearing type wear testing machines and corresponding environmental conditions for test.

The orbital bearing hip wear simulator is widely used and has been successful in evaluating the wear properties of total hip arthroplasty (THA) articulating surfaces. It is mentioned that these test results have been shown to correlate well with clinical experience. Importance of standardization of the parameters is described.

The motion of this simulator is simpler and less anatomic than the motion described in ISO 14242-1. Here, simplified two-peak loading and simplified angular movement for abduction/adduction and extension/flexion are applied.

In 2019, the inverted (head above liner) and anatomical (liner above head) positions were prepared. Lubricant was replaced as calf serum diluted with deionized water in accordance with ISO 3696. The fluid test medium was filtered through a 2 μm filter and had a protein mass concentration of 30 g/L ±2 g/L.

For special condition, the following standard was applied.

ISO 14242-4:2018

Implants for surgery – wear of total hip-joint prostheses – Part 4: Testing hip prostheses under variations in component positioning which results in direct edge loading.

The development of knee simulator testing for total knee replacement was recently reviewed by Maag et al. (2021). Since the start of total knee replacements, various knee simulators with physiological loading and motions for wear testing have been used. In Leeds knee simulator as a pioneer single station hydraulic machine for tibiofemoral joint, the tibial axis compressive (three peaks) and anterior-posterior forces were synchronized with flexion/extension motion for a steady walking cycle. The varus/valgus and internal/external rotation were allowed to freely rotate. The wear and creep deformation could be estimated by dual-index holographic contouring (Dowson, Gillis, & Atkinson, 1985, pp. 215–228).

Various knee simulators are divided to two types of force control (ISO 14243-1) and displacement control (ISO 14243-3). The load control mode is recommended to be used for highly conforming/congruent, constraining and medial-pivot/deep dish designs, while the displacement control mode could be used for low conformity/congruent, low internal/external constraint designs and flat tibial plateaus. In case of mobile bearing designs both control modes can be applied (Maag et al., 2021). For testing of patellofemoral joint, two standards: ISO 14243-4 and ISO 14243-5 are used.

Currently the knee joint simulators have been applied to simulate other daily living activities such as stair ascent, stair descent, jogging, sitting and standing up from chair, and so on. In addition, new functions to accommodate customized biomechanics tests using cadaver were applied. For further improvement of knee simulators, Hirokawa et al. (2015) indicated the importance of reproduction of physical active knee motion (in Section 2.6.4).

4.4.4.3 Computer simulation for wear evaluation

Theoretical modeling of wear in hip implants has been developed (Mattei et al., 2011). Usually, wear is evaluated with Archard's wear law for the volumetric wear U_v for ball-in-socket geometry in walking and other daily activities.

$$U_v = kW_n s \tag{4.10}$$

where, k: a dimensional wear coefficient (wear factor), W_n: normal load, s: Relative surface sliding distance. In the wear estimation, it is important to appropriately select the value of k on the basis of the corresponding experimental data and tribological conditions for hip prostheses. In these modeling, not only wear depth/volume and wear rate but related contact stress contour and heat generation are expected to be predicted including long-term phenomena.

In wear model of metal-on-polyethylene total knee replacement, Gao et al. (2018) developed the combined lubrication and wear modeling using wear factor as an exponential function of the "Lambda ratio" in mixed lubrication modeling. In this study, lubrication and wear in medial and lateral knee compartments during walking in two million cycles were evaluated. It is observed that the wear stage changes from running-in wear to steady state wear in medial compartment. The volumetric wear

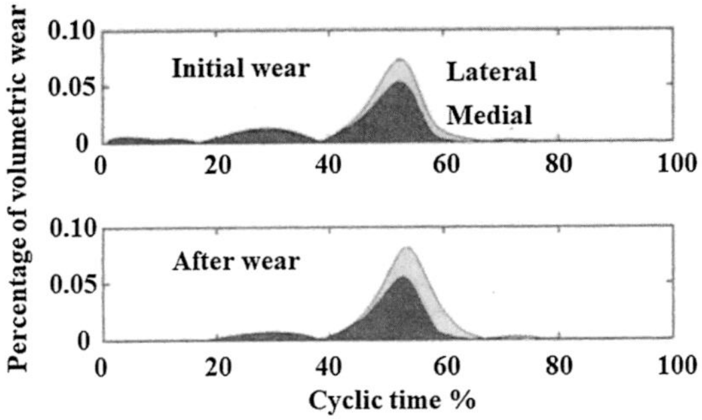

Figure 4.28 Volumetric wear percentage distribution in one walking cycle. Lateral wear is plotted on top of medial.
Modified from Gao, L., Hua, Z., Hewson, R., Andersen, M. S., & Jin, Z. (2018). Elastohydrodynamic lubrication and wear medelling of the knee joint replacements with surface topography. *Biosurface and Biotribology*, *4*(1), 18–23. https://doi.org/10.1049/bsbt.2017.0003

during the first and the last gate cycles of the total wear simulation is presented in Fig. 4.28, where wear behaviors at individual walking phases can be observed. The maximum pressure and minimum film thickness in the medial compartment are compared in the first and last wear cycles in Fig. 4.29. With progression of wear the pressure distribution becomes flattened and the fluid film thickness is increased. These lubrication and wear modeling is expected to contribute to the optimization of knee prosthesis design.

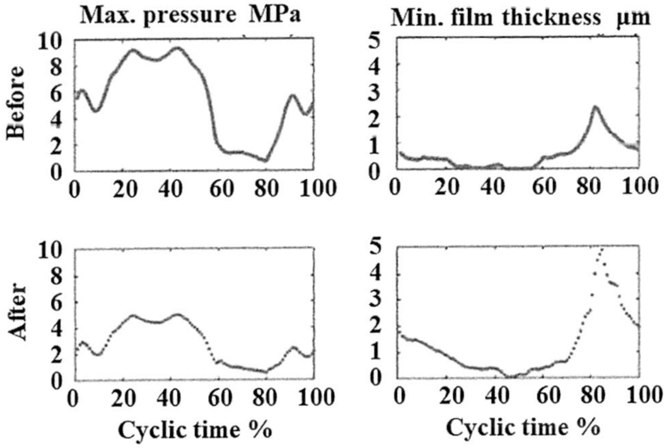

Figure 4.29 Maximum pressure and minimum film thickness at the first and last wear cycles (in medial). With progression of wear the pressure distribution becomes flattened and the fluid film thickness is increased.
Modified from Gao, L., Hua, Z., Hewson, R., Andersen, M. S., & Jin, Z. (2018). Elastohydrodynamic lubrication and wear medelling of the knee joint replacements with surface topography. *Biosurface and Biotribology*, *4*(1), 18–23. https://doi.org/10.1049/bsbt.2017.0003

4.5 Improvement of tribological properties in artificial joints

To extend the durability of artificial joints, the material structure and properties have been improved. However, for clinical application of new materials, unanticipated failures can occur. In the previous clinical application of the carbon fiber-reinforced UHMWPE, the mechanical strength of reinforced polyethylene was improved but unexpected wear and fatigue failure occurred. It is suggested that the addition of carbon fibers decreased the resistance of the polyethylene to fatigue crack initiation and propagation, probably because stiff but poorly bonded carbon fibers did little to reinforce the ductile polyethylene matrix (Connelly et al., 1984). In clinical application of Hylamer with high mechanical strength and high crystallinity as UHMWPE prepared with hot isostatically pressed (HIP) treatment, high wear including cracking, delamination and pitting occurred. It is suggested that Hylamer liners had greater property change with oxidation degradation by gamma sterilization in air (Collier et al., 1998).

4.5.1 Improvement of wear resistance in UHMWPE

4.5.1.1 Cross-linking for UHMWPE

UHMWPE with extremely high molecular weight as simple chemical composition for $[-CH_2-CH_2-]_n$ exhibits superior mechanical toughness, chemical inertness, and wear resistance, and has been used as a bearing material in artificial joints since successful clinical application by Charnley in 1962. Wear rate of UHMWPE was about one-thirtieth of that of PTFE. In longer clinical usage, however, UHMWPE wear has been criticized to cause wear-debris-induced osteolysis and eventual loosening of joint prostheses. Therefore, it was required to improve wear resistance of UHMWPE. UHMWPE is composed of crystalline and amorphous regions. On wear mechanism of UHMWPE, Wang (Wang, 2001) indicated that multidirectional motion imposes a higher wear rate than reciprocating linear motion. In multidirectional motion, the UHMWPE wear surface experiences both shear and tensile stresses in multiple directions, and UHMWPE molecules rupture more easily perpendicular to their principle alignment. Therefore, the cross-linking by gamma-ray or electron beam irradiation was considered as strategy to improve wear resistance, material structure, and properties of UHMWPE. Oonishi, et al. (2001), (Oonishi, Clarke, et al., 2001) applied high dose gamma irradiation at 1000 kGy to UHMWPE socket for 1971 to 1978 and confirmed the superior wear resistance without oxidative degradation in retrieved sockets in long-term clinical usage.

Many years later with various improvements of material properties, the radiation in an inert environment and thermal treatment have been applied. To reduce wear and the incidence of revision resulting from osteolysis, highly cross-linked and thermally treated polyethylenes (HXLPEs) were clinically introduced starting in 1998 (first-generation) (Kurtz et al., 2011; Kurtz & Patel, 2016). These materials are processed with a total dose (including sterilization) ranging from 50 to 105 kGy, depending on manufacturer. As thermal treatment after irradiation, annealing or remelting is selected.

In annealing as thermal processing below the crystalline melt transition (~137°C) to preserve crystallinity and mechanical properties, when combined with terminal gamma sterilization in an inert environment, this method brought the elevation of residual free radicals which have the potential to oxidize in vivo. In remelting as thermal processing above the melt transition, this treatment resulted in a material with undetectable free radicals but at the expense of reduced crystallinity and lower material properties. Although there are issues to be improved, the first-generation highly cross-linked polyethylene technologies have brought about reduction in wear and osteolysis in total hip arthroplasty.

In the second-generation HXLPE starting around 2005, various technologies have been applied to improve the further wear resistance and mechanical properties without oxidative degradation for total hip replacements with larger head and thinner UHMWPE acetabular cup, which are favorable to prevent the dislocation and to extend the range of motion. Suitable application of HXLPE to total knee replacements had been expected. For example, the sequential irradiation and annealing treatments (Manley, 2016), optimized irradiation and melting (Muratoglu & Bragdon, 2016), addition of vitamin E and the grafting of phospholipid polymer to HXLPE have been applied.

4.5.1.2 Addition of vitamin E

The addition of vitamin E (α-tocopherol) by blending or diffusing to UHMWPE could improve wear resistance and mechanical properties significantly and reduce oxidative degradation. The former addition treatment is the blending of the liquid antioxidant vitamin E with UHMWPE resin powder, consolidating the mixture by compression molding and irradiating the consolidated blend for cross-linking, and the latter one is the diffusing of vitamin E into already consolidated and radiation cross-linked UHMWPE.

Tomita et al. (1999) examined the effect of blending of vitamin E on the crack formation and/or propagation at grain boundaries in UHMWPE using 2-dimensional sliding fatigue testing in U-shaped motion and micro indenter testing. In gamma-irradiated UHMWPEs without vitamin E under multidirection loading, severe flaking-like destruction was exhibited, but in gamma-irradiated vitamin-E-containing UHMWPEs, no subsurface crack formation and no-flaking-like destruction were observed. The addition of vitamin E reduced the dynamic hardness at grain boundaries. The presence of vitamin E appeared to prevent crack propagation partly due to reduced hardness at grain boundaries.

The vitamin-E-diffused, irradiated UHMWPE as a second-generation cross-linked bearing surface improved the mechanical and fatigue strength of cross-linked UHMWPE without sacrificing wear and oxidative stability (Oral & Muratoglu, 2011). The vitamin-E-containing, irradiated polyethylene has superior oxidative stability due to the reaction of vitamin E in uniform concentration with the free radicals on the polyethylene chains and also free radicals resulting from their reaction with oxygen.

4.5.2 Improvement for wear reduction and prevention of osteolysis by phospholipid polymer grafted cross-linked polyethylene

To prevent wear-debris-induced periprosthetic osteolysis and subsequent aseptic loosening in artificial joints, Moro et al. (2004, 2017) developed a new type of hip prosthesis with surface modification by grafting of biocompatible phospholipid polymer onto the surface of artificial joint material, which has a surface structure similar to natural articular cartilage. Natural articular cartilage has a hydration gel layer and phospholipid layer on its surface. These surface layers contribute to the excellent lubrication property of natural synovial joint. Therefore, they have developed a new artificial joint system with a 100-nm thick surface hydration layer formed by poly(2-methacryloyloxyethyl phosphorylcholine (MPC)) (PMPC) (Ishihara et al., 1990) grafted onto the surface of cross-linked polyethylene (PMPC-grafted CLPE) (Moro et al., 2004), mimicking the surface layer of natural articular cartilage (Fig. 4.30). MPC polymers possess good biocompatibility, high lubricity, low friction, hydrophilicity, antiprotein adsorption, and a cell membrane-like surface. Highly cross-linked polyethylene (CLPE) liners were manufactured from compression-molded PE sheet stock (GUR 1020 resin) and treated with a total dose of 75 kGy gamma-ray irradiation (25 kGy: for sterilization). PMPC grafting at the surface of the CLPE liner was performed using a photoinduced polymerization technique.

As exhibited in Fig. 4.31 (Moro et al., 2014), wear test in hip joint simulator test under ISO 14242-3 conditions indicates PMPC-grafted CLPE has extremely low wear compared with CLPE. Furthermore, it is noticed that the effect of PMPC grafting was sustained up to 70 million cycles in hip joint simulator test. Gravimetric, surface, and wear particle analyses revealed that PMPC grafting onto the CLPE liner surface markedly decreased the production of wear particles and showed that the effect of PMPC grafting was maintained through 70 million cycles. It is considered that PMPC grafting can significantly improve the wear resistance of artificial hip joints.

In addition, the PMPC-grafted particles were biologically inert and did not cause subsequent bone-resorptive responses (Moro et al., 2004). In the follow-up study of

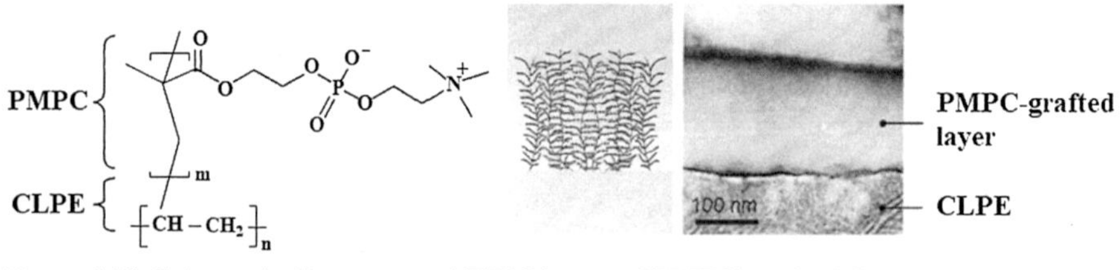

Figure 4.30 Schematic diagrams and TEM image of PMPC-grafted CLPE sample. PMPC-grafted layer mimics the surface layer of natural articular cartilage which acts in hydration lubrication mechanism.
Modified from Uehara, S., Yarimitsu, S., Moro, T., Kyomoto, M., Watanabe, K., Tanaka, S., Ishihara, K., & Murakami, T. (2016). Evaluation of wear suppression for phospholipid polymer-grafted ultra-high molecular weight polyethylene at concentrated contact. *Tribology International*, *101*, 264–272. https://doi.org/10.1016/j.triboint.2016.04.026

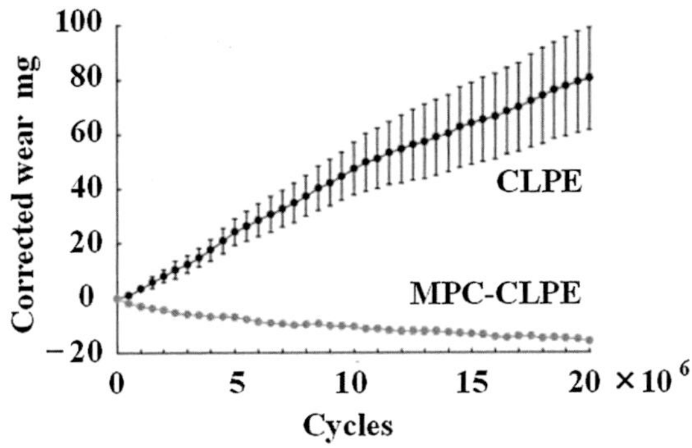

Figure 4.31 Wear amounts of CLPE liners with or without PMPC grafting. The wear of the CLPE liner increased in a cycle-dependent manner, with a final total weight loss of 80.3 ± 18.7 mg (mean ± standard deviation) after 20 million cycles. In contrast, MPC-CLPE liners continued to gain weight even after correction for water absorption, showing a total weight gain of 16.1 ± 0.9 mg. This weight gain might be due in part to absorption of fluid by the liners during soaking.
Modified from Moro, T., Takatori, Y., Kyomoto, M., Ishihara, K., Hashimoto, M., Ito, H., Tanaka, T., Oshima, H., Tanaka, S., & Kawaguchi, H. (2014). Long-term hip simulator testing of the artificial hip joint bearing surface grafted with biocompatible phospholipid polymer. *Journal of Orthopaedic Research*, *32*(3), 369–376. https://doi.org/10.1002/jor.22526

clinical applications of PMPC-grafted CLPE (Moro et al., 2017), neither periprosthetic osteolysis nor femoral component migration in the radiographic evaluation and the linear wear rate of PMPC-grafted CLPE was extremely smaller than other conventional CLPEs. Mean bedding-in penetration (penetration during the first year: including creep and/or plastic deformation) was 0.225 ± 0.107 mm, but the mean steady-state wear rate was very low as 0.002 ± 0.023 mm/year (between 1 and 5 years postoperatively). On clinical evaluation, the mean Japanese Orthopedic Association (JOA) score for pain, range of motion, gait, activities of daily living had increased significantly at post-operation. The relief from pain was remarkable and both the mean flexion and abduction range had improved significantly.

Thus, PMPC-grafted CLPE should contribute dramatically toward the improved longevity of artificial hip joint. In this system, even if some wear might occur under severer or accidental operating conditions where the hydration lubrication becomes ineffective around polymer brush layer, the underlying cross-linked polyethylene will be able to maintain good wear resistance as a fail-safe system.

Kyomoto et al. (2022) investigated the effect of a hydrated PMPC graft layer for all-polymer poly(ether-ether-ketone) (PEEK)-on-CLPE bearing on wear resistance and metal ion release. It is noticed that PEEK-on-CLPE showed lower wear and less metal ion release than Co-Cr-Mo alloy-on-CLPE, and the wear resistance of the all-polymer hip bearings was further improved with hemi- or both-surface modified with a hydrated PMPC layer.

The surface gel hydration lubrication using a hydrated gel layer formed by hydrophilic polymer on the rubbing surface becomes important in maintaining low friction

(Ishikawa et al., 2001; Sasada, 2000). In aqueous solution, the hydrophilic macromolecules on the bearing surfaces attract water molecules and form a water-rich hydrated gel layer. If a load is applied to the surface, it is mostly supported by the fluid pressure in the hydrated gel layer. In the case of sliding motion, water molecules are sheared with low stress and low friction is maintained until the water is discharged by loading from the hydrated gel layer in hydration lubrication (Ikeuchi, 2007; Klein, 2013). Excellent lubricity and durability of the PMPC-grafted CLPE are exhibited, owing to this hydration lubrication. Therefore, it is important to elucidate the details pertaining to hydration lubrication of the PMPC layer in the bearing interface for further improvement of the longevity of artificial joints.

Yarimitsu et al. (2015) investigated the effects of dehydration and rehydration by loading-unloading procedure on lubrication properties of PMPC-grafted CLPE using various lubricants. An in situ visualization technique was used in simplified reciprocating friction tests to track changes in the surface of PMPC-grafted CLPE.

For a flat-ended, conical/cylindrical pin of untreated CLPE or PMPC-grafted CLPE (tip diameter of 2 mm) and a flat single-crystal sapphire plate for in situ visualization, the reciprocating tests were performed at vertical load of 7.84 N (mean contact pressure was 2.5 MPa), a sliding speed of 10 mm/s, and a reciprocating stroke of 10 mm. The total sliding distance was 2.0 m. Four lubricants were used in this study: pure water, 0.5 mass% hyaluronic acid (HA, $M_w = 9.2 \times 10^5$) aqueous solution, 30 vol% fetal bovine serum (FBS; Invitrogen Corporation, Carlsbad, CA, USA) solution, and a mixture of 30 vol% FBS and 0.5 mass% HA.

Before the reciprocating test with or without rehydration procedure, the PMPC-grafted CLPE pin specimens were soaked in each lubricant for 10 min for hydration of PMPC layer. In rehydration procedure, the reciprocating test was initiated immediately after applying a load, and the reciprocating motion continued for five cycles (5 s). After the reciprocating cycles, the upper pin specimen was separated from the counterface and was kept with an unloading for 60 s in a lubricant. Then, the friction test was restarted after loading, and these procedures were repeated until a total sliding distance of 2.0 m was reached. The PMPC layer fluorescently labeled with rhodamine 6G on the sliding surface was imaged using fluorescence microscopy at definite sliding distances.

The frictional behaviors of untreated and PMPC-grafted CLPE with/without rehydration in pure water are shown in Fig. 4.32. The start-up friction at the first cycle of PMPC-grafted CLPE was significantly lower than that of CLPE, regardless of rehydration. With the rehydration procedures, the PMPC-grafted CLPE maintained a lower friction during the 2.0 m test. The recovery and gradual increase of friction coefficient for each rehydration procedure was observed both in cases of untreated and PMPC-grafted CLPE. In contrast, the steady state friction coefficient of PMPC-grafted CLPE without rehydration was nearly identical with that of CLPE. Fluorescent images of PMPC layers indicated partial damage for rehydration and severely damage for without rehydration in the convex region of machine marks on PMPC-grafted CLPE surfaces (Fig. 4.34A).

In HA solution, the start-up friction of PMPC-grafted CLPE was lower than that of CLPE regardless of rehydration. However, there was no significance in the friction

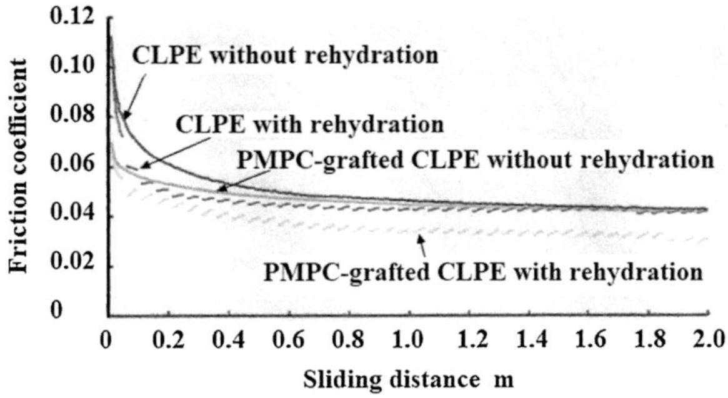

Figure 4.32 Transient of friction coefficient in pure water. As average time-depending trend, friction gradually decreases during running-in process, but friction with rehydration shows the recovery and gradual increase.

Modified from Yarimitsu, S., Moro, T., Kyomoto, M., Watanabe, K., Tanaka, S., Ishihara, K., & Murakami, T. (2015). Influence of dehydration and rehydration on the lubrication properties of phospholipid polymer-grafted cross-linked polyethylene. *Proceedings of the Institution of Mechanical Engineers, Part H: Journal of Engineering in Medicine, 229*, 506–514.

coefficient at steady state between untreated and PMPC-grafted CLPE. Friction lowering and gradual increase were observed after rehydration similarly to in water. The damage of PMPC layer was slight in HA solution and the rehydration procedure significantly suppressed the damage.

In 30 vol.% FBS solution, the start-up friction of PMPC-grafted CLPE was lower than that of CLPE, but friction level was higher than that in water. In FBS solution, friction increase and gradual lowering were observed after rehydration for CLPE and PMPC-grafted CLPE. FBS solution contains proteins and phospholipids, which appear to form adsorbed films on rubbing surfaces. It is considered that immediately after reloading, the excessively adsorbed molecules interact with high friction although a hydrated PMPC layer prevents protein adsorption. Thereafter, during continuous rubbing, the adsorbed films are likely to be reoriented by frictional force. Thus, after reloading, friction increases and gradually decreases to the definite level. Steady state friction coefficient of PMPC-grafted CLPE was lowest. The damage of PMPC layer was limited to the convex region of the machine marks on the surface of PMPC-grafted CLPE with rehydration. In contrast, the damage of PMPC layer ranged from convex to concave regions without rehydration. In FBS solution, the damage of PMPC layer without rehydration was accelerated compared with that of water or HA lubricants.

In a mixture of FBS and HA solution, both the start-up and steady state friction of the PMPC-grafted CLPE was lower than that of CLPE, regardless of rehydration (Fig. 4.33). In rehydration process, friction increased and gradually decreased at definite level, similarly in FBS solution. Damages of the PMPC layer on the convex region on the PMPC-grafted CLPE surface were suppressed with rehydration (Fig. 4.34b). Considerable low friction of PMPC-grafted CLPE without rehydration may have been brought by the remaining PMPC layer within and around the concave region of the machine mark (Fig. 4.34b (vi) and Fig. 4.35).

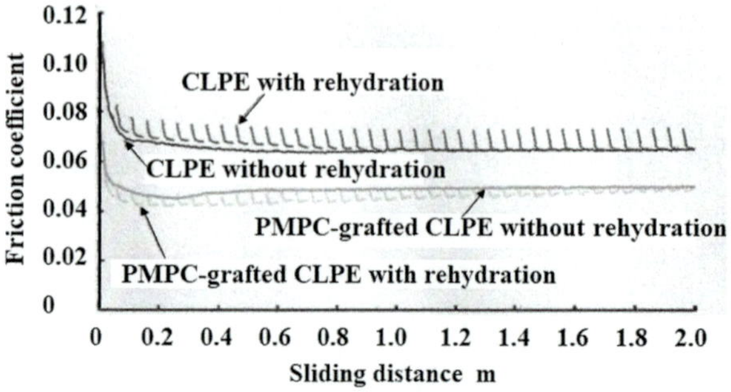

Figure 4.33 Transient of friction coefficient in a mixture of FBS and HA solution. Both the start-up and steady state friction of the PMPC-grafted CLPE were lower than that of CLPE. Modified from Yarimitsu, S., Moro, T., Kyomoto, M., Watanabe, K., Tanaka, S., Ishihara, K., & Murakami, T. (2015). Influence of dehydration and rehydration on the lubrication properties of phospholipid polymer-grafted cross-linked polyethylene. *Proceedings of the Institution of Mechanical Engineers, Part H: Journal of Engineering in Medicine*, *229*, 506–514.

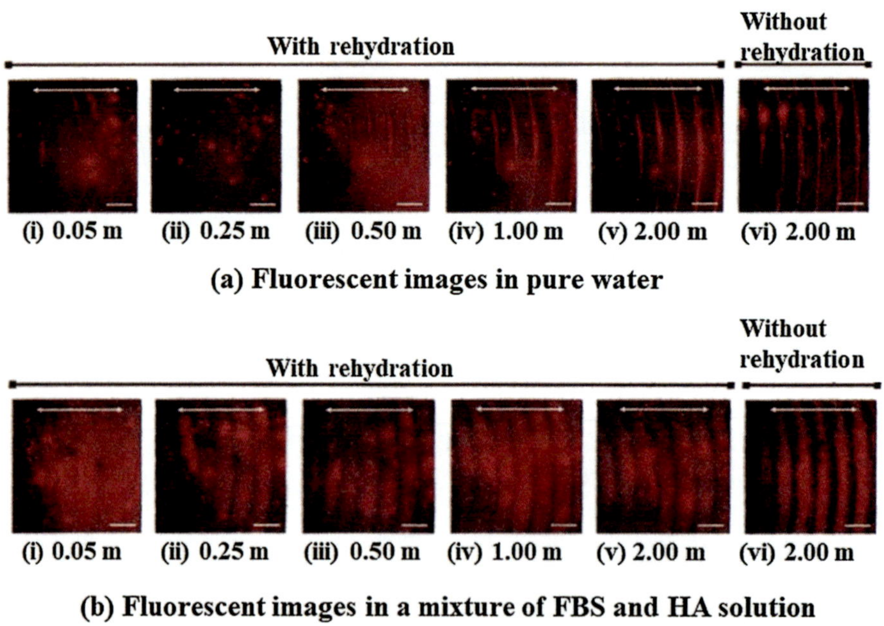

Figure 4.34 Fluorescence microscopy images of the PMPC layer lubricated in pure water and in a mixture of FBS 30 vol.% and HA 0.5 mass% solution. Numbers in the subcaption indicate sliding distance. bars: 50 μm, arrows: sliding direction.
Modified from Yarimitsu, S., Moro, T., Kyomoto, M., Watanabe, K., Tanaka, S., Ishihara, K., & Murakami, T. (2015). Influence of dehydration and rehydration on the lubrication properties of phospholipid polymer-grafted cross-linked polyethylene. *Proceedings of the Institution of Mechanical Engineers, Part H: Journal of Engineering in Medicine*, *229*, 506–514.

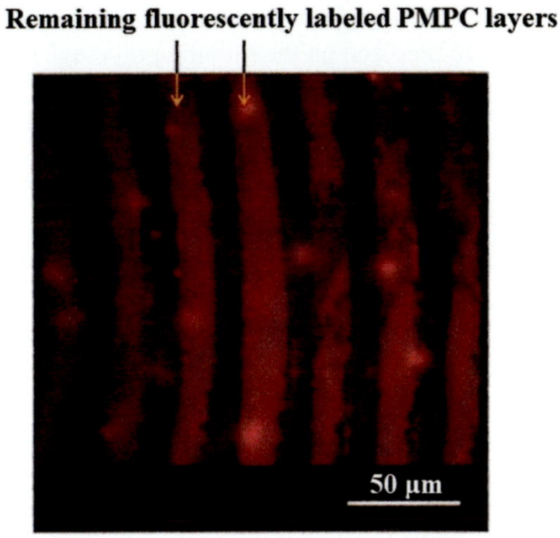

Figure 4.35 Remaining fluorescently labeled PMPC layer within concave regions of the machine marks. Enlarged image of Fig. 43B (vi) at 2.0 m sliding without rehydration. Modified from Uehara, S., Yarimitsu, S., Moro, T., Kyomoto, M., Watanabe, K., Tanaka, S., Ishihara, K,. & Murakami, T. (2016). Evaluation of wear suppression for phospholipid polymer-grafted ultra-high molecular weight polyethylene at concentrated contact. *Tribology International*, *101*, 264–272. https://doi.org/10.1016/j.triboint.2016.04.026

As described above, the start-up friction of PMPC-grafted CLPE was reduced, and the damage of PMPC layer was suppressed by rehydration in water or hyaluronic acid (HA) solutions. In contrast, the start-up friction of PMPC-grafted CLPE increased in fetal bovine serum (FBS) solution, and the damage for PMPC layer was quite noticeable for without rehydration. Interestingly, the start-up friction of PMPC-grafted CLPE was reduced in FBS solution containing HA, and the damage of PMPC layer was suppressed. These results indicate that the rehydration by unloading and HA are elemental in maximizing the lubrication effect of hydrated PMPC layer. It was confirmed that the PMPC layer exhibited superior hydration lubrication ability in a mixture of FBS and HA solution as the simulated synovial fluid like that of articular cartilage in synovial fluid.

Uehara et al. (2016) evaluated the effectiveness of the application of PMPC grafting to CLPE with concentrated contact load such as total knee prostheses. In this study, the roller-on-flat tests were performed for sliding pair of a polycrystalline alumina ceramic roller with a spherical surface of 30 mm in radius and 10 mm in width ($R_a = 0.01$ μm) and untreated CLPE or PMPC-grafted CLPE rectangular flat plate (60×20, 4 mm thickness, $R_a = 0.2$ μm) at 20 N load (initial Hertzian contact stress: $p_{max} = 19.5$ MPa, $p_{mean} = 13.0$ MPa) at sliding speed of 94.2 mm/s in 0.5 mass% HA solution containing 30 vol.% FBS. In this study to examine very low wear loss less than 0.01 mg, the indentation technique was applied to evaluate the degree of wear by comparing the projected areas of trigonal pyramid indents in wear zone. To quantify the depth of wear scars, arrays of indents were prepared on the rubbing polyethylene surfaces, and wear tests were performed.

The wear scars after the wear tests on the arrays of indents for CLPE and PMPC-grafted CLPE are shown in Fig. 4.36. The local wear depths were estimated from changes in projected area of the indents. It is noticed that the wear scar depth became greater for higher contact pressure and longer testing times and was almost more than twice or thrice deep for CLPE compared with PMPC-grafted CLPE at all contact pressure zones. Fig. 4.37 shows laser microscopic images and surface profiles of the indents in nonworn or worn zone on the surfaces of CLPE and PMPC-grafted CLPE. For CLPE, the significant wear around indent and roughening in slope area of indent were confirmed. In contrast, mild wear around indent and sustaining of smooth valley slope faces of indent were observed in PMPC-grafted CLPE.

In this study, the effectiveness of PMPC-grafting to CLPE in wear suppression was clearly indicated for simplified rubbing condition at concentrated contact at 3.0–7.0 MPa.

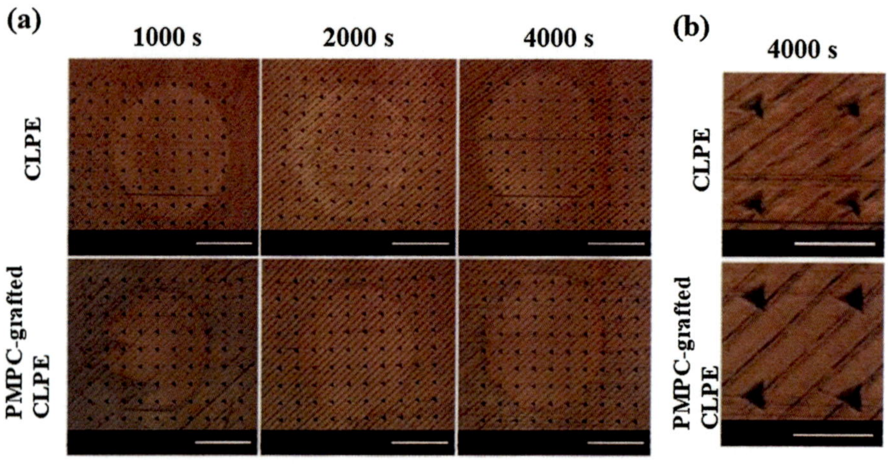

Figure 4.36 Optical microscopic images of the wear scars after the wear tests on the arrays of indents for CLPE and PMPC-grafted CLPE. (a) Images of the entire wear scars after the testing times of 1000 s, 2000 s, and 4000 s. Scale bar, 1 mm. (b) Enlarged images of the indents, which received the contact pressure of 6.0–7.0 MPa at a testing time of 4000 s. Scale bar, 250 μm.

Modified from Uehara, S., Yarimitsu, S., Moro, T., Kyomoto, M., Watanabe, K., Tanaka, S., Ishihara, K., & Murakami, T. (2016). Evaluation of wear suppression for phospholipid polymer-grafted ultra-high molecular weight polyethylene at concentrated contact. *Tribology International*, *101*, 264–272. https://doi.org/10.1016/j.triboint.2016.04.026

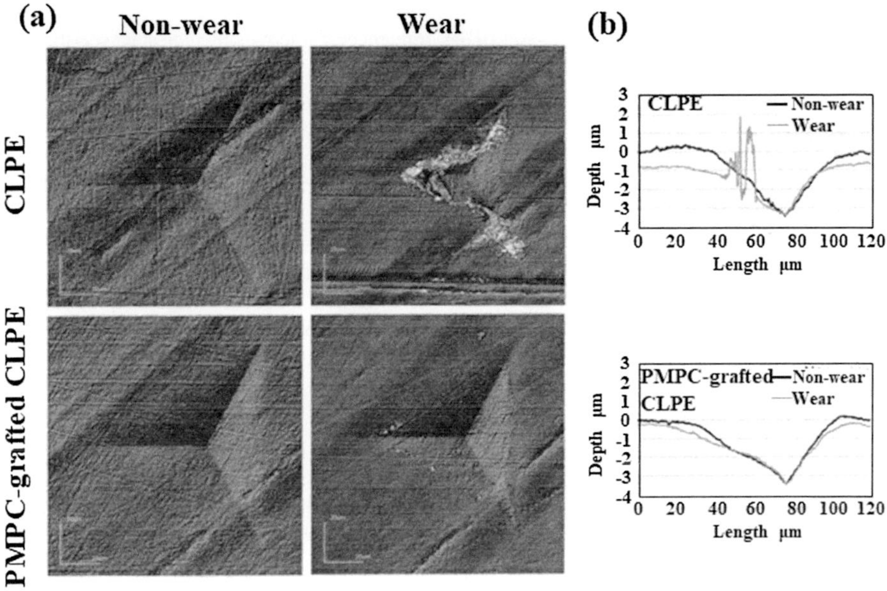

Figure 4.37 Laser microscopic images and surface profiles of the indents on the surfaces of CLPE and PMPC-grafted CLPE. The indents are shown at not worn zone and at worn zone with 6.0−7.0 MPa for 4000 s. (a) Laser microscopic images of the indents. Scale bar: 20 μm. (b) Surface profiles along left ridge lines of the indents.
Modified from Uehara, S., Yarimitsu, S., Moro, T., Kyomoto, M., Watanabe, K., Tanaka, S., Ishihara, K., & Murakami, T. (2016). Evaluation of wear suppression for phospholipid polymer-grafted ultra-high molecular weight polyethylene at concentrated contact. *Tribology International, 101*, 264−272. https://doi.org/10.1016/j.triboint.2016.04.026

References

Amstutz, H. C., Campbell, P., Kossovsky, N., & Clarke, I. C. (1992). Mechanisms and clinical significance of wear debris-induced osteolysis. *Clinical Orthopaedics and Related Research, 276*, 7−18. https://doi.org/10.1097/00003086-199203000-00003

Auger, D. D., Dowson, D., & Fisher, J. (1995). Cushion form bearings for total knee joint replacement part 1: Design, friction and lubrication. *Proceedings of the Institution of Mechanical Engineers, Part H: Journal of Engineering in Medicine, 209*(2), 73−81. https://doi.org/10.1243/PIME_PROC_1995_209_323_02

Bergmann, G., Deuretzbacher, G., Heller, M., Graichen, F., Rohlmann, A., Strauss, J., & Duda, G. N. (2001). Hip contact forces and gait patterns from routine activities. *Journal of Biomechanics, 34*(7), 859−871. https://doi.org/10.1016/S0021-9290(01)00040-9

Brockett, C. L., Harper, P., Williams, S., Isaac, G. H., Dwyer-Joyce, R. S., Jin, Z., & Fisher, J. (2008). The influence of clearance on friction, lubrication and squeaking in large diameter metal-on-metal hip replacements. *Journal of Materials Science: Materials in Medicine, 19*(4), 1575−1579. https://doi.org/10.1007/s10856-007-3298-9

Catelas, I., Wimmer, M. A., & Utzschneider, S. (2011). Polyethylene and metal wear particles: Characteristics and biological effects. *Seminars in Immunopathology, 33*(3), 257−271. https://doi.org/10.1007/s00281-011-0242-3

Charnley, J. (1979). *Low friction arthroplasty of the hip*. Springer-Verlag.

Cho, C.-H., Murakami, T., & Sawae, Y. (2010). Influence of microscopic surface asperities on the wear of ultra-high molecular weight polyethylene in a knee prosthesis. *Proceedings of the Institution of Mechanical Engineers, Part H: Journal of Engineering in Medicine, 224*(4), 515–529. https://doi.org/10.1243/09544119jeim690

Cho, C., Murakami, T., & Sawae, Y. (2016). Influence of elastic recovery of machining surface asperities on the contact stress and plastic strain of polyethylene tibial insert in virgin knee prosthesis. *Tribology Online, 11*(2), 320–325. https://doi.org/10.2474/trol.11.320

Cho, C. H., Murakami, T., Sawae, Y., Sakai, N., Miura, H., Kawano, T., & Iwamoto, Y. (2004). Elasto-plastic contact analysis of an ultra-high molecular weight polyethylene tibial component based on geometrical measurement from a retrieved knee prosthesis. *Proceedings of the Institution of Mechanical Engineers, Part H: Journal of Engineering in Medicine, 218*(4), 251–259. https://doi.org/10.1243/0954411041561018

Clarke, I. C., & Gustafson, A. (2000). Clinical and hip simulator comparisons of ceramic-on-polyethylene and metal-on-polyethylene wear. *Clinical Orthopaedics and Related Research, 379*, 34–40. https://doi.org/10.1097/00003086-200010000-00006

Collier, J. P., Bargmann, L. S., Currier, B. H., Mayor, M. B., Currier, J. H., & Bargmann, B. C. (1998). An analysis of hylamer and polyethylene bearings from retrieved acetabular components. *Orthopedics, 21*(8), 865–871. https://doi.org/10.3928/0147-7447-19980801-09

Collier, J. P., Sperling, D. K., Currier, J. H., Sutula, L. C., Saum, K. A., & Mayor, M. B. (1996). Impact of gamma sterilization on clinical performance of polyethylene in the knee. *Journal of Arthroplasty, 11*(4), 377–389. https://doi.org/10.1016/S0883-5403(96)80026-X

Connelly, G. M., Rimnac, C. M., Wright, T. M., Hertzberg, R. W., & Manson, J. A. (1984). Fatigue crack propagation behavior of ultrahigh molecular weight polyethylene. *Journal of Orthopaedic Research, 2*, 119–125. https://doi.org/10.1002/jor.1100020202

Costa, L., Luda, M. P., Trossarelli, L., Brach Del Prever, E. M., Crova, M., & Gallinaro, P. (1998). Oxidation in orthopaedic UHMWPE sterilized by gamma-radiation and ethylene oxide. *Biomaterials, 19*(7–9), 659–668. https://doi.org/10.1016/S0142-9612(97)00160-9

Dowson, D., Diab, M. M. E.-H., Gillis, B. J., & Atkinson, J. R. (1985). Influence of counterface topography on the wear of ultra high molecular weight polyethylene under wet and dry conditions. In *Polymer wear and its control* (pp. 171–187). American Chemical Society.

Dowson, D., Gillis, B. J., & Atkinson, J. R. (1985). Penetration of metallic femoral components into polymer tibial components observed in a knee joint simulator. *Polymer Wear and Its Control* (pp. 215–228). American Chemical Society (ACS).

Dowson, D., McNie, C. M., & Goldsmith, A. A. J. (2000). Direct experimental evidence of lubrication in a metal-on-metal total hip replacement tested in a joint simulator. *Proceedings of the Institution of Mechanical Engineers, Part C: Journal of Mechanical Engineering, 214*(1), 75–86. https://doi.org/10.1243/0954406001522822

Dowson, D., Unsworth, A., Cooke, A. F., & Gvozdanovic, D. (1981). Lubrication of joints. In D. Dowson, & V. Wright (Eds.), *Introduction to the biomechanics of joints and joint replacement* (pp. 120–145). Mechanical Engineering Publications.

Dumbleton, J. H. (1981), *Tribology Series, 3, Tribology of natural and artificial joints*. Elsevier Scientific Publishing Company.

Dwyer-Joyce, R. S., Drinkwater, B. W., & Donohoe, C. J. (2003). The measurement of lubricant-film thickness using ultrasound. *Proceedings of the Royal Society A: Mathematical, Physical and Engineering Sciences, 459*(2032), 957–976. https://doi.org/10.1098/rspa.2002.1018

Dwyer-Joyce, R. S., Harper, P., & Drinkwater, B. W. (2004). A method for the measurement of hydrodynamic oil films using ultrasonic reflection. *Tribology Letters, 17*(2), 337–348. https://doi.org/10.1023/B:TRIL.0000032472.64419.1f

Gallo, J., Goodman, S. B., Konttinen, Y. T., Wimmer, M. A., & Holinka, M. (2013). Osteolysis around total knee arthroplasty: A review of pathogenetic mechanisms. *Acta Biomaterialia, 9*(9), 8046–8058. https://doi.org/10.1016/j.actbio.2013.05.005

Gallo, J., Vaculova, J., Goodman, S. B., Konttinen, Y. T., & Thyssen, J. P. (2014). Contributions of human tissue analysis to understanding the mechanisms of loosening and osteolysis in total hip replacement. *Acta Biomaterialia, 10*(6), 2354–2366. https://doi.org/10.1016/j.actbio.2014.02.003

Gao, L., Hua, Z., Hewson, R., Andersen, M. S., & Jin, Z. (2018). Elastohydrodynamic lubrication and wear medelling of the knee joint replacements with surface topography. *Biosurface and Biotribology, 4*(1), 18–23. https://doi.org/10.1049/bsbt.2017.0003

Gao, L., Wang, F., Yang, P., & Jin, Z. (2009). Effect of 3D physiological loading and motion on elastohydrodynamic lubrication of metal-on-metal total hip replacements. *Medical Engineering and Physics, 31*(6), 720–729. https://doi.org/10.1016/j.medengphy.2009.02.002

Glyn-Jones, S., McLardy-Smith, P., Gill, H. S., & Murray, D. W. (2008). The creep and wear of highly cross-linked polyethylene A three-year randomised, controlled trial using radiostereometric analysis. *The Journal of Bone and Joint Surgery (Br), 90-B*, 556–561. https://doi.org/10.1302/0301-620X.90B5.20545

Greenbaum, E. S., Burroughs, B. B., Harris, W. H., & Muratoglu, O. K. (2004). Effect of lipid adsorption on wear and compressive properties of unirradiated and highly crosslinked UHMWPE: An in vitro experimental model. *Biomaterials, 25*(18), 4479–4484. https://doi.org/10.1016/j.biomaterials.2003.11.049

Hamrock, B. J., & Dowson, D. (1978). Elastohydrodynamic lubrication of elliptical contacts for materials of low elastic modulus: 1 fully flooded conjunction. *Journal of Lubrication Technology, 100*(1), 236–245.

Hirokawa, S., Murakami, T., Kiguchi, K., & Fukunaga, M. (2015). Development of the novel simulator that reproduces physiological and active knee motion. *Transactions of the JSME (In Japanese), 81*(824), 1–14. https://doi.org/10.1299/transjsme.14-00488

Huang, C.-H., Liau, J.-J., & Cheng, C.-K. (2007). Fixed or mobile-bearing total knee arthroplasty. *Journal of Orthopaedic Surgery and Research, 2*(1). https://doi.org/10.1186/1749-799X-2-1

Ikeuchi, K. (2007). Origin and future of hydration lubrication. *Proceedings of the Institution of Mechanical Engineers, Part J: Journal of Engineering Tribology, 221*(3), 301–305. https://doi.org/10.1243/13506501JET214

Ingham, E., & Fisher, J. (2005). The role of macrophages in osteolysis of total joint replacement. *Biomaterials, 26*(11), 1271–1286. https://doi.org/10.1016/j.biomaterials.2004.04.035

Ishihara, K., Ueda, T., & Nakabayashi, N. (1990). Preparation of phospholipid polymers and their properties as polymer hydrogel membrane. *Polymer Journal, 22*(5), 355–360. https://doi.org/10.1295/polymj.22.355

Ishikawa, Y., Sasada, T., & Ikeuchi, K. (2001). Surface gel hydration lubrication – proposal of a new idea for joint lubrication. In JAST (Ed.), *2. Proc. ITC Nagasaki 2000* (pp. 1491–1496). JAST.

Jagatia, M., & Jin, Z. M. (2001). Elastohydrodynamic lubrication analysis of metal-on-metal hip prostheses under steady state entraining motion. *Proceedings of the Institution of Mechanical Engineers, Part H: Journal of Engineering in Medicine, 215*(6), 531–541. https://doi.org/10.1243/0954411011536136

Jalali-Vahid, D., Jagatia, M., Jin, Z. M., & Dowson, D. (2001). Prediction of lubricating film thickness in UHMWPE hip joint replacements. *Journal of Biomechanics, 34*(2), 261–266. https://doi.org/10.1016/S0021-9290(00)00181-0

Jalali-Vahid, D., & Jin, Z. M. (2002). Transient elastohydrodynamic lubrication analysis of ultra-high molecular weight polyethylene hip joint replacements. *Proceedings of the Institution of Mechanical Engineers, Part C: Journal of Mechanical Engineering Science, 216*(4), 409–420. https://doi.org/10.1243/0954406021525205

Jin, Z. M., Dowson, D., & Fisher, J. (1993). Fluid film lubrication in natural hip joints. *Tribology Series, 25*(C), 545–555. https://doi.org/10.1016/S0167-8922(08)70410-3

Jin, Z. M., Dowson, D., & Fisher, J. (1997). Analysis of fluid film lubrication in artificial hip joint replacements with surfaces of high elastic modulus. *Proceedings of the Institution of Mechanical Engineers, Part H: Journal of Engineering in Medicine, 211*(3), 247–256. https://doi.org/10.1243/0954411971534359

Johnson, K. L., Greenwood, J. A., & Poon, S. Y. (1972). A simple theory of asperity contact in elastohydrodynamic lubrication. *Wear, 19*(1), 91–108. https://doi.org/10.1016/0043-1648(72)90445-0

Klein, J. (2013). Hydration lubrication. *Friction, 1*(1), 1–23. https://doi.org/10.1007/s40544-013-0001-7

Kurtz, S. M. (Ed.). (2016). *UHMWPE biomaterials Handbook: Ultra-high molecular weight polyethylene in total joint replacement and medical devices* (3rd ed.). Elsevier.

Kurtz, Steven M., Gawel, H. A., & Patel, J. D. (2011). History and systematic review of wear and osteolysis outcomes for first-generation highly crosslinked polyethylene. *Clinical Orthopaedics & Related Research, 469*(8), 2262–2277. https://doi.org/10.1007/s11999-011-1872-4

Kurtz, S. M., & Patel, J. D. (2016). The clinical performance of highly cross-linked UHMWPE in hip replacements. In S. M. Kurtz (Ed.), *UHMWPE biomaterials Handbook: Ultra high molecular weight polyethylene in total joint replacement and medical devices* (3rd ed., pp. 57–71). Elsevier Inc.

Kyomoto, M., Moro, T., Yamane, S., Watanabe, K., Hashimoto, M., Tanaka, S., & Ishihara, K. (2022). Efficacy of hydrated phospholipid polymer interfaces between all-polymer bearings for total hip arthroplasty. *Journal of Biomedical Materials Research - Part B Applied Biomaterials, 110*(1), 89–102. https://doi.org/10.1002/jbm.b.34892

Maag, C., Metcalfe, A., Cracaoanu, I., Wise, C., & Auger, D. D. (2021). The development of simulator testing for total knee replacements. *Biosurface and Biotribology, 7*(2), 70–82. https://doi.org/10.1049/bsb2.12001

Mak, M. M., & Jin, Z. M. (2002). Analysis of contact mechanics in ceramic-on-ceramic hip joint replacements. *Proceedings of the Institution of Mechanical Engineers, Part H: Journal of Engineering in Medicine, 216*(4), 231–236. https://doi.org/10.1243/09544110260138718

Mak, M., Jin, Z., Fisher, J., & Stewart, T. D. (2011). Influence of acetabular cup rim design on the contact stress during edge loading in ceramic-on-ceramic hip prostheses. *Journal of Arthroplasty, 26*(1), 131–136. https://doi.org/10.1016/j.arth.2009.10.019

Manley, M. T. (2016). Highly cross-linked and annealed UHMWPE. In S. M. Kurtz (Ed.), *UHMWPE Biomaterials Handbook, Ultra-high molecular weight polyethylene in total joint replacement and medical devices* (3rd ed., pp. 274–292). Elsevier Inc.

Masaoka, T. (2017). Wear, oxidation and osteolysis. In K. Yamamoto (Ed.), *Biomaterials for artificial hip joints (In Japanese)* (pp. 242–249). Medical View.

Mattei, L., Di Puccio, F., Piccigallo, B., & Ciulli, E. (2011). Lubrication and wear modelling of artificial hip joints. *Tribology International, 44*(5), 532–549. https://doi.org/10.1016/j.triboint.2010.06.010

Medley, J. B., Dowson, D., & Wright, V. (1984). Transient elastohydrodynamic lubrication models for the human ankle joint. *Engineering in Medicine, 13*(3), 137−151. https://doi.org/10.1243/EMED_JOUR_1984_013_035_02

Medley, J. B., Murakami, T., & Jin, Z. (2021). My personal stories and anecdotes with the late Professor Duncan Dowson. *Biosurface and Biotribology, 7*(2), 43−52. https://doi.org/10.1049/bsb2.12013

Meng, Q., Liu, F., Fisher, J., & Jin, Z. (2013). Contact mechanics and lubrication analysis of ceramic-on-metal total hip replacements. *Tribology International, 63*, 51−60. https://doi.org/10.1016/j.triboint.2012.02.012

Mongkolwongrojn, M., Wongseedakaew, K., & Kennedy, F. E. (2010). Transient elastohydrodynamic lubrication in artificial knee joint with non-Newtonian fluids. *Tribology International, 43*(5−6), 1017−1026. https://doi.org/10.1016/j.triboint.2009.12.041

Moro, T., Takatori, Y., Ishihara, K., Konno, T., Takigawa, Y., Matsushita, T., Chung, U. I. L., Nakamura, K., & Kawaguchi, H. (2004). Surface grafting of artificial joints with a biocompatible polymer for preventing periprosthetic osteolysis. *Nature Materials, 3*(11), 829−836. https://doi.org/10.1038/nmat1233

Moro, T., Takatori, Y., Kyomoto, M., Ishihara, K., Hashimoto, M., Ito, H., Tanaka, T., Oshima, H., Tanaka, S., & Kawaguchi, H. (2014). Long-term hip simulator testing of the artificial hip joint bearing surface grafted with biocompatible phospholipid polymer. *Journal of Orthopaedic Research, 32*(3), 369−376. https://doi.org/10.1002/jor.22526

Moro, T., Takatori, Y., Tanaka, S., Ishihara, K., Oda, H., Kim, Y. T., Umeyama, T., Fukatani, E., Ito, H., Kyomoto, M., Oshima, H., Tanaka, T., Kawaguchi, H., & Nakamura, K. (2017). Clinical safety and wear resistance of the phospholipid polymer-grafted highly cross-linked polyethylene liner. *Journal of Orthopaedic Research, 35*(9), 2007−2016. https://doi.org/10.1002/jor.23473

Murakami, T. (1990). The lubrication in natural synovial joints and joint prostheses. *JSME International Journal. Ser. 3, 33*(4), 465−474. https://doi.org/10.1299/jsmec1988.33.465

Murakami, T., & Ohtsuki, N. (1987). Lubricating film formation in knee prostheses under walking conditions. *Tribology Series, 11*, 387−392.

Murakami, T., Ohtsuki, N., & Higaki, H. (1993). Transient elastohydrodynamic lubrication in artificial knee joint with non-Newtonian fluids. *Tribology Series, 25*(C), 673−682. https://doi.org/10.1016/S0167-8922(08)70420-6

Muratoglu, O. K., & Bragdon, C. R. (2016). Highly cross-linked and melted UHMWPE. In *UHMWPE biomaterials handbook: Ultra high molecular weight polyethylene in total joint replacement and medical devices* (3rd ed., pp. 264−273). Elsevier Inc. https://doi.org/10.1016/B978-0-323-35401-1.00015-6

Myant, C., & Cann, P. (2013). In contact observation of model synovial fluid lubricating mechanisms. *Tribology International, 63*, 97−104. https://doi.org/10.1016/j-triboint.2012.04.029

Nečas, D., Sadecká, K., Vrbka, M., Galandáková, A., Wimmer, M. A., Gallo, J., & Hartl, M. (2021). The effect of albumin and γ-globulin on synovial fluid lubrication: Implication for knee joint replacements. *Journal of the Mechanical Behavior of Biomedical Materials, 113*. https://doi.org/10.1016/j.jmbbm.2020.104117

OECD. (1969). In *Friction, wear and lubrication = tribology =, glossary of terms and definitions in the field of friction, wear and lubrication = tribology -*. OECD.

Oonishi, H., Clarke, I. C., Masuda, S., & Amino, H. (2001). Study of retrieved acetabular sockets made from high-dose, cross-linked polyethylene. *Journal of Arthroplasty, 16*(8), 129−133. https://doi.org/10.1054/arth.2001.28371

Oonishi, H., Kadoya, Y., & Masuda, S. (2001). Gamma-irradiated cross-linked polyethylene in total hip replacements - analysis of retrieved sockets after long-term implantation. *Journal of Biomedical Materials Research, 58*(2), 167–171. https://doi.org/10.1002/1097-4636(2001)58:2<167::AID-JBM1003>3.0.CO;2-A

Oral, E., & Muratoglu, O. K. (2011). Vitamin E diffused, highly crosslinked UHMWPE: A review. *International Orthopaedics, 35*(2), 215–223. https://doi.org/10.1007/s00264-010-1161-y

Saikko, V. (2003). Effect of lubricant protein concentration on the wear of ultra-high molecular weight polyethylene sliding against a CoCr counterface. *Journal of Tribology, 125*(3), 638–642. https://doi.org/10.1115/1.1537751

Sakoda, H., Osaka, Y., Uetsuki, K., Okamoto, Y., & Haishima, Y. (2019). Evaluating the durability of UHMWPE biomaterials used for articulating surfaces of joint arthroplasty using delamination tests. *Journal of Biomedical Materials Research Part B: Applied Biomaterials, 107B*(1), 65–72. https://doi.org/10.1002/jbm.b.34095

Sasada, T. (2000). Lubrication of human joints – nature of joint friction and "surface gel hydration lubrication"-(In Japanese). *Japanese Journal of Clinical Biomechanics, 21*, 17–22.

Sawae, Y. (2009). Effects of physiological factors on wear of UHMWPE for joint prosthesis. In S. K. Shinha, & B. J. Briscoe (Eds.), *Polymer tribology* (pp. 195–226). Imperial College Press.

Sawae, Y., & Murakami, T. (1999). Wear evaluation of UHMWPE wear in multi-directional tests. In *Preprint of JAST annual meeting* (pp. 91–94).

Sawae, Y., Murakami, T., & Chen, J. (1998). Effect of synovia constituents on friction and wear of ultra-high molecular weight polyethylene sliding against prosthetic joint materials. *Wear, 216*(2), 213–219. https://doi.org/10.1016/S0043-1648(97)00196-8

Sawae, Y., Yamamoto, A., & Murakami, T. (2008). Influence of protein and lipid concentration of the test lubricant on the wear of ultra high molecular weight polyethylene. *Tribology International, 41*(7), 648–656. https://doi.org/10.1016/j.triboint.2007.11.010

Scholes, S. C., & Unsworth, A. (2000). Comparison of friction and lubrication of different hip prostheses. *Proceedings of the Institution of Mechanical Engineers, Part H: Journal of Engineering in Medicine, 214*(1), 49–57. https://doi.org/10.1243/0954411001535237

Stevenson, H., & Cann, P. M. (2021). Protein content of model synovial fluid and CoCrMo wear. *Biotribology, 26*. https://doi.org/10.1016/j.biotri.2021.100172

Stewart, T., Tipper, J., Streicher, R., Ingham, E., & Fisher, J. (2001). Long-term wear of HIPed alumina on alumina bearings for THR under microseparation conditions. *Journal of Materials Science: Materials in Medicine, 12*(10–12), 1053–1056. https://doi.org/10.1023/A:1012802308636

Tallian, T. E. (1967). On competing failure modes in rolling contact. *ASLE Transactions, 10*(4), 418–439. https://doi.org/10.1080/05698196708972201

Tomita, N., Kitakura, T., Onmori, N., Ikada, Y., & Aoyama, E. (1999). Prevention of fatigue cracks in ultrahigh molecular weight polyethylene joint components by the addition of vitamin E. *Journal of Biomedical Materials Research, 48*(4), 474–478. https://doi.org/10.1002/(SICI)1097-4636(1999)48:4<474::AID-JBM11>3.0.CO;2-T

Uehara, S., Yarimitsu, S., Moro, T., Kyomoto, M., Watanabe, K., Tanaka, S., Ishihara, K., & Murakami, T. (2016). Evaluation of wear suppression for phospholipid polymer-grafted ultra-high molecular weight polyethylene at concentrated contact. *Tribology International, 101*, 264–272. https://doi.org/10.1016/j.triboint.2016.04.026

Unsworth, A., Pearcy, M. J., White, E. F. T., & White, G. (1987). Soft layer lubrication of artificial hip joints. In *Proceedings of the Institution of Mechanical Engineers*,

International Conference on Tribology - Friction, lubrication and wear, fifty years on, (London: July 1–3, 1987), II(1987–5) (pp. 715–724).

Wang, A. (2001). A unified theory of wear for ultra-high molecular weight polyethylene in multi-directional sliding. *Wear, 248*(1–2), 38–47. https://doi.org/10.1016/S0043-1648(00)00522-6

Wang, A., Essner, A., Polineni, V. K., Stark, C., & Dumbleton, J. H. (1998). Lubrication and wear of ultra-high molecular weight polyethylene in total joint replacements. *Tribology International, 31*(1–3), 17–33. https://doi.org/10.1016/S0301-679X(98)00005-X

Willert, H. G., & Semlitsch, M. (1977). Reaction of the articular capsule to wear products of artificial joint prostheses. *Journal of Biomedical Materials Research, 11*(2), 157–164. https://doi.org/10.1002/jbm.820110202

Yan, Y., Neville, A., & Dowson, D. (2006). Biotribocorrosion – An appraisal of the time dependence of wear and corrosion interactions – Part I: The role of corrosion. *Journal of Physics D: Applied Physics, 39*(15), 3200–3205. https://doi.org/10.1088/0022-3727/39/15/S10

Yan, Y., Neville, A., Hesketh, J., & Dowson, D. (2013). Real-time corrosion measurements to assess biotribocorrosion mechanisms with a hip simulator. *Tribology International, 63*, 115–122. https://doi.org/10.1016/j.triboint.2012.08.006

Yarimitsu, S., Moro, T., Kyomoto, M., Watanabe, K., Tanaka, S., Ishihara, K., & Murakami, T. (2015). Influence of dehydration and rehydration on the lubrication properties of phospholipid polymer-grafted cross-linked polyethylene. *Proceedings of the Institution of Mechanical Engineers, Part H: Journal of Engineering in Medicine, 229*, 506–514.

Yarimitsu, Seido, Nakashima, K., Sawae, Y., & Murakami, T. (2007). Study on the mechanism of wear reduction of artificial cartilage through in situ observation on forming protein boundary film. *Tribology Online, 2*(4), 114–119. https://doi.org/10.2474/trol.2.114

Zhang, L., Sawae, Y., Yamaguchi, T., Murakami, T., & Yang, H. (2015). Effect of radiation dose on depth-dependent oxidation and wear of shelf-aged gamma-irradiated ultra-high molecular weight polyethylene (UHMWPE). *Tribology International, 89*, 78–85. https://doi.org/10.1016/j.triboint.2014.12.011

Biotribology of artificial joints with artificial cartilage

5

5.1 Effectiveness of compliant artificial cartilage

The natural synovial joints with extremely low friction and minimum wear are likely to operate in not single lubrication mode but various lubrication modes from fluid film lubrication to boundary lubrication (Dowson, 1966–67; Unsworth et al., 1975). As described in detail in Chapter 3, this lubrication mechanism is called the adaptive multimode lubrication mechanism (Murakami, 1990; Murakami, Yarimitsu, Sakai, Nakashima, Yamaguchi, & Sawae, 2017), in which both the compliant articular cartilage and synovial fluid play important roles, and various lubrication mechanisms become synergistically effective depending on the severity of various daily activities. Particularly, it is recognized that the elastic deformation of articular cartilage is favorable in forming the fluid film as soft-EHL (Medley et al., 1984) or micro-EHL (Dowson & Jin, 1986) and reducing the contact pressure. Furthermore, natural articular cartilage is a kind of natural hydrogel with high water content of 70%–80% as biphasic material composed of extracellular matrix of proteoglycan reinforced by collagen (mainly type II) fibers and chondrocytes. Therefore, even at slow-speed conditions, the interstitial fluid pressurization based on biphasic property and fiber-reinforced structure in articular cartilage appear to contribute to the superior load-carrying capacity with low friction. In addition, the surface lubricity of articular cartilage is effective in protection of rubbing surfaces and sustaining of low friction even at local direct contact areas.

In contrast, most of the artificial joints, such as total hip and total knee replacements, are composed of biocompatible artificial materials of UHMWPE and mating anticorrosive metal or bioinert ceramics. UHMWPE has Young's modulus of 0.5–1.5 GPa, which is stiffer at two or three order than that of articular cartilage, and the minimum fluid film thickness is estimated to be thinner than 0.1 μm for total hip and knee prostheses during the stance phase of walking (Auger et al., 1995a; Jalali-Vahid et al., 2001). The fluid film thickness is usually smaller than the surface roughness of UHMWPE. Therefore, mixed or boundary lubrication mode is the main lubrication mode, and thus, local direct contact occurs, resulting in medium friction and considerable wear, as pointed out in Chapter 4. In cases with significant wear, considerable wear debris was produced in joint prostheses and resulted in joint loosening due to particle-derived osteolysis (Ingham & Fisher, 2005; Willert & Semlitsch, 1977).

To extend the durability of artificial joints, the improvement of material structure and properties has been conducted. The crosslinking (Kurtz et al., 2011; Oonishi et al., 2001) by gamma-ray or electron beam irradiation with annealing or remelted treatment and the addition of vitamin E (Oral & Muratoglu, 2011; Tomita et al., 1999) by blending or diffusion to UHMWPE could improve significantly wear

Biotribology of Natural and Artificial Joints. https://doi.org/10.1016/B978-0-12-823669-7.00004-1

resistance and reduce oxidative degradation. However, some wear still occurred at severe loading conditions. Moro et al. (2004, 2017) developed a new type of hip prosthesis with poly (2-methacryloyloxyethyl phosphorylcholine) (PMPC)-grafted cross-linked polyethylene, which could reduce wear and the risk of osteolysis, and improve the durability. Although some wear may occur under severer operating conditions where the hydration lubrication becomes ineffective around polymer brush layer, the underlying cross-linked polyethylene will be able to maintain good wear resistance as a fail-safe system.

For hard-on-hard hip prostheses such as metal-on-metal, ceramic-on-ceramic, and ceramic-on-metal with high precision and high wear resistance, the possibility of fluid film lubrication is expected to be actualized under the normal walking condition. However, some problems such as stripe wear, streaking, breakage of ceramic component, taper failure with modular stem designs, adverse biological reaction to metallic ions, and others occurred particularly at severe operating conditions under edge loading and/or at malpositioning (Mihalko et al., 2014; Zywiel et al., 2011).

In order to fundamentally improve the lubrication modes in artificial joints, the applications of various compliant artificial cartilage materials with low elastic moduli as cartilage replacement have been developed. The application of appropriate compliant artificial cartilage materials with properties similar to articular cartilage is expected to duplicate the superior load-carrying capacity and tribological properties of natural synovial joints. The fluid film formation is enhanced, and the contact stress level is reduced due to the elastic deformation effect of low-modulus materials in joint prostheses with compliant artificial cartilage. The changes in fluid film thickness with different elastic modulus were evaluated by numerical analyses based on elastohydrodynamic lubrication theory and the experimental measurement of fluid film formation by means of electric resistance method for electric conductive surfaces, and/or estimated by frictional behaviors as Stribeck curves.

Some examples on tribological evaluations are shown here. To evaluate the actual fluid film formation in knee prostheses with soft layer or UHMWPE tibial components in knee joint simulators under walking condition, the electric resistance method for conductive femoral and tibial components was applied by using the electric circuit as shown in Fig. 4.3 (Murakami & Ohtsuki, 1987; Murakami, 1990; Murakami et al., 1993). To indicate the effectiveness of fluid film formation, the degree of separation is defined as the ratio of measured voltage to the applied electric voltage as mentioned in Chapter 4. Here, the degree of separation = 1 means full separation, and 0 means intimate local contact.

In Fig. 5.1, the experimental results on fluid film formation in two kinds of knee prostheses with different tibial components are shown during walking condition in knee joint simulators with electric circuit to measure the degree of separation (Murakami et al., 1993). As lubricants, nonconductive lubricants of silicone oils S-100 or S-10,000 (numeral means kinematic viscosity in mm^2/s at 20°C) different in viscosity were used. In knee prosthesis composed of metallic femoral and UHMWPE tibial component, significant intimate contacts occurred except extremely slight fluid film formation even in lubricated condition with very viscous lubricant, where UHMWPE surface was coated with Pt. In contrast, it was shown that the

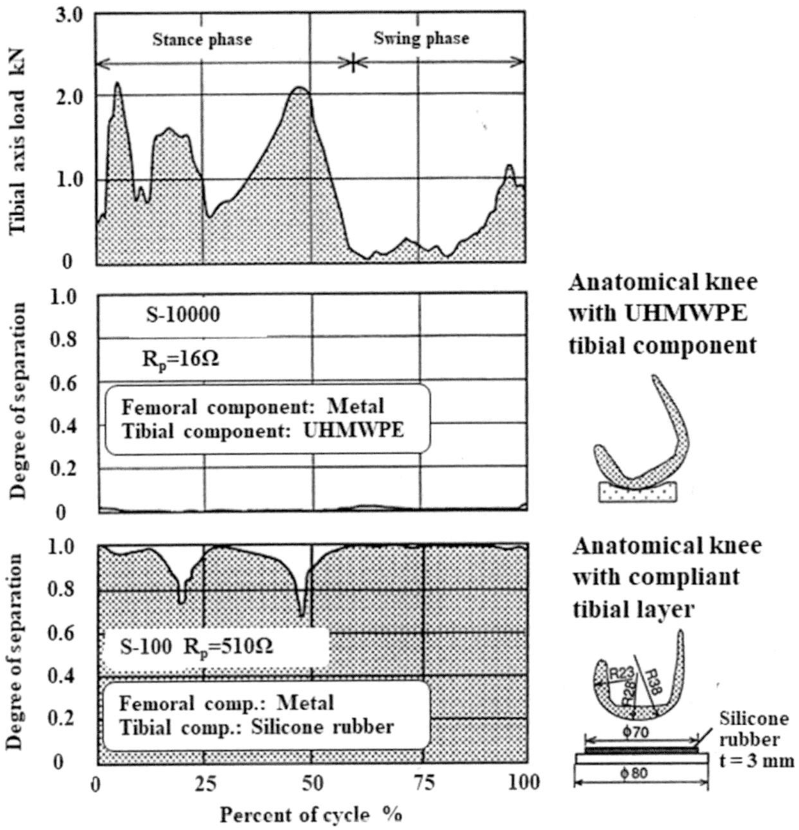

Figure 5.1 Fluid film formation in knee prostheses with UHMWPE or silicone rubber tibial component in simulator tests. Little fluid film is formed in knee prosthesis with UHMWPE component, but significant fluid film is formed in knee prosthesis with compliant silicone rubber layer during walking.
Modified from Murakami, T., Ohtsuki, N., & Higaki, H. (1993). The adaptive multimode lubrication in knee prostheses with compliant layer during walking motion. *Tribology Series*, *25*(C), 673–682. https://doi.org/10.1016/S0167-8922(08)70420-6.

significant elastohydrodynamic film is formed except short drops at stroke ends even during the loading stance phase in a knee prosthesis with a compliant conductive silicone rubber tibial component with elastic modulus of 9.1 MPa lubricated with medium viscosity oil. These results indicate the effectiveness of elastic deformation of compliant layer on the fluid film formation as soft-EHL mechanism.

An example of comparison of fluid film thickness in simplified artificial knee joint models of cylindrical component with soft layer as 2 mm thickness with different elastic modulus and rigid flat plate is shown in Fig. 5.2 (Murakami et al., 1997). For soft layers, two kinds of materials of 1.0 and 40 MPa as elastic moduli were selected. The equivalent radius 30 mm for the femoral and tibial surfaces of knee joint model is simplified as the radius of the nondeformed cylindrical surface.

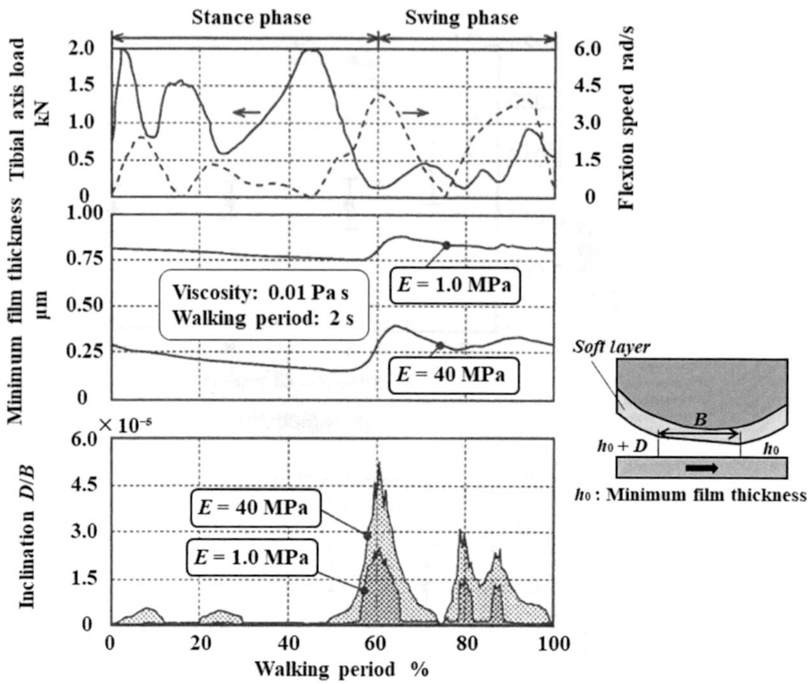

Figure 5.2 Fluid film formation in knee prosthesis models with artificial cartilage as soft layer during walking. Minimum fluid film thickness and inclination of inclined plane are shown corresponding to tibial axis load and flexion speed during walking.
Modified from Murakami, T., Sawae, Y., Higaki, H., Ohtsuki, N., & Moriyama, S. (1997). The adaptive multimode lubrication in knee prostheses with artificial cartilage during walking. *Tribology Series*, *32*, 371–382. https://doi.org/10.1016/s0167-8922(08)70466-8.

In this numerical analysis for soft-EHL, the simplified model as plane inclined surface (PI) model proposed by Medley and Dowson (2008), Gladstone and Medley (1990) was applied considering entraining and squeeze actions. In this model, the load-bearing area is assumed as elastically deformed plane inclined surface with bearing length of *B* and inclination *D*/*B*. In this walking condition with period of 2 s (slow walking), the maximum loads during stance phase are about 2.0 kN and the flexion-extension motions occur twice during walking. In Fig. 5.2, the results of numerical analysis for the minimum film thickness as h_0 and the inclination of inclined plane *D*/*B* are shown for two kinds of soft layers. The lubricant viscosity is assumed as 0.01 Pa·s.

As shown in the graph for minimum film thickness versus walking period (Fig. 5.2), the considerable fluid film is formed corresponding to the walking phase in knee joints with a soft layer. The fluid film gradually decreases during the stance phase at higher load and slow speed and then recovers to thicker level during low load and higher speed swing phase in artificial knee joint with a soft layer. The hydrodynamic action as the wedge effect in the converging narrow clearance by the appropriate inclination of plane surface to enhance the fluid film thickness is observed at high-speed and low-load conditions during swing phase. In contrast, at high-load and low-speed conditions during the stance phase, the fluid film thickness decreases due to insufficient wedge effect caused by the infinitesimal inclination, but the squeeze film action at loading appears to contribute to preserve fluid film.

If this result is applied in comparison between soft layer-A (elastic modulus: $E = 1$ MPa) and soft layer-B ($E = 40$ MPa), the layer-A shows twofold to threefold thicker film thickness than the layer-B, although the inclination of the layer-B is larger than that of the layer-A. It is considered that larger bearing area with elastic deformation in the layer-A is probably effective to form thicker fluid film in soft-EHL mechanism. If both elastomeric rubbing materials have similar surface roughness, the effectiveness of fluid film lubrication becomes better for the layer-A with lower elastic modulus.

In clinical application of knee prostheses composed of soft layer and smooth rigid (metallic or ceramic) surface, the influence of surface roughness of soft layer surface on the possibility of full fluid film lubrication should be evaluated. The values for maximum height of surface roughness of both soft layers are larger than 1 μm as usual case, which is larger than minimum fluid film thickness in Fig. 5.2. Therefore, local direct contact is likely to occur between rubbing surfaces. However, Dowson and Jin (1986) indicated that the significant flattening of compliant surface asperities of articular cartilage in natural synovial joints enables to avoid direct contact between rubbing surfaces during walking in micro-EHL mechanism. Similar mechanism is expected in artificial joints with compliant artificial cartilage. The actual lubrication mode in these cases should be evaluated based on the detailed micro-EHL analysis or experimental measurement of contact situation in lubricated conditions, as discussed below.

If the sufficient fluid film is formed to prevent the direct contact between rubbing surfaces, the low friction and low wear are expected to be established under mild operating conditions. However, nonnegligible wear occurred in some joint prostheses with artificial cartilage under severer conditions. Therefore, to attain the longer durability in clinical application, the sufficient strength and superior tribological properties of artificial cartilage are required. The optimum geometric and material design to reduce the severity of rubbing may be effective.

Next, the evaluation of lubrication mode in hip prostheses with compliant artificial cartilage is introduced. In Fig. 5.3 as frictional characteristics in Stribeck curves, the soft-layered artificial hip joint shows remarkable low frictional property in comparison with the traditional hip joint with UHMWPE cup in simulator testing (Unsworth et al., 1987). In hip prosthesis composed of metallic femoral head and acetabular cup lined with polyurethane of appropriate compliance and thickness, the fluid film lubrication can be achieved even with low-viscosity lubricants in experimental simulator tests in lubricants of different viscosity from 0.002 to 0.071 Pa·s. It is noticed that the soft-layered hip joint exhibits the superiority of fluid film lubrication as indicated by low friction for hollow circle marks in darkened area in Fig. 5.3 even for low viscosity conditions. In contrast, the hip joint

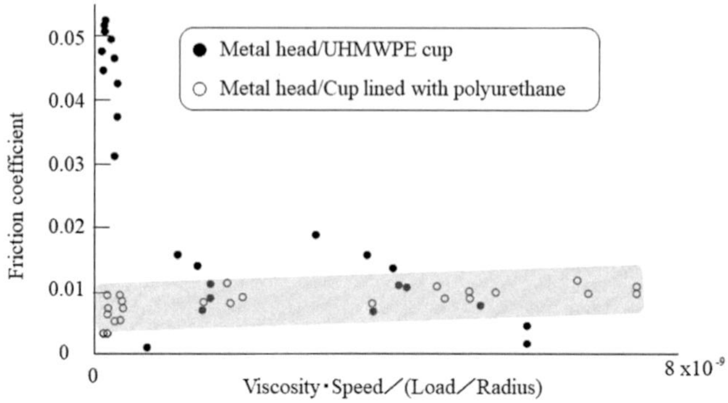

Figure 5.3 Friction characteristics in hip prostheses with UHMWPE or polyurethane acetabular cup in simulator testing. In hip prosthesis of metal head/UHMWPE cup, significant increase in friction is observed in lowering zone of lubrication parameter (Zommerfeld number): viscosity·speed/(load/radius). In hip one of metal head/cup lined with polyurethane, low friction coefficient almost less than 0.01 is observed in darkened area in wide operating range.

Modified from Unsworth, A., Pearcy, M. J., White, E. F. T., & White, G. (1987). Soft layer lubrication of artificial hip joints. In: *Proceeding IMechE international conference on tribology - friction, lubrication and wear, 50 years on, (London, U.K.: July. 1–3, 1987)*, (Vol. II, pp. 715–724).

with UHMWPE cup (solid marks) appears to operate in mixed lubrication mode as indicated by friction increase for low viscosity conditions.

As shown above, the lubrication mode for rubbing pairs of hard material with UHMWPE appears to be mixed or boundary lubrication where some local direct contacts may occur at severe contact conditions. Dowson (1989) discussed the adequacy of the current joint replacement materials and indicated the important feature that the dominant form of lubrication in current prostheses such as UHMWPE-on-metal and UHMWPE-on-ceramic is mixed lubrication, while the dominant mode in natural synovial joint is fluid film lubrication including micro-EHL. He introduced the term "cushion joints" from the viewpoint that the application of low elastic modulus linings as artificial cartilage with similar properties to natural articular cartilage is expected to solve the friction and wear problems by promotion of the fluid film lubrication.

Dowson et al. (1991) proposed the design concept as cushion form bearings, which comprise low elastic modulus layers on the articulating surfaces for artificial hip joints. It is shown in cushion form bearings with appropriate elastic modulus, for example, for polyurethane layer of 5–20 MPa as Young's modulus, that in a similar way to natural synovial joints the soft-EHL film is formed by larger contact areas and lower contact pressure, and the significant deformation of surface asperities reduces the effective surface roughness and maintains effective fluid film in micro-EHL mechanism. The comparison of lubricating properties between Muller artificial hip joint (UHMWPE-on-Co-Cr-Mo alloy) and cushion bearing (2 mm thick polyurethane-on-metal), which have the same femoral head diameter of 32 mm, the same radial clearance of 0.25 mm, i.e., the same reduced radius of 1.04 m (as sphere on plane) is shown in Table 5.1 under

Table 5.1 Analysis for current artificial hip joint and cushion bearing.
The cushion hip joint exhibits thicker elastohydrodynamic minimum film and thicker squeeze film than Muller hip joint.

	Contact half- width	**Minimum film thickness**	**Maximum contact pressure**	**Squeeze film thickness at 1 s**
	a **mm**	h_{min} **μm**	p_{max} **MPa**	h_s **μm**
Muller hip joint	10.56	0.075	10.7	0.111
Cushion bearing	15	0.26	8.2	0.22

Modified from Dowson, D., Fisher, J., Jin, Z. M., Auger, D. D., & Jobbins, B. (1991). Design considerations for cushion form bearings in artificial hip joints. *Proceedings of the Institution of Mechanical Engineers, Part H: Journal of Engineering in Medicine, 205*(2), 59–68. https://doi.org/10.1243/PIME_PROC_1991_205_269_02.

a condition of a maximum load of 2.5 kN, an entraining velocity of 75 mm/s and lubricant viscosity of 0.001 Pa s. Here, Young's modulus E and Poisson's ratio ν for these material combinations are as follows:

Muller hip (UHMWPE/Co-Cr–Mo): $E = 1.4$ GPa/210 GPa, $\nu = 0.4/0.3$.

Cushion bearing (polyurethane/metal): $E = 20$ MPa/100 GPa, $\nu = 0.5/0.3$.

In this analysis, the minimum film thickness for Muller hip prosthesis was predicted by Hamrock and Dowson (1978) formula for an equivalent elastic sphere and a rigid plane. In contrast, the minimum film thickness for cushion bearing was estimated based on the formula for the elastohydrodynamic lubrication for circular contact of thin layer of compliant material upon rigid substrate (Dowson et al., 1991). Detailed analysis for elastohydrodynamic lubrication film thickness of soft-layered solids at elliptical contacts is given by Dowson and Yao (1994). Squeeze film thickness at 1 s is predicted by Higginson formula (Higginson, 1978). As shown in Table 5.1, cushion hip joint exhibits thicker elastohydrodynamic minimum film and thicker squeeze film than Muller hip joint but has smaller thickness than the surface roughness of soft layer. So, the effect of the elastic deformation of the asperities (δ) for a roughness modeled as a sinusoidal form in micro-EHL mechanism was examined on the basis of the following formula:

$$\delta = \frac{(1-\nu^2)\lambda p_a}{\pi E} \tag{5.1}$$

where.

λ = wave-length of sinusoidal surface for soft layer

p_a = pressure perturbation (determine by surface roughness and film thickness).

The deformation δ of the asperities for $\lambda = 20$ μm and $p_a = 1$ MPa is estimated from Eq. (5.1).

$\delta = 0.0034$ μm (in UHMWPE cup of Muller hip joint).

$\delta = 0.25$ μm (in soft layer of cushion form bearing).

As shown here, the low elastic modulus material of the cushion form bearing provides better conditions for micro-EHL than the UHMWPE cup. In cushion form bearing, pressure perturbation produced by micro-EHL can flatten surface asperities, effectively producing smoother surfaces and hence maintaining the integrity of thin fluid film. It is found that the main advantage of using a cushion bearing with low elastic modulus layer is associated with the micro-EHL mechanism. The elastic modulus of the soft layer should not be lower than necessary to produce effective micro-EHL, as an excessive reduction in modulus will produce increased shear strains, which may reduce the durability of the soft layer.

In both total hip prostheses with the UHMWPE cup and soft-layered cup, similar elastohydrodynamic film is formed if both cups are designed to have similar contact areas. Auger et al. (1993) performed the friction measurement for these hip prostheses in a pendulum simulator and compared frictional behaviors. It was observed that friction coefficient in the cushion cup with the polyurethane layer showed much lower values (0.003−0.009) than in the UHMWPE cup (0.017−0.042). It is indicated that the effective fluid film lubrication was preserved in the cushion form joint by micro-EHL action.

Scholes et al. (2005) performed contact analyses and friction tests on compliant layer hip joints (metal-on-polyurethane). It was shown in Stribeck plots for four biocompatible polyurethane materials that significantly low friction was produced within full fluid film lubrication regime by all joints with a hardness of 6 MPa and less compared with the conventional metal-on-UHMWPE joints. They gave an essential comment for incompressible compliant material that the radial clearance should be above 0.08 mm to prevent the acetabulum from grabbing the femoral component. The importance of the compliant layer thickness for minimum fluid film thickness was indicated.

On start-up friction, which increases with loading time in friction testing for the hip prosthesis, which mimics the natural joint (Scholes et al., 2006), it is noticed that high start-up friction quickly reduced to very low level in full-fluid-film lubrication within less than half a walking cycle after preloading for 20 min.

In the design of the cushion form hip joint, maximizing the contact area simultaneously maximizes the theoretical fluid film thickness except very low radial clearance. In the design of cushion form knee joint, different consideration is recommended as described below.

Most of artificial knee joints composed of the hard material (metal or ceramic) and UHMWPE operate in mixed or boundary lubrication regime and consequently produce wear debris, which can ultimately cause loosening and failure of the prostheses. To promote the fluid film lubrication, Auger et al. (1995a) proposed the cushion form bearings for total knee joint replacement. In knee prostheses to require a large range of flexion, it is necessary to have very little conformity compared with hip prostheses. The low elastic modulus layer in cushion knee joint can deform macroscopically to produce larger contact areas and reduce stress levels. Furthermore, the compliant layer allows microscopic elastic deformation of the surface asperities which is enabled to form micro-EHL film. In friction test of cushion knee joint with a flat polyurethane tibial surface in pendulum simulator, remarkably low levels of friction (friction coefficient less than 0.005) were observed for polyurethane layer of 6 MPa modulus within the mixed lubrication regime with high percentage of fluid film lubrication in water and water-based lubricants. The flattening of surface asperities by micro-EHL mechanism appears to prevent substantial surface-to-surface contact and thus to keep friction low. For polyurethane layer of 20 MPa modulus, higher friction than 6 MPa polyurethane was observed in mixed lubrication regime at low Sommerfeld number.

At the next step, the wear and creep performance of cushion knee joints was investigated under loading and sliding conditions similar to those encountered in vivo (Auger et al., 1995b). As specimens, flat polyurethane–layered tibial components were constructed to articulate against standard condylar metallic femoral components. The cushion knee joints produce nearly equivalent or larger contact areas and therefore have equivalent or thicker elastohydrodynamic lubricating film thickness than an existing UHMWPE knee prosthesis. The wear tests were performed in the knee joint simulator under dynamic walking condition at 1 Hz and at maximum load of 2230 N (about three x body weight). The articulating surfaces of knee joint were immersed in a deionized water bath. The creep tests were conducted under a sinusoidal compressive loading (0–2.0 kN) at 1 Hz with no articulation. In wear tests for polyurethane layer of 6 MPa modulus, which gave improved lubrication with low friction and low stress (Auger et al., 1995a), severe material failure occurred both at the articulating surface and the interface with the substrate probably due to larger strains at less than 20, 000 cycles. It was considered that low modulus 6 MPa polyurethane material has insufficient structural integrity. In contrast, it was confirmed that the cushion knee with 20 MPa polyurethane layer in thickness of 5 mm generate very little wear debris after 5.0 million cycles. It is pointed out that debonding the polyurethane layer from the substrate should be prevented for longer durability. In creep tests, it was found that the rate of penetration of femoral component into the polyurethane layer decreases and essentially stabilizes after approximately 1.0 million cycles of compressive loading. If the polyurethane layer is not extruded and fluid entrainment is not impaired in walking motion, it is expected that the creep will act to increase the area of contact, thus reducing stress and enhancing the fluid film thickness.

Jones et al. (2008) performed to optimize the design parameters to establish optimum conditions for compliant layer knee joints. It is indicated on selection of modulus of polyurethane that the optimization considering the trade-off relation between lubricating fluid film thickness and the deformation or strain is required. Lower modulus enhances thick fluid film resulting in low friction but produces excessive creep deformation, which increases the shear stress and shear strain at the interface making it more susceptible to bonding and fatigue. In their evaluation, 19.2 MPa material was the optimum choice. Effects of soft layer thicknesses (2 mm, 3 and 4 mm) and bearing conformities (low (a flat tibial bearing), medium (concave), and high (concave)) with convex femoral component on the performance were evaluated. The increasing soft-layer thickness for a given modulus resulted in increases in both the contact area and the theoretical fluid film thickness. For medium conformity design, all specimens different in layer thickness showed similar friction levels in cases where the fluid film lubrication is predominant. All the three bearing conformities performed similar friction performance. Optimum design on soft-layer thickness and conformity was suggested in considering other requirements including manufacture and reproducibility.

As described above, in the cushion form bearing with appropriate design under normal walking motion, the lubricating fluid film can be maintained and hence wear can be minimized. In various daily activities, however, more severe operating conditions can occur in artificial joints under adverse lubrication conditions. Stewart et al. (1997) performed Stribeck assessment in pendulum simulator to investigate the tribological conditions leading to fluid film breakdown in composite cushion bearings for total knee joint replacement in severe cyclic loading, reduced sliding velocity, reduced stroke length, and start-up after a period of constant loading. In the friction test, the cushion knee joint composed of the polished metallic cylinder femoral component and the polyurethane flat composite (4 mm thick soft layer of 20 MPa modulus and 4 mm thick stiff layer of 1000 MPa modulus, manufactured by injection molding) as the tibial component was used lubricated with four lubricants different in viscosity from 0.000,845 to 0.80 Pa·s. Experimental results were evaluated in Stribeck assessment by considering the lubrication modes, which is predicted by the surface separation ratio (the film parameter, lambda ratio) Λ, defined as the ratio of the central film thickness to the combined surface roughness. Peak friction coefficients with water and bovine serum as lubricants rose from about 0.02 during normal walking to about 0.15 in conditions with severe cyclic loading and decreased sliding velocities in mixed lubrication or boundary lubrication regime ($\Lambda < 3$). In contrast, Peak friction coefficients with high viscosity lubricants maintained low friction coefficients of around $0.01 - 0.005$ in condition with severe cyclic loading and decreased sliding velocities in fluid film lubrication regime with thick elastohydrodynamic fluid film ($\Lambda > 3$). In reduced stroke length, friction increased probably due to a decrease in the amount of fluid entrainment resulting in thinning of lubricating fluid. It was confirmed that friction coefficient at start-up in water and bovine serum significantly increased as a function of stationary loading time before movement.

Thus, in cushion form bearing as joint prostheses with compliant articular layer, the fluid film lubrication can be attained by effective soft-EHL and micro-EHL mechanisms in normal activities such as walking.

As mentioned above, it is expected that fluid film lubrication is attained in cushion form bearings for activities with hydrodynamic action during normal walking and running motions. Under other daily activities under adverse lubrication conditions, however, sufficient fluid film may not be sufficiently formed. For very thin film conditions with significant surface-to-surface contact under adverse lubrication conditions at slower motion and for activities at little movement, starting and stopping, alternate lubrication mechanisms with surface lubricity are required. Furthermore, superior biocompatibility and durability for longer usage are required. For clinical application, it should be taken into consideration that the appropriate fixation of compliant artificial cartilage in such as acetabular component and tibial component is another important feature.

As a medical treatment for patients with joint diseases in which only one side of the synovial joint is damaged, the hemiarthroplasty is an attractive alternative to total joint replacement particularly for the young active patient. The partial replacements of articular cartilage with artificial cartilage such as for hip femoral head or for localized cartilage lesions in knee joint as hemiarthroplasty have been developed, and the tribological performances have been evaluated. At the present stage, except specified application, the extensive clinical application of artificial cartilage implantation has not yet been established due to several causes of insufficient strength, wear-resistance, friction property, and fixation-stability, but the various research have been promoted to improve the performance of artificial cartilage materials or devices.

The in vitro tribological evaluation was conducted for the knee hemiarthroplasty model of polyurethane tibial plates with different moduli (1.4 MPa, 6.5 and 22 MPa) against femoral cartilage in the pendulum friction simulator (Luo et al., 2010). It was demonstrated that the lower-modulus polyurethane caused lowering of contact stress and frictional shear stress and thus resulted in very little or no visible wear of opposing articular cartilage, whereas higher-modulus (22 MPa) polyurethane and stainless steel tibial plate showed higher frictional shear stress and higher wear. These results suggest the possibility of knee hemiarthroplasty with polyurethane tibial plate as clinical application.

Next, some examples of tribological evaluations for cushion form prostheses with compliant layer in clinical application are shown.

The polycarbonate urethane acetabular component was clinically applied for total hip replacement, and the retrieved (in short term) acetabular component showed lower wear on articulating surfaces (1.4 mm^3/year and 15 mm^3/year, in micro-CT wear measurement) probably due to micro-EHL effect than the corresponding UHMWPE acetabular cup (Kurtz, Siskey, & Shah, 2008; Kurtz, Siskey, Wippermann, et al., 2008). Further examination for long-term retrievals is required.

5.2 Effectiveness of hydrogel artificial cartilage

Hydrogels are three-dimensional crosslinked hydrophilic polymer networks swollen by water. Crosslinks, which bond the polymers of a hydrogel, are divided into physically and chemically ones. Hydrogels have biphasic properties, which show both solid- and liquid-like behaviors.

The articular cartilage in natural synovial joints is considered as one of natural hydrogel containing high water content, and thus the hydrogel material is a possible candidate material to alternate the articular cartilage as cartilage replacement. The application of compliant artificial cartilage of hydrogel as cartilage replacement is expected to mimic the natural articular cartilage and to improve lubrication by combined effect of soft elastohydrodynamic lubrication, biphasic lubrication, hydration lubrication, and boundary lubrication.

Biocompatible and hydrophilic poly(vinyl alcohol) (PVA) hydrogel is one of candidate hydrogels for artificial cartilage. Sasada (1988), Sasada et al. (1988) developed the hip prosthesis containing artificial cartilage layer of PVA hydrogel prepared by repeated freezing-thawing method (Nambu, 1982) with 85–90 mass% water content on inner surface of the acetabular cup. The new hip prosthesis showed quite similar low frictional behavior to natural synovial joint and further improvement in frictional behavior with hyaluronate solution. However, it had insufficient durability for longer clinical application.

In contrast, Oka et al. (2000) prepared the PVA hydrogel through other synthetic process with lower water content, which had better wear resistance in unidirectional pin (PVA)-on-disc (alumina) test. In this PVA hydrogel, the squeeze film effect was enhanced, and better shock-absorbing ability than traditional UHMWPE was confirmed. The γ-ray irradiation with doses of over 50 kGy improved the wear resistance; however, considerable wear was confirmed in reciprocating test against alumina plate including thin film conditions at stroke ends. Therefore, this hydrogel was used as a counter material for articular cartilage in hemiarthroplasty because of its low friction property against articular cartilage. In the in vivo animal study of hemiarthroplasty, the composite osteochondral device (COD) composed of PVA hydrogel and titanium fiber mesh implanted in dog femoral chondyle showed that the menisci and tibial cartilage were intact, whereas in tests with alumina and titanium femoral specimens, the ulceration and cartilage loss were apparent at 8 weeks after implantation. The superior biocompatibility of this PVA hydrogel was confirmed at in vivo tests as implantation of plugs into osteochondral defect and introduction of small particles into animal knee joints. For this COD, the firm attachment of the material to the underlying bone was confirmed with abundant new bone ingrowth into the pores of fiber mesh (Fig. 5.4). These results indicated the possibility of clinical application in partial surface replacement. As recent study, the application of biodegradable fiber mesh is explored.

At that time, the role of interstitial fluid pressure in hydrogel was not examined for this PVA hydrogel with lower water content (20%−45%), but the superiority in fluid film formation and shock-absorbing ability was clearly shown with good fixation to bone.

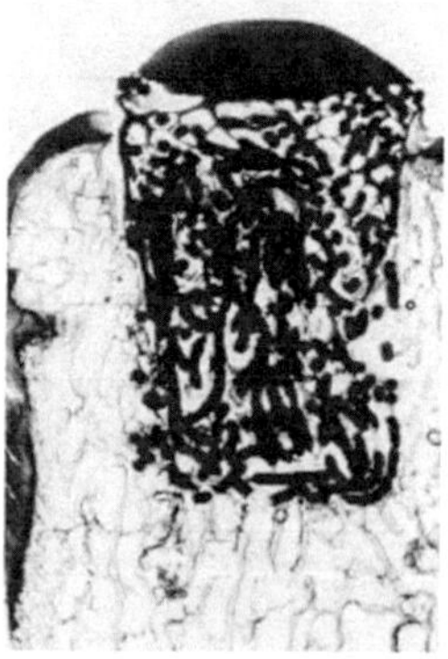

Figure 5.4 Photomicrograph of the artificial osteochondral composite, 26 weeks after implantation. The composite osteochondral device (COD) was attached to underlying bone by bone ingrowth into the titanium fiber mesh (TFM). The dark blunt area at the top of the TFM shows PVA hydrogel (Giemsa surface stain).
Modified from Oka, M., Ushio, K., Kumar, P., Ikeuchi, K., Hyon, S. H., Nakamura, T., & Fujita, H. (2000). Development of artificial articular cartilage. *Proceedings of the Institution of Mechanical Engineers, Part H: Journal of Engineering in Medicine*, *214*(1), 59−68. https://doi.org/10.1243/0954411001535246.

Next, an important case on the differences in frictional behaviors of simplified knee prosthesis models with soft layer in knee joint simulator is shown in Fig. 5.5 (Murakami et al., 1997). As femoral component, the cylindrical stainless steel specimen of 30 mm radius and surface roughness of 0.02 μm (R_{rms}) was used. As tibial component, 2 mm thick flat soft layer of PVA hydrogel or polyurethane supported by acrylic plate was used. PVA specimen was prepared by repeated freezing-thawing method (Nambu, 1982). Average degree of polymerization is 2000, and average degree of saponification is 98.5–99.4 mol%. PVA hydrogel of high water content (79%) has elastic modulus E of 1.1 MPa and surface roughness of 0.5 μm

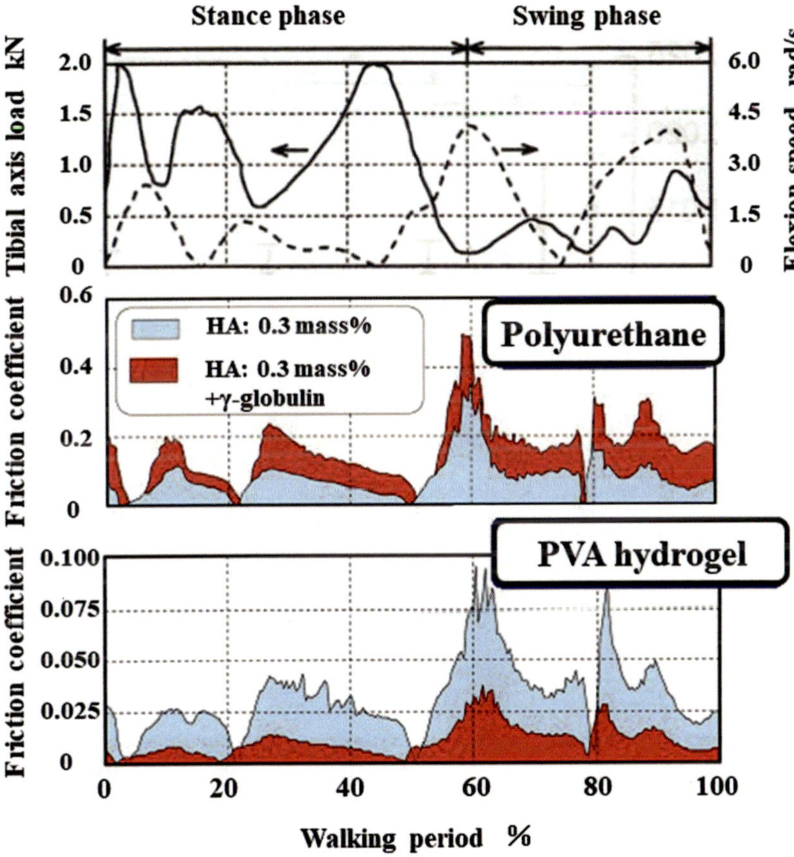

Figure 5.5 Frictional behavior of knee prostheses with soft layer in knee joint simulator. With HA (molecular weight: 1.95×10^6) solution as lubricant, PVA hydrogel shows lower friction than polyurethane. Addition of serum protein to HA solution reduces friction for PVA hydrogel but increases friction for polyurethane.

Modified from Murakami, T., Sawae, Y., Higaki, H., Ohtsuki, N., & Moriyama, S. (1997). The adaptive multimode lubrication in knee prostheses with artificial cartilage during walking. *Tribology Series*, *32*, 371–382. https://doi.org/10.1016/s0167-8922(08)70466-8.

(R_{rms}). As polyurethane specimen, medical-grade poly(ether)-polyurethane (Pellethane, E = 37 MPa) with surface roughness of 0.4 − 0.5 μm (R_{rms}) was used.

For both knee prosthesis models with the soft layer, the minimum fluid film thickness can be estimated by the previous analysis for elastic soft layer. As shown in Fig. 5.2 for lubricant of 0.01 Pa·s, the compliant layer with elastic modulus of 1.0 MPa corresponding to PVA hydrogel shows thicker fluid film thickness than 0.75 μm during walking, while the polyurethane layer with elastic modulus of about 40 MPa exhibits thinner film than 0.2 μm.

The possibility of fluid film lubrication is evaluated by estimation of the surface separation ratio (lambda ratio) Λ as the ratio of minimum fluid film thickness to composite surface roughness.

For PVA hydrogel: $\Lambda = 0.75/0.5 = 1.5$.

For polyurethane: $\Lambda = 0.2/0.45 = 0.44$.

Therefore, the lubrication regime for PVA hydrogel specimen is considered as the mixed lubrication because $1 < \Lambda < 3$. For polyurethane, $\Lambda < 1$, and this situation indicates significant direct contacts between rubbing surfaces in boundary or severe mixed lubrication regime. For compliant articulating materials, however, the flattening of surface asperities due to local pressure perturbations caused by surface roughness itself in conjunction zone facilitates the full fluid film lubrication in micro-elastohydrodynamic lubrication mechanism (Dowson & Jin, 1986), as expected in human synovial joint during walking. So, it should be examined about actual frictional behaviors of these knee prostheses with the soft layer.

As shown in Fig. 5.5, PVA hydrogel exhibits lower friction than polyurethane in walking condition lubricated with hyaluronate (HA) solution. This difference in friction is attributable to the difference in effectiveness of fluid film lubrication in mixed lubrication as discussed above.

It is further noted that the addition of γ-globulin as serum protein to HA solution remarkably reduced friction coefficient (around 0.01 or less for stance phase) and attenuated stick-slip during stance phase. In contrast, the addition of γ-globulin increased friction for polyurethane. The addition of albumin brought similar changes in friction for both PVA hydrogel and polyurethane. Synovial fluid contains not only hyaluronic acid and serum proteins but phospholipids and lubricin and other constituents. It should be kept in mind that PVA hydrogel exhibited low friction in synovial fluid similar to friction in HA solution with γ-globulin.

To investigate the influence of adsorbed film on stainless steel on frictional behaviors, the adsorbed film formation on stainless steel flat surfaces was observed in saline by AFM as fluid tapping mode after immersion in test fluid for 5 min. AFM images in Fig. 5.6 show the changes in adsorbed film formation on stainless steel plate with addition of γ-globulin to HA solution. Globular molecules of γ-globulin appears to adsorb on stainless steel surface, although some hyaluronate molecules may adsorb on surface. These images correspond to adsorbed film on noncontact zone. The rubbing process probably changes the adsorbed state, but it is considered that adsorbed films similar to images in Fig. 5.6 can induce the difference in frictional behaviors between PVA hydrogel and polyurethane, depending upon difference in lubrication mode. In mild mixed lubrication with some surface-to-surface contact for PVA hydrogel, the

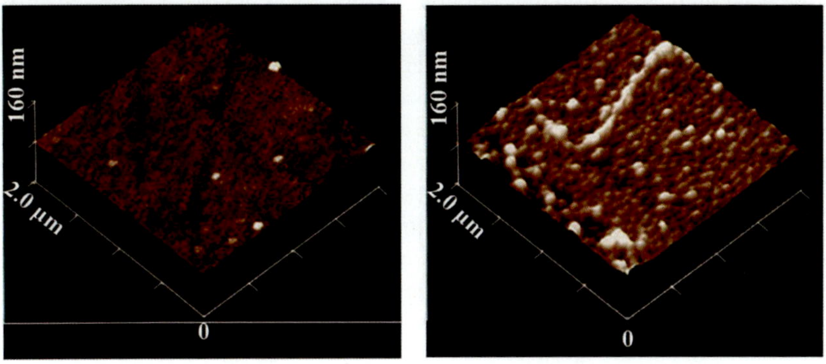

Figure 5.6 AFM images of adsorbed films on stainless steel plate. (a) On the stainless steel surface in HA solution, little adsorbed molecules are observed. (b) The addition of γ-globulin to HA solution, globular molecules of γ-globulin appear to adsorb on stainless steel surface, although some hyaluronate molecules may adsorb on surface.
Modified from Murakami, T., Sawae, Y., Higaki, H., Ohtsuki, N., & Moriyama, S. (1997). The adaptive multimode lubrication in knee prostheses with artificial cartilage during walking. *Tribology Series*, *32*, 371–382. https://doi.org/10.1016/s0167-8922(08)70466-8.

adsorbed films can protect the rubbing surfaces and reduce the shearing resistance between adsorbed films. In contrast, in severe mixed lubrication with significant surface-to-surface contact or boundary lubrication mode for polyurethane, the interaction between adsorbed molecules on rubbing surfaces can enhance frictional resistance. Therefore, it is suggested that PVA hydrogels show superior frictional behavior in walking condition.

Detailed features of adsorbed films are mentioned in 5.4.

Under various daily activities, artificial hydrogel cartilage may have significant surface-to-surface contact. For such severe conditions, the strength and tribological performance should be improved. An improving method for properties of PVA hydrogel is to control the nanoscopic structure, to reinforce the structure with fibrous element, and thus to enhance biphasic properties for efficient lubrication and load-carrying property, which is described in Section 5.3.3. The PVA hydrogel based on hydrogen bonding network as physical bonding prepared by simple methods without any additives or irradiative stimulus, such as repeated freezing-thawing method and cast-drying method has superiority in safety in human body. The further improvement of key properties of PVA hydrogel is expected. The low permeability with compaction effect (strain-depending compaction) is another requirement for superior interstitial fluid pressurization in biphasic hydrogel.

As other method to enhance the strength of hydrogel materials, the chemical crosslinking treatments were applied. Various semi-interpenetrating network (SIPN) hydrogel materials were synthesized by polymerizing a mixture of lightly crosslinked, vinyl monomers around a reinforcing polymer. SIPN hydrogel of N-vinyl-pirrolidone (NVP) and methylmethacrylate (MMA) was developed as candidate materials with similar strength to natural articular cartilage (Corkhill et al., 1990). The friction and wear properties were evaluated for flat specimens of different reinforcing polymers of polyurethane (pellethane) and cellulose acetate (CA), i.e., NVP-MMA-pellethane (Young's modulus: $E = 21$ MPa, water content: 40 wt%) and NVP-MMA-CA ($E = 23$ MPa, water content: 50 wt%) (Murakami et al., 2000). SIPN hydrogel showed higher friction than PVA hydrogel with high water content in simulator tests. In reciprocating tests at high-load condition, SIPN exhibited higher wear than PVA hydrogel with a high water content. These results suggest the existence of optimum conditions of elastic modulus and water content for low friction and minimum wear. Furthermore, the importance of surface lubricity is noticed.

High-strength candidate material is double network (DN) hydrogel composed of two kinds of hydrophilic polymers, which have higher strength but high water content (Gong et al., 2003). DN hydrogel composed by the combination of stiff and brittle first network and soft and ductile second network. The DN hydrogel materials have 0.4−0.9 MPa elastic modulus and exhibit the compressive fracture strength as high as a few to several tens of MPa, despite of containing 90 mass% water. The high strength of the DN gels is not due to a linear combination of two component networks, like the common IPN or fiber-reinforced hydrogels, but due to a nonlinear effect of binary structure. An example is shown in Fig. 5.7 for PAMPS poly-(2-acrylamido-2-methylpropane sulfonic acid)-1 (molar monomer concentration) −4 (crosslinker concentration, mol%)/PAAm (poly(acrylamide)) -2-0.1 DN gel. Furthermore, DN gels exhibit very low friction coefficient as 10^{-3}, when two gel surfaces have a repulsive interaction each other.

The DN gel composed of PAMPS/PDMAAm (poly-(N,N′-dimethyl acrylamide)) was applied for localized lesions in animal knee joints as hemiarthroplasty (Arakaki

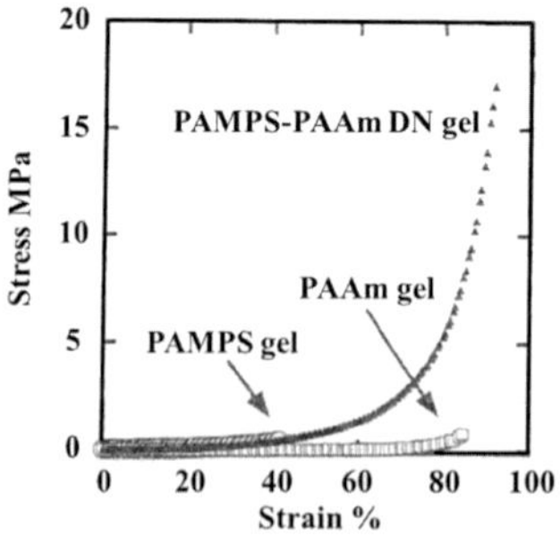

Figure 5.7 Stress-strain curves for hydrogels under uniaxial compression. *Circles*: PAMPS-one to four SN gel (Water content: 90 mass%), Squares: PAAM-2-0.1 SN gel (Water content: 90 mass%), *Triangles*: PAMPS-1-4/PAAm-2-0.1 DN gel (Water content: 90 mass%). *SN*: single net.
Modified from Gong, J. P., Katsuyama, Y., Kurokawa, T., & Osada, Y. (2003). Double-network hydrogels with extremely high mechanical strength. *Advanced Materials*, *15*(14), 1155−1158. https://doi.org/10.1002/adma.200304907.

et al., 2010). This showed the possibility of successful application with little or no visible wear of articular cartilage, although the evaluations for longer mechanical and biological durability under various operating conditions are required.

The various applications of compliant artificial cartilage as articulating surface in total hip or knee prostheses, or hemiarthroplasty have been developed.

5.3 Role of biphasic lubrication in hydrogel artificial cartilage

Hydrogels such as natural articular cartilage and PVA hydrogels with high water content exhibit a time-dependent biphasic behavior due to the simultaneous coexistence of solid and liquid phases under compressive loading or deformation (Mow et al., 1980; Mow et al., 1990). In usual biphasic models, it is assumed that the interstitial fluid flow follows Darcy's law as shown below:

$$Q = \kappa \, \mathrm{d}p/\mathrm{d}h \tag{5.2}$$

where Q is volume flow rate per unit cross-sectional area, κ is permeability, p is pressure, and h is distance. Thus, the pressure differences in the fluid phase cause flow.

In biphasic materials, the permeability and pressure gradient control the interstitial fluid pressurization, fluid flow, and deformation. Articular cartilage with low permeability can considerable fluid pressurization to contribute to load support.

In this section, some important cases on biphasic lubrication in artificial hydrogel cartilage are shown.

5.3.1 *Measurement of interstitial fluid pressure*

The superior lubrication performances in PVA hydrogels with high water content were shown from the viewpoints of soft-EHL effect in Fig. 5. 2 (Murakami et al., 1997) and effective adsorbed film formation in mixed lubrication in Figs. 5.5 and 5.6 (Murakami et al., 1997). Subsequently the lubrication mechanisms of PVA hydrogel were discussed on the basis of biphasic finite element analyses that the fluid load support by biphasic PVA hydrogels can control their frictional behaviors (Murakami et al., 2014; Murakami, Sakai, et al., 2015; Murakami, Yarimitsu, et al., 2015), but the in situ measurement of interstitial fluid pressure had not yet been performed. In this section, the results using microelectromechanical system (MEMS) pressure sensor by Sakai et al. (2018) are shown.

5.3.1.1 *Pressure measurement*

In this experiment for sliding pair of cylindrical layered PVA hydrogel (repeated freeze-thawing gel, 2 mm thickness layer adhered on impermeable semi-cylindrical substrate, outer radius: 18 mm, width: 20 mm) moving specimen and acrylic flat stationary specimen embedded with a small MEMS pressure sensor (FISO technologies

Inc., Canada: FOP-M), friction and fluid pressure in contact area were simultaneously measured in wide reciprocating speeds from 1.56 mm/s in the boundary lubrication mode to 400 mm/s in fluid film lubrication one. The pressure measurement hole for the pressure sensor of 1.6 mm diameter was 0.6 mm diameter at the acrylic sliding surface. Normal load of 1.8 N was applied and then the average contact pressure was estimated as 0.03 MPa for contact width of 3 mm. Stroke was 38.6 mm. As lubricant, 3.3 mass% PVA solution was used, and its viscosity was 20 mPa·s at 23°C.

Examples of measured pressure distribution at low- and high-speed conditions are shown in Fig. 5.8. The fluid pressure in Fig. 5.8a at 3.125 mm/s is considered to be derived from the interstitial fluid pressure, and the position of the pressure peak shifts

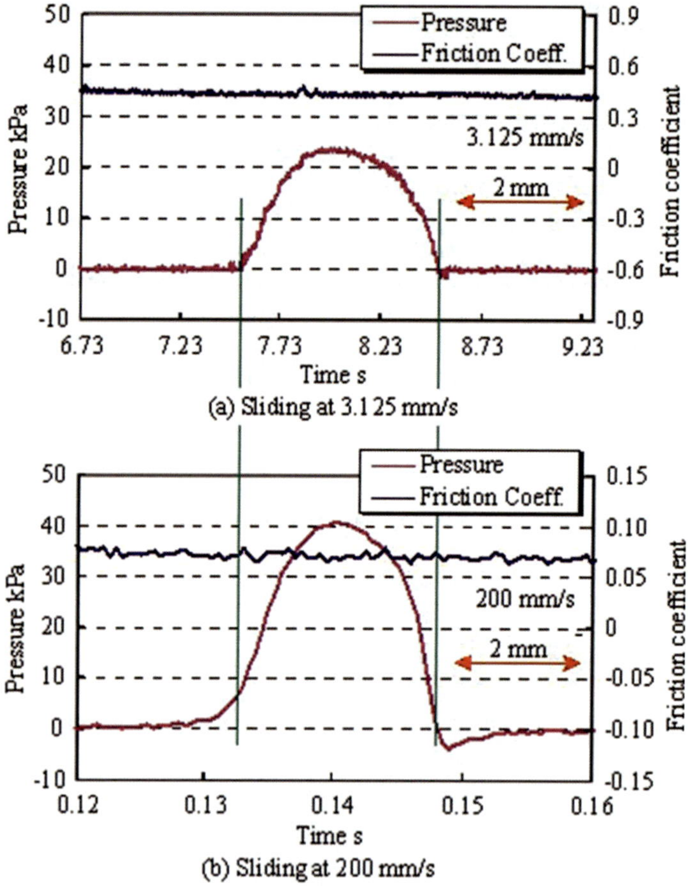

Figure 5.8 Pressure transition of the fluid in the same position and length at different speeds and friction coefficient. (a) Pressure at 3.125 mm/s is generated by the interstitial fluid pressurization. (b) Pressure at 200 mm/s is generated by the hydrodynamic fluid pressurization. The widths of the graphs were adjusted by the same 8 mm. The absolute positions were arranged by the position encorder in the experimental apparatus.
Modified from Sakai, N., Yarimitsu, S., Sawae, Y., Komori, M., & Murakami, T. (2018). Transitional behavior between biphasic lubrication and soft elastohydrodynamic lubrication of poly(vinyl alcohol) hydrogel using microelectromechanical system pressure sensor. *Biosurface and Biotribology*, *4*(1), 24–33. https://doi.org/10.1049/bsbt.2018.0001.

a little to the forward position in the contact area which is indicated by two green vertical lines. On the contrary, in Fig. 5.8b at higher speed of 200 mm/s, the pressure generation by the entraining effect in the leading zone and the negative pressure in the trailing zone are observed as a feature of hydrodynamic lubrication modes. Thus, the features of pressure profiles by in situ pressure measurement indicate the interstitial fluid pressurization at low speed and the hydrodynamic fluid pressurization at high speed.

5.3.1.2 Biphasic fluid load support and estimation of friction

Experimental results for friction coefficient and the proportion of fluid load support to total load and calculated friction coefficient by load partitioning of biphasic lubrication are shown in Fig. 5.9. The effective friction coefficient μ_{eff} can be estimated for μ_{eq} as equilibrium coefficient of friction achieved when the interstitial fluid pressure has subsided, using the following formula by Ateshian (2009).

$$\mu_{eff} = \mu_{eq}(1 - (1 - \phi)W^p / W) \tag{5.3}$$

where W is the total load, W^p is the load support by fluid pressure and ϕ is the solid-to-solid contact area fraction in microscopic view.

As shown in Fig. 5.9a. the friction increases with decreasing sliding speed at lower speed region < 100 mm/s. The friction level at lower speed range is considerably high compared to the previous reports on PVA hydrogels (Murakami et al., 1997, 2000). In this study special sliding pair with high friction property (solid-to-solid friction coefficient: around 1.0) (Stoimenov et al., 2013) for PVA hydrogel and acrylic specimens were prepared to highlight the function of the biphasic lubrication mode near hydrodynamic lubrication. Acrylic resin plate caused very high adhesion to the PVA hydrogel.

To examine the cause of sliding speed dependency of friction, the proportion of the fluid load support to total load was calculated by the integration of measured pressure (Fig. 5.9b). In the higher speed region than 200 mm/s, it is confirmed that almost all of the load is supported by the fluid pressure as indicated with the value of 1.0 for the ratio of fluid load support force to total load. In lower speed region than 10 mm/s, the proportion of fluid load support ranged from 0.4 to 0.8. It is considered that the fluid load support at lower speed region is generated by interstitial fluid pressure in biphasic lubrication mechanism.

Next, the effective friction coefficient was estimated by Formula (5.3) with the data for proportion of fluid load support in Fig. 5.9b. As shown in Fig. 5.9c, larger fluid load support results in lower friction coefficient. It is noticed that the calculated effective friction coefficient shows good agreement with the measured data in Fig. 5.9a, although the calculated effective friction coefficient does not include the friction drag of Couette flow in hydrodynamic lubrication.

It is shown in this reciprocating test for PVA hydrogel and acrylic flat plate embedded with MEMS pressure sensor that the significant part of load is supported by the interstitial fluid pressure of biphasic surface in thin film lubrication regime at a low sliding speed. In mixed lubrication regime at lower sliding speed <100 mm/s, the fluid pressure for supporting the load is considered to be generated by the

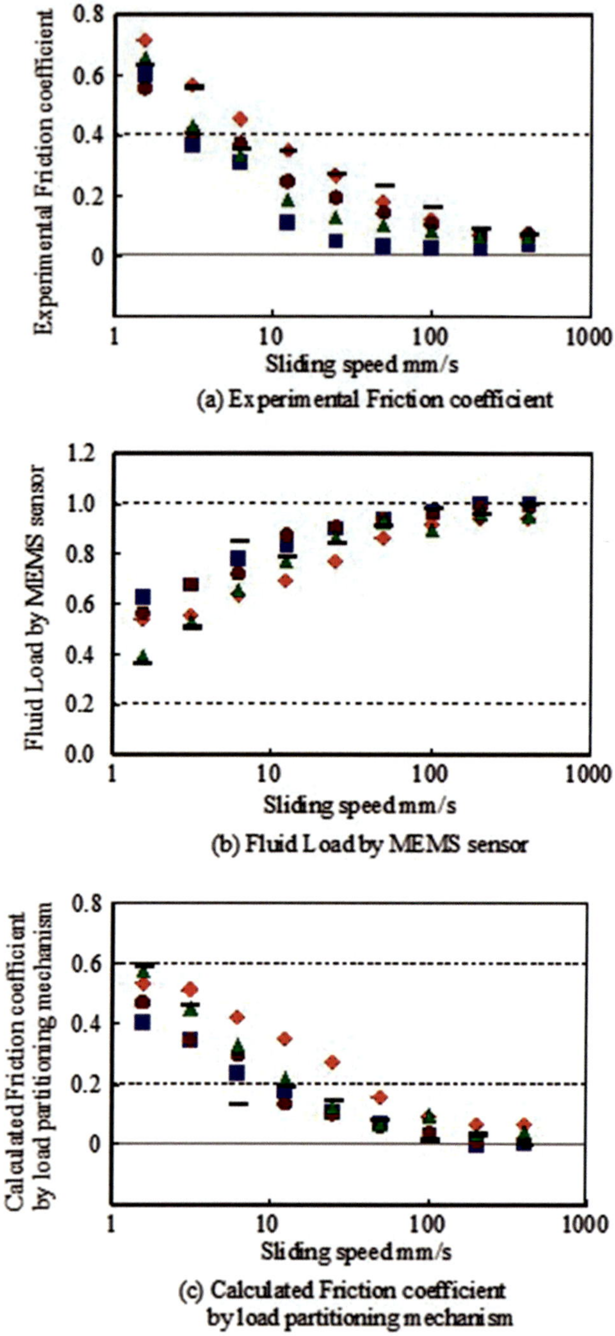

Figure 5.9 Sliding velocity dependency of friction coefficient, proportion of fluid load support to total contact load and calculated friction coefficient based on load partitioning. (a) Friction coefficient in experiment (b) Proportion of fluid load support to total contact load (c) Calculated friction coefficient based on load partitioning mechanism of biphasic mechanism in N = 5. The same marks correspond to the same experimental specimen. Modified from Sakai, N., Yarimitsu, S., Sawae, Y., Komori, M., & Murakami, T. (2018). Transitional behavior between biphasic lubrication and soft elastohydrodynamic lubrication of poly(vinyl alcohol) hydrogel using microelectromechanical system pressure sensor. *Biosurface and Biotribology*, *4*(1), 24–33. https://doi.org/10.1049/bsbt.2018.0001.

coexistence of both the interstitial fluid pressure of biphasic lubrication and the hydrodynamic lubrication pressure.

In their study (Sakai et al., 2018), a similar test was performed for alternate specimen of the impermeable surface model in which the sliding surface of PVA hydrogel was covered by a low-density polyethylene film of 50 μm thickness. The impermeable specimens exhibited lower fluid load support ratio but lower friction than biphasic permeable PVA specimens in the middle sliding speed from around 10 to 100 mm/s in mixed lubrication regime. Although further investigation is required for the clarification of flowing behaviors of fluid between rubbing surfaces in both cases, the selection of optimum permeability and surface lubricity of hydrogels are suggested for improvement of friction property.

5.3.2 Role of biphasic properties in PVA hydrogels different in manufacturing methods and structures

5.3.2.1 PVA hydrogel specimens and method of reciprocating tests

It is expected that the application of artificial hydrogel cartilage with similar properties to natural articular cartilage brings appropriate solution to solve the friction and wear problems by improvement of lubrication mechanism. In this section, as artificial cartilage specimens, four kinds of PVA hydrogels different in manufacturing methods and structures were prepared. Friction and wear behaviors for these hydrogels were examined in reciprocating test, and the differences in friction were evaluated on the basis of the biphasic finite element (FE) analysis (Murakami, Sakai, et al., 2015; Murakami, Yarimitsu, Sakai, Nakashima, Yamaguchi, Sawae, & Suzuki, 2017).

Four kinds of PVA hydrogel sheet specimens were prepared via repeated freeze-thawing (FT), cast-drying (CD), and hybrid (a combination of FT and CD) methods as described below.

1. Repeated freeze-thawing (FT) gel (Fig. 5.10a)

PVA powder (type A) (polymerization degree: 2000, average degree of saponification: 98.5-99.4 mol%, Kishida Chemical Co. Ltd., Japan) at 20 mass% was dissolved in pure water and heated at 95°C for more than 3 h in autoclave and cast in a mold and then gelled by repeated freeze-thawing (for 10 h at −20°C to for 20 h at 4°C) method (Nambu, 1982). Number of freeze-thawing cycles was 5 times as optimum condition to establish maximum tensile strength and stiffness (Nakashima et al., 2001). The elastic modulus of PVA hydrogel is 1.2 MPa and equivalent water content is 79%. As shown in Fig. 5.10a, PVA FT gel is opaque, and has porous, fibrous, and rough surface (arithmetic average roughness: $R_a = 0.39$ μm).

2. Cast-drying (CD) gel (Fig. 5.10b).

This method was developed by Otsuka and Suzuki (2009a,b) and Otsuka et al. (2010). An aqueous solution of PVA was obtained by dissolving 15.0 mass% of the PVA powder (type B) (polymerization degree: 1700, average degree of

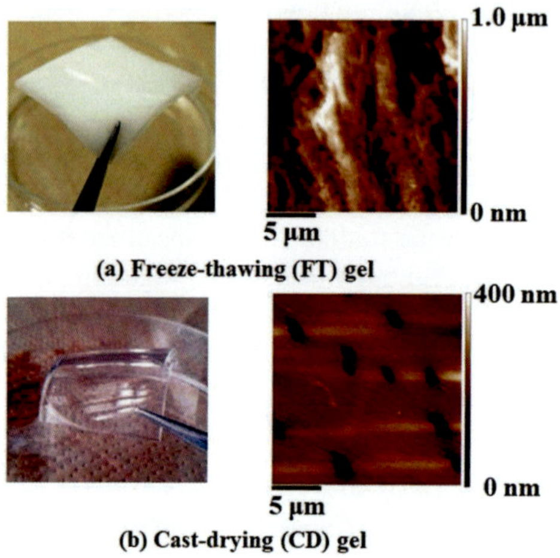

Figure 5.10 Appearances and AFM images of FT gel and CD gel. PVA FT gel is opaque and has porous, fibrous and rough surface. CD gel is transparent and has smooth surface.
From Murakami, T., Yarimitsu, S., Sakai, N., Nakashima, K., Yamaguchi, T., Sawae, Y., & Suzuki, A. (2017). Superior lubrication mechanism in poly(vinyl alcohol) hybrid gel as artificial cartilage. *Proceedings of the Institution of Mechanical Engineers, Part J: Journal of Engineering Tribology*, 231(9), 1160–1170. SAGE Publications Ltd. https://doi.org/10.1177/1350650117712881.

saponification: 98–99 mol%, Kuraray Co. Ltd. Japan) into deionized and distilled water at 90-95°C for more than 2 h, and PVA solution was cooled to the room temperature. The PVA solution was poured into a plastic dish of polystyrene and left in air at room temperature (ca. 25°C). The solution became very viscous within several days and solidified within a week or more. As indicated by Otsuka and Suzuki (2009a,b) and Otsuka et al. (2010) on the basis of the measurements using Fourier transform infrared spectroscopy (FT-IR) with attenuated total reflection (ATR) method and X-ray diffraction (XRD), the crosslinks due to hydrogen bonds and micro-crystals were formed during the dehydration process of the PVA solution. After the drying process, specimen was put into the pure water to be swollen. The equivalent water content is 73% to 78%. PVA CD gel is transparent, which indicates its uniform network structure. As shown in AFM image, it has smooth surface ($R_a = 0.013$ μm).

Schematic models for PVA gels at nano-scale (Otsuka & Suzuki, 2009a) and FT and CD gels at micro- and milli-scale are shown in Fig. 5.11, which were proposed by Suzuki and Sasaki (2015).

3. Hybrid gel as FT on CD or CD on FT

As shown in Fig. 5.11, FT gel has heterogeneous network structure composed of crystalline and amorphous regions, and thus it has porous structure and high permeability. In contrast, CD gel has uniform structure and low permeability. In hybrid method (Suzuki et al., 2013; Suzuki & Sasaki, 2015; Suzuki & Murakami, 2016), the double-layered laminated structure composed of FT and CD gel layers could be prepared.

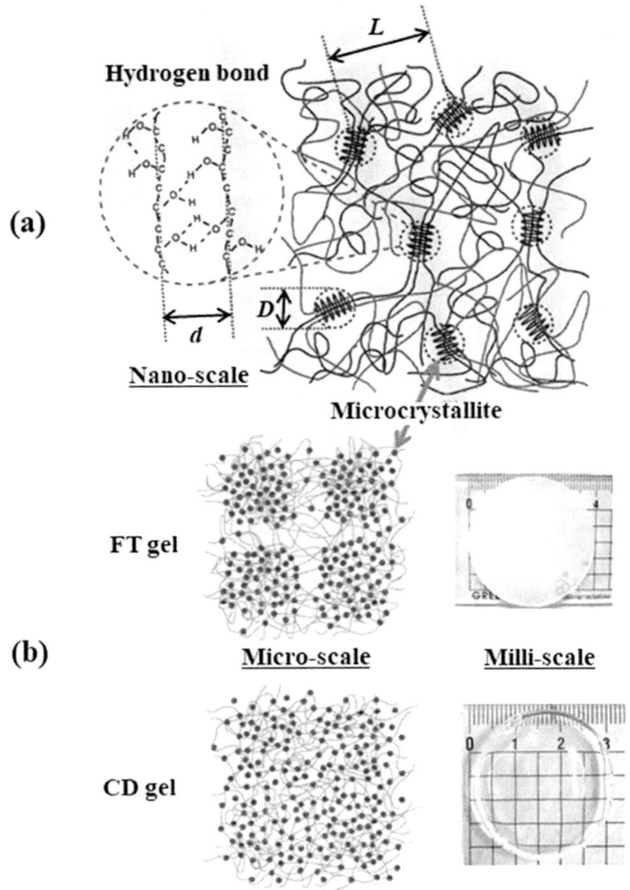

Figure 5.11 Schematic models for PVA gel structures in nano-, micro- and milli-scale. (a) Network structure of PVA gel in nano-scale: microcrystallites are physically crosslinked with hydrogen bonds. In enlarged view, single hydrogen is omitted. Three characteristic lengths, the distance between main chains in microcrystals: *d*, the size of microcrystals: *D* and the distance between microcrystals: *L* were deduced from the diffraction peaks of X-ray diffraction (XRD) spectra. (**b**) Schematic models for two kinds of hydrogel structures in micro- and milli-scales: FT gel has heterogeneous network structure composed of crystalline and amorphous regions. CD gel has uniform structure.
Modified from Suzuki, A., & Sasaki, S. (2015). Swelling and mechanical properties of physically crosslinked poly(vinyl alcohol) hydrogels. *Proceedings of the Institution of Mechanical Engineers, Part H: Journal of Engineering in Medicine*, *229*(12), 828–844. https://doi.org/10.1177/0954411915615469

First, FT gel sheet was prepared from aqueous PVA solution for 15 mass% of PVA powder (type B) by repeating 4 times freeze-thawing (for 8 h at −20°C to for 16 h at 4°C) procedures. The second process as cast-drying method was started by pouring 15 mass% PVA (type B) solution on the FT gel sheet. Then, the specimen was dried in the refrigerator. Yarimitsu, Yoshida, et al. (2016) indicated that the two-stage drying process with optimum temperature and relative humidity improved the tribological properties of hybrid gel. Therefore, hybrid specimens were prepared by two-stage drying process with the following optimum conditions. That is, temperature and relative humidity of first stage in drying process were 8°C and 50% RH for 7 days, and those in second process were 20°C and 40%RH. After the drying process, specimen was put into the pure water to be swollen. In reciprocating test, the CD layer or FT layer was arranged as a rubbing surface, while the bottom layer was fixed to the specimen holder with cyanoacrylate adhesive. Therefore, FT on CD and CD on FT hybrid specimens were prepared as shown in Fig. 5.12. The hybrid gel is semi-transparent, and surface roughness levels are different for FT and CD surfaces. The hybrid (FT on CD) gel has a little smoother surface ($R_a = 0.212$ μm) than that of FT gel, and the hybrid (CD on FT) gel has the smoothest surface ($R_a = 0.007$ μm).

The reciprocating tests were conducted for sliding pair of an upper specimen composed of a 2 mm thick, soft layer of PVA hydrogel adhered on acrylic ellipsoid (40 × 25 mm, in diameter) and glass plate. The sliding speed, stroke, and load were 20 mm/s, 35 mm, and 2.94 N, respectively. Total sliding distance was 140 m. Lubricant was saline.

Table 5.2 shows material properties estimated from the stress relaxation tests of cuboid under uni-axial compression at definite deformation (Yamaguchi & Murakami, 2012; Murakami et al., 2014).

5.3.2.2 Experimental results of reciprocating test

In saline solution, four kinds of hydrogels exhibited highly different frictional behaviors as hybrid (CD on FT) < CD < FT < hybrid (FT on CD), as shown in Fig. 5.13. It is noteworthy that the hybrid (CD on FT) gel maintained extremely

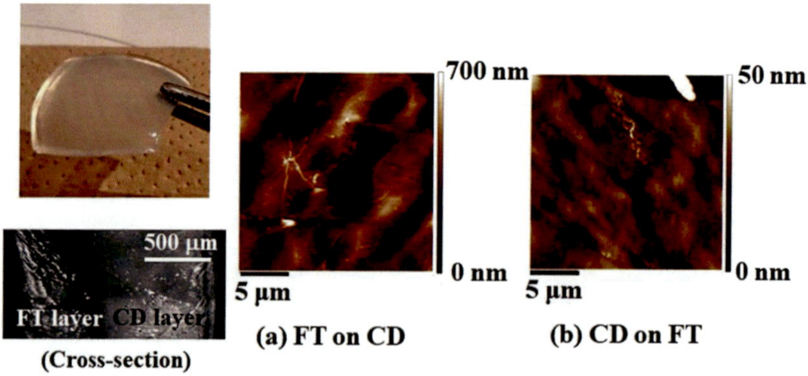

Figure 5.12 Hybrid gel and AFM images of surfaces as FT on CD and CD on FT (a) FT on CD has rougher surface than (b) CD on FT.
From Murakami, T., Yarimitsu, S., Sakai, N., Nakashima, K., Yamaguchi, T., Sawae, Y., & Suzuki, A. (2017). Superior lubrication mechanism in poly(vinyl alcohol) hybrid gel as artificial cartilage. *Proceedings of the Institution of Mechanical Engineers, Part J: Journal of Engineering Tribology*, 231(9), 1160–1170. SAGE Publications Ltd. https://doi.org/10.1177/1350650117712881.

Table 5.2 Material properties for PVA hydrogels.
CD gel has significantly lower permeability than FT gel.

Preparation method for PVA hydrogel	**Permeability**	**Young's modulus**	**Poisson's ratio**
	m⁴/Ns	**kPa**	
Freeze-thawing (FT)	2.0×10^{-13}	110	0
Cast-drying (CD)	2.4×10^{-15}	190	0.41

From Murakami, T., Yarimitsu, S., Nakashima, K., Yamaguchi, T., Sawae, Y., Sakai, N., & Suzuki, A. (2014). Superior lubricity in articular cartilage and artificial hydrogel cartilage. *Proceedings of the Institution of Mechanical Engineers, Part J: Journal of Engineering Tribology*, *228*(10), 1099−1111. https://doi.org/10.1177/1350650114530273.

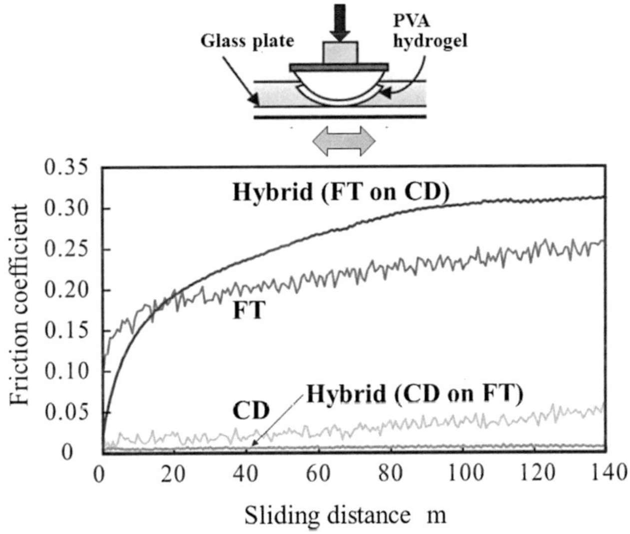

Figure 5.13 Frictional behaviors of PVA hydrogels in reciprocating test. Clear differences in time-dependent friction level are observed as hybrid (CD on FT) < CD < FT < hybrid (FT on CD).
From Murakami, T., Yarimitsu, S., Sakai, N., Nakashima, K., Yamaguchi, T., Sawae, Y., & Suzuki, A. (2017). Superior lubrication mechanism in poly(vinyl alcohol) hybrid gel as artificial cartilage. *Proceedings of the Institution of Mechanical Engineers, Part J: Journal of Engineering Tribology*, 231(9), 1160−1170. SAGE Publications Ltd. https://doi.org/10.1177/1350650117712881.

low friction coefficient 0.006 until the end of test. The CD gel showed low but slightly higher friction coefficient with gradual increase. Meanwhile, the FT gel showed initial medium friction and a gradual increase over 0.25 at final stage. Hybrid (FT on CD) gel exhibited the highest friction, although initial friction was low.

The extent of wear was evaluated by comparison of worn surfaces observed by optical microscope after the wear test with intact surfaces before wear test, as shown in Fig. 5.14. FT and CD gels in saline showed some scratches on rubbing surfaces. Minimal wear was observed on hybrid (CD on FT) gel. On the contrary, the highest wear was noticed in hybrid (FT on CD) gel.

5.3.2.3 Evaluation of frictional behaviors based on biphasic analysis

To elucidate the cause for these differences in friction and wear behaviors, the biphasic analysis was conducted for simplified model as shown in Fig. 5.15. Two-dimensional biphasic FE analysis for 1.5 mm thick PVA hydrogel layer fixed on the cylindrical

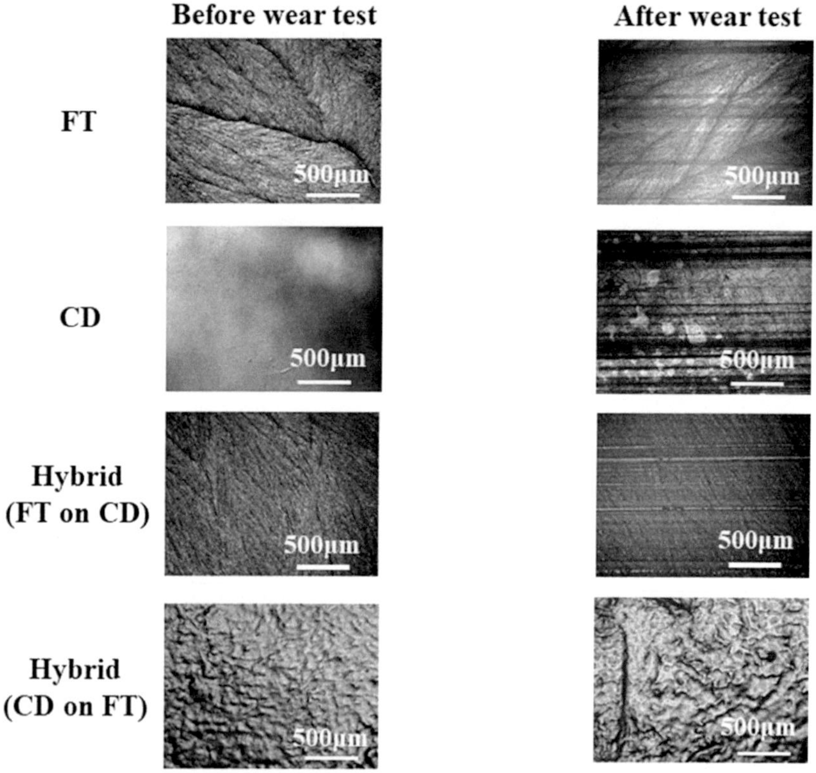

Figure 5.14 Changes in rubbing surfaces of PVA hydrogels in reciprocating test. Minimal wear was observed on hybrid (CD on FT) gel, but the highest wear was noticed in hybrid (FT on CD) gel.
From Murakami, T., Yarimitsu, S., Sakai, N., Nakashima, K., Yamaguchi, T., Sawae, Y., & Suzuki, A. (2017). Superior lubrication mechanism in poly(vinyl alcohol) hybrid gel as artificial cartilage. *Proceedings of the Institution of Mechanical Engineers, Part J: Journal of Engineering Tribology*, 231(9), 1160–1170. SAGE Publications Ltd. https://doi.org/10.1177/1350650117712881.

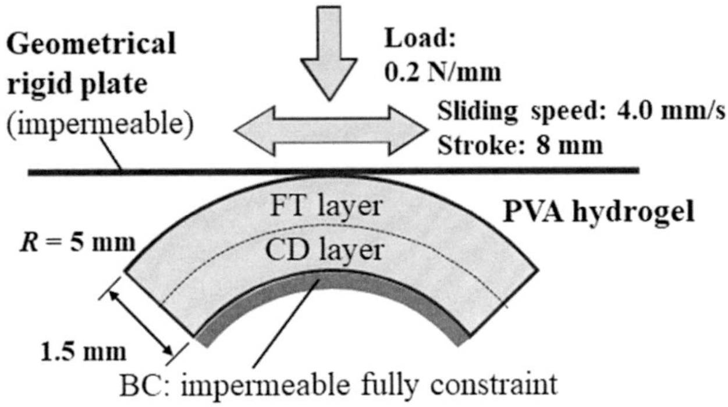

Figure 5.15 Biphasic finite element model for cylindrical PVA hydrogel. Simplified two-dimensional biphasic analysis was conducted for reciprocating test.
From Murakami, T., Yarimitsu, S., Sakai, N., Nakashima, K., Yamaguchi, T., Sawae, Y., & Suzuki, A. (2017). Superior lubrication mechanism in poly(vinyl alcohol) hybrid gel as artificial cartilage. *Proceedings of the Institution of Mechanical Engineers, Part J: Journal of Engineering Tribology*, 231(9), 1160−1170. SAGE Publications Ltd. https://doi.org/10.1177/1350650117712881.

rigid body of 3.5 mm radius was conducted under continuous constant loading by impermeable rigid plate in reciprocating motion, where biphasic FE analysis method for articular cartilage in reciprocating motion by Sakai, Hagihara, et al. (2012) was modified for PVA hydrogels. The commercial ABAQUS (6.8−4) package was applied with biphasic elements. A load of 0.2 N/mm was applied with a ramp time of 1 s by impermeable rigid plate and held constant for further 200 s during reciprocating motion. The biphasic tissue was modeled by CPE4RP (four-node bilinear displacement and pore pressure, reduced integration with hour glass control) elements and a mesh was chosen as 0.1 mm thick with 0.1 mm on surface. For hybrid gels, the thickness of FT layer was 0.8 mm and that of CD layer was 0.7 mm. The bottom of the model was fixed and impermeable. The sliding speed was 4 mm/s for a stroke of 8 mm with a period of 4 s, where the hydrodynamic action diminishes. The changes in interstitial fluid pressure and stress in solid phase during reciprocating motion were examined. The friction coefficient for solid-to-solid contact μ_{solid} between the geometrical rigid plate and the solid phase of PVA gel was set to 0.006 to 0.3 according to measured values at equilibrium in experiment as described below. The mechanical properties (permeability, Young's modulus, and Poisson's ratio) of FT and CD PVA hydrogels are shown in Table 5.1. It is noteworthy that CD hydrogel has lower permeability as two orders than FT hydrogel. The permeability of CD hydrogel has similar level to that of natural articular cartilage. For hybrid gels, corresponding material properties as constant values were used to CD and FT layers, respectively. The elastic moduli have the uniform values without depth-dependency. The permeability as a constant value was used without strain dependency.

As shown in Fig. 5.16, friction coefficients were estimated from the horizontal surface force divided by normal load in simplified two-dimensional biphasic FE analysis, in which friction coefficients for solid-to-solid contact μ_{solid} for hybrid (FT on CD),

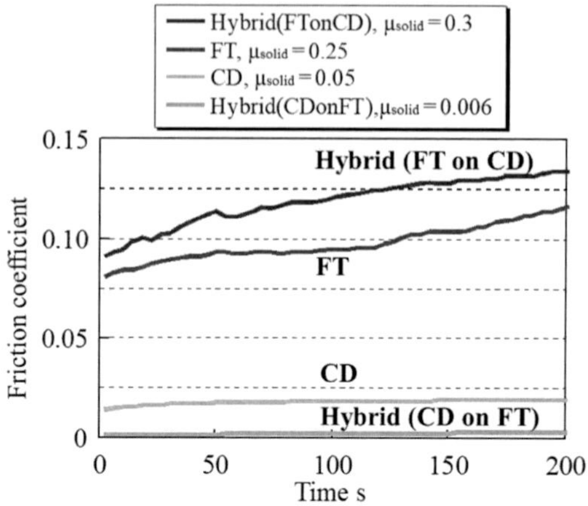

Figure 5.16 Changes in frictional behaviors for PVA hydrogels estimated by biphasic FE analysis. Time-dependent frictional behaviors estimated by biphasic FE analysis correspond to experimental behaviors in Fig. 5.13.
Modified from Murakami, T., Yarimitsu, S., Sakai, N., Nakashima, K., Yamaguchi, T., Sawae, Y., & Suzuki, A. (2017). Superior lubrication mechanism in poly(vinyl alcohol) hybrid gel as artificial cartilage. *Proceedings of the Institution of Mechanical Engineers, Part J: Journal of Engineering Tribology*, 231(9), 1160–1170. SAGE Publications Ltd. https://doi.org/10.1177/1350650117712881.

FT, CD, and hybrid (CD on FT) were set to 0.3, 0.25, 0.05, and 0.006, respectively, according to measured values at equilibrium in the experiment. The comparison of estimated time-dependent frictional behaviors based on FE analysis (Fig. 5.16) with experimental results (Fig. 5.13) clearly indicates the good correspondence, although some shifts in phase and friction level are observed.

To clarify these differences, the changes in the interstitial fluid pressure and load support by fluid phase in biphasic FE analysis were examined for immediately after loading and after 50–200 s running. As shown in Fig. 5.17, all hydrogels have similar pressurization immediately after loading. With repetition of reciprocating motion, the fluid pressure distribution changed. The CD hydrogel with low permeability maintains similar pressure distribution even after 50 s, but in FT gel with high permeability, the fluid pressure was remarkably reduced, and the loading area became larger accompanied with significant exudation of fluid from the periphery of conjunction of hydrogel surface. In hybrid gels, high pressure is maintained in each CD layer with low permeability of FT on CD and CD on FT. The difference in flow patterns of four kinds of PVA hydrogels after 50 s running is shown in Fig. 5.18, which have mutual relation with the changes in the interstitial fluid pressurization in Fig. 5.17.

The changes in load supports by fluid and solid phases are exhibited in Fig. 5.19. Immediately after loading, all gels showed similar load support by fluid phase from 70.1% to 73.9%, but after 50 s, the load support by fluid phase gradually decreased in three kinds of gels except hybrid (CD on FT) gel. In hybrid (CD on FT) gel, the load support by fluid phase was almost completely maintained until 50 s. After more than 50 s, its fluid load support was remarkably reduced

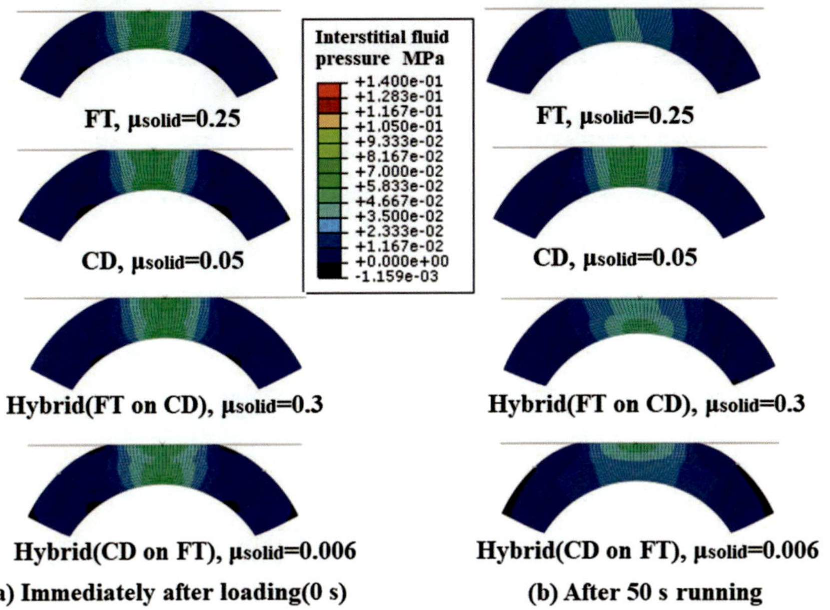

Figure 5.17 Changes in interstitial fluid pressure of PVA hydrogels in biphasic FE analysis. All PVA hydrogels show similar interstitial fluid pressurization immediately after loading, but the differences in permeability of FT and CD gels change the corresponding interstitial fluid pressurization in repetition of reciprocating motion.
Modified from Murakami, T., Yarimitsu, S., Sakai, N., Nakashima, K., Yamaguchi, T., Sawae, Y., & Suzuki, A. (2017). Superior lubrication mechanism in poly(vinyl alcohol) hybrid gel as artificial cartilage. *Proceedings of the Institution of Mechanical Engineers, Part J: Journal of Engineering Tribology*, 231(9), 1160–1170. SAGE Publications Ltd. https://doi.org/10.1177/1350650117712881.

As artificial hydrogel cartilage candidate materials containing high water content, four kinds of PVA gels showed quite different time-dependent frictional behaviors and wear properties in reciprocating tests in saline (Figs. 5.13 and 5.14). The differences in time-dependent frictional behaviors estimated on the basis of biphasic FE analysis (Fig. 5.16) showed quite similar properties to experiment (Fig. 5.13). As indicated in Formula (5.3) for friction by Ateshian (2009), both high partitioning of load support by fluid phase and low equilibrium friction coefficient: μ_{eq} (= friction coefficient for solid-to-solid contact: μ_{solid}) contribute to lowering in friction. Although time-dependent increase in friction corresponded to decreasing of fluid load support, the friction level was not only by load support by fluid phase but by the level of surface friction coefficient μ_{solid} for contact of solid phase. Therefore, it is pointed out that both biphasic property and surface lubricity are important to maintain superior lubrication.

The time-dependent friction increase from the initial medium level of FT gel is considered to be due to biphasic properties with high permeability (Table 5.1). The interstitial fluid pressure in FT gel gradually diminished as shown in Fig. 5.17, and

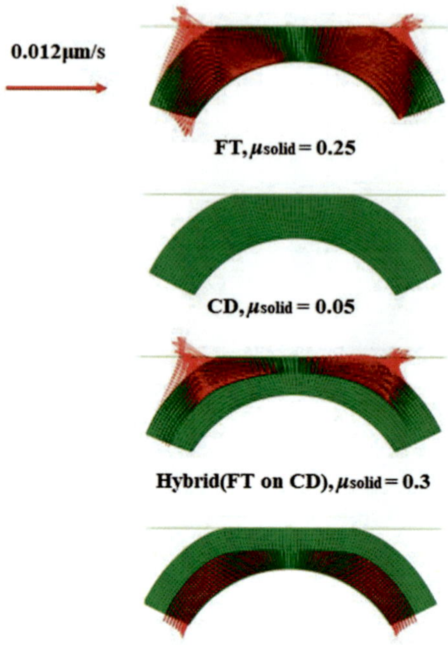

Figure 5.18 Changes in fluid flow for PVA hydrogels after 50 s running in biphasic FE analysis. Large fluid flow occurs in gel zone driven along higher pressure gradient. Modified from Murakami, T., Yarimitsu, S., Sakai, N., Nakashima, K., Yamaguchi, T., Sawae, Y., & Suzuki, A. (2017). Superior lubrication mechanism in poly(vinyl alcohol) hybrid gel as artificial cartilage. *Proceedings of the Institution of Mechanical Engineers, Part J: Journal of Engineering Tribology*, 231(9), 1160−1170. SAGE Publications Ltd. https://doi.org/10.1177/1350650117712881.

therefore, the fluid load support became small as 54.1% after 200 s from initial 71.9% (Fig. 5.19). The friction in saline is expected to increase gradually as indicated by the curve for FT gel with $\mu_{solid} = 0.25$ (Fig. 5.16). In contrast, CD gel with low permeability showed slower decreasing of load support by fluid phase to 61.4% after 200 s from initial 72.5% (Fig. 5.19). The estimated friction for CD gel increases gradually from initial low friction but is limited at low friction level because of low surface friction $\mu_{solid} = 0.05$ for solid-to-solid contact.

For hybrid gels, CD on FT gel showed extremely low friction and minimal wear. In contrast, hybrid (FT on CD) gel showed very high friction and high wear, which has similar friction and wear properties to FT gel but become worse, irrespective of smoother surface roughness. For common properties of FT and hybrid (FT on CD) gels, FT gels of porous and heterogeneous network structure have rough surface and high permeability, which may enhance interaction between rubbing surfaces in saline in mixed lubrication and decreases fluid load support. FT surfaces appear to have lower surface lubricity including hydration lubrication ability. Further research is

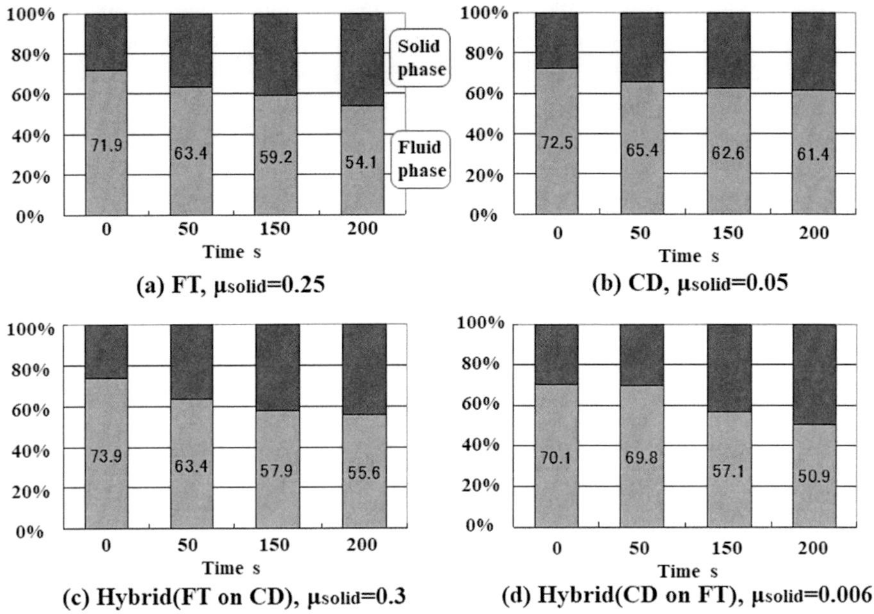

Figure 5.19 Changes in load support by fluid phase for PVA hydrogels in biphasic FE analysis. The values of fluid load support immediately after loading are similar in all PVA hydrogels, but they decrease differently with repetition of reciprocating motion.
Modified from Murakami, T., Yarimitsu, S., Sakai, N., Nakashima, K., Yamaguchi, T., Sawae, Y., & Suzuki, A. (2017). Superior lubrication mechanism in poly(vinyl alcohol) hybrid gel as artificial cartilage. *Proceedings of the Institution of Mechanical Engineers, Part J: Journal of Engineering Tribology*, 231(9), 1160–1170. SAGE Publications Ltd. https://doi.org/10.1177/1350650117712881.

required to clarify the reason why the hybrid (FT on CD) gel showed higher friction coefficient as 0.3 and higher wear than those of FT gel.

On low friction coefficient as 0.05 of CD gel and extremely low friction coefficient as 0.006 of hybrid (CD on FT) gel even after lowering of fluid load support, it is expected that another lubrication mechanism as surface lubricity of hydrogels enables to reduce friction remarkably (Fig. 5.13). In this reciprocating test in saline without lubricating synovia constituents, the effective role by boundary lubrication is not taken into consideration. Therefore, as other lubrication mechanism, hydration lubrication mechanism and/or gel film lubrication is likely to reduce friction. The transparency of CD hydrogel indicates its uniform network structure. Furthermore, CD gel has low permeability, and the fluid retained inside of gel with slow flows out, and thus the interstitial fluid pressure maintains for longer time. As shown in this study, hybrid (CD on FT) gel exhibited lower friction coefficient of 0.006 and lower wear than CD gel. In measurement by AFM, CD surface of hybrid (CD on FT) gel has smoother surface ($R_a = 0.007$ μ m) than usual CD gel ($R_a = 0.013$ μ m) and probably superior hydration lubrication ability derived from ingenious lubricity at micro- and nano-levels as discussed by Yarimitsu, Yoshida, et al. (2016). CD layer with smooth rubbing surface in hybrid (CD on FT) gel prepared by two-stage drying process with optimum temperature and relative humidity, which has non-cross-linked protruded PVA molecular chains with optimum extent of cross-linking, may contribute to extremely low friction and minimum wear. It appears that the optimized gel structure can contribute to the sustaining of the superior lubrication properties.

5.3.3 Is fiber-reinforcement effective?

It is predicted for natural articular cartilage that the collagen fiber network and its arrangement are effective in improvement of interstitial fluid pressurization and reduction of friction by a biphasic lubrication mechanism (Sakai, Hagihara, et al., 2012; Sakai, Hosoda, et al., 2012; Sakai et al., 2016). From the biomimetic viewpoint, Sakai et al. (2019) exhibited the effectiveness of fiber reinforcement in FT-PVA hydrogel on interstitial fluid pressurization and friction reduction by sliding FE simulation and reciprocating ball-on-plate test.

A PVA hydrogel sheet was prepared by a repeated freeze-thawing method. Fibre-reinforced (FR) PVA specimen was prepared by pouring PVA solution into acrylic casting mold in which PVA fibers had been arranged with 1 mm mesh interval at either specified one layer or two layers. Material properties in compression test of PVA specimen were identified by FE analysis for model with CPE4RP elements as a commercial package ABAQUS V6. Normal biphasic PVA model with viscoelastic solid phase without strain-dependent permeability was selected. The fiber layer was represented by SPRINGA element (axial spring between two nodes). The stiffness of PVA fiber was estimated by fitting the fiber-reinforced FE model to the time-depending experimental results of the FR-PVA specimen. Fig. 5.20 shows the curve-fitting result of the fiber-reinforced FE model and the experimental compression test result of the FR-PVA hydrogel sheet of 2 mm thickness. Additionally, it is indicated that fiber reinforcement is effective to enhance the stiffness to compression load and interstitial fluid pressurization in comparison with those for normal PVA without fiber.

A two-dimensional FE sliding analysis was conducted under migrating contact condition for the reciprocating impermeable geometrical rigid cylinder of 13 mm radius on PVA hydrogel flat plate of 2 mm thickness at load of 0.2 N/mm, at 8 mm/s and 25 mm stroke. The contours of von Mises stress and interstitial fluid pressure for no-fiber PVA model and two FR-PVA models at 60% phase of the second forward sliding are shown in Fig. 5.21.

It is noticed that the fiber reinforcement enhances the interstitial fluid pressurization, and the middle two layers model (Fig. 5.21b) generated a higher interstitial fluid pressure over the whole depth compared with the single surface layer model (Fig. 5.21c). The peak fluid pressure at the surface for the single surface layer model is a little lower, but the contact area is larger than the middle two layers model, therefore, attains the highest load support of 88.4% and the lowest effective friction coefficient μ_{eff} of 0.029. The values of μ_{eff} for the middle two layers and no-fiber models are 0.047 and 0.063, respectively. Here, the effective friction coefficient μ_{eff} was calculated by formula (5.3) as $\mu_{\text{eq}} - \mu_{\text{solid}} = 0.25$ and $\phi = 0$. The small difference between

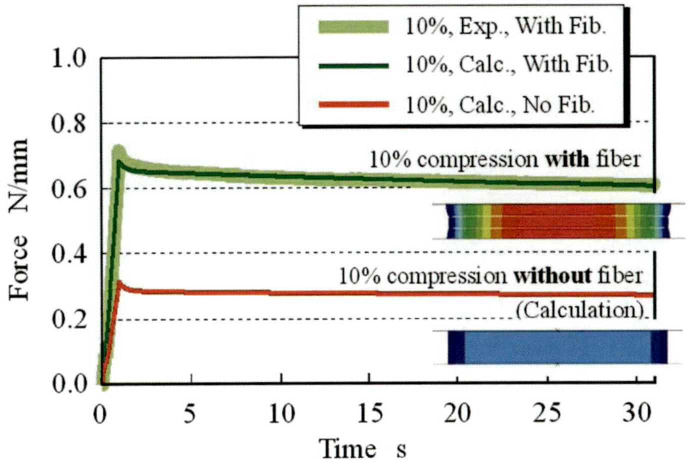

Figure 5.20 Curve-fitting results to estimate the fiber stiffness of FR-PVA hydrogel. 10% compression was given in 1 s. Force-time curve for normal PVA hydrogel without fiber and contours for pore pressure for both hydrogels are added.
From Sakai, N., Yarimitsu, S., Sawae, Y., Komori, M., & Murakami, T. (2019). Biomimetic artificial cartilage: Fibre-reinforcement of PVA hydrogel to promote biphasic lubrication mechanism. *Biosurface and Biotribology*, *5*(1), 13−19. https://doi.org/10.1049/bsbt.2018.0031.

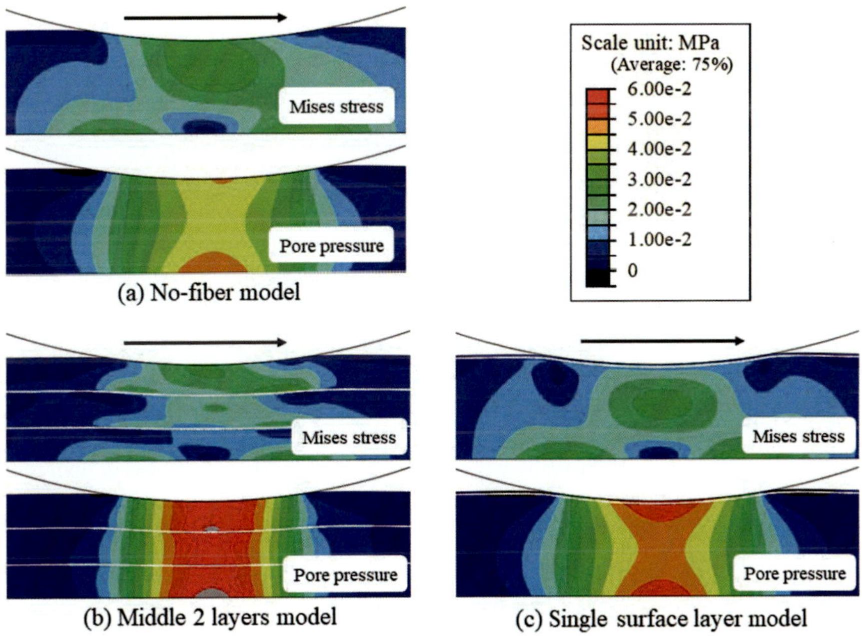

Figure 5.21 Contour of von Mises stress and fluid pore pressure in migrating sliding of three kinds of FT-PVA models. (a) No-fiber model (b) Middle two layers model (depth of fiber mesh as *white line*: 0.65 and 1.35 mm from the surface) (c) single surface layer model (depth of fiber mesh as *white line*: 0.05 mm from the surface) Mises stress as the solid phase of single surface layer model is lower than that of middle two layers model. Superficial pore pressures of the single fiber layer is a little lower than that of middle two layers model, but the contact length is larger.
Modified from Sakai, N., Yarimitsu, S., Sawae, Y., Komori, M., & Murakami, T. (2019). Biomimetic artificial cartilage: Fibre-reinforcement of PVA hydrogel to promote biphasic lubrication mechanism. *Biosurface and Biotribology*, *5*(1), 13−19. https://doi.org/10.1049/bsbt.2018.0031.

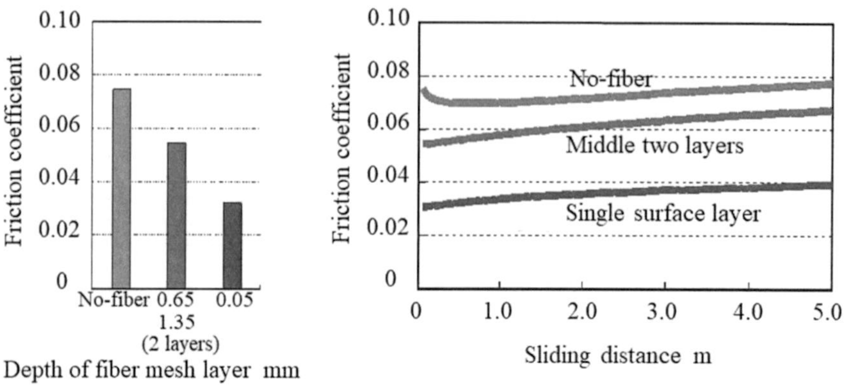

(a) Sliding biphasic FE simulation **(b) Experimental sliding test in pure water**

Figure 5.22 Comparison of friction coefficients obtained by FE analysis and experiment. (a) Sliding FE analysis, (b) experimental sliding test in pure water In the prediction by biphasic FE analysis and experimental sliding test, the fiber reinforcement of PVA hydrogels reduced friction coefficient. Single surface fiber specimen showed lowest friction.
Modified from Sakai, N., Yarimitsu, S., Sawae, Y., Komori, M., & Murakami, T. (2019). Biomimetic artificial cartilage: Fibre-reinforcement of PVA hydrogel to promote biphasic lubrication mechanism. *Biosurface and Biotribology, 5*(1), 13−19. https://doi.org/10.1049/bsbt.2018.0031.

the friction coefficients obtained from the sliding FE simulation (Fig. 5.22a), and the calculated effective friction coefficient could be from the plowing friction in migrating contact condition. The von Mises stress of the single surface layer model was lowest because the horizontal strain was constrained by the surface horizontal fiber.

The results of sliding biphasic FE simulation and experimental sliding test were shown in Fig. 5.22. The friction coefficient in sliding FE analysis (Fig. 5.22a) is a little higher than μ_{eff} because it includes plowing term in migrating sliding. The frictional behaviors in experimental sliding test as alumina ceramic ball of 13 mm radius on PVA flat specimen at load of 6 N, at 8 mm/s and 25 mm stroke in pure water are shown in Fig. 5.22b. Friction coefficients in experiment gradually increase with sliding distance, but friction levels correspond to those in biphasic FE analysis. Effectiveness of fiber reinforcement for PVA hydrogel predicted by sliding FE analysis was confirmed in experimental sliding test. Fiber reinforcement of PVA hydrogel successfully reduced friction.

In this study, with a single layer model, the surface fiber model showed the highest reduction of friction in sliding FE simulation. The experimental sliding test confirmed the prediction of FE analysis. In production process of fiber-reinforced PVA hydrogel specimen, it should be considered that it is necessary to prevent initial relaxation and preserve the initial tension of embedded fiber.

5.4 Roles of boundary lubrication

On hydrogel surfaces rubbed in the mixed or boundary lubrication regime, mild or severe direct contacts occur depending on the rubbing severity. To reduce friction and to protect the compliant hydrogel surfaces, adsorbed films or some lubricating layers should be formed on the surfaces. The synovial fluid as a lubricant in synovial joint is regarded as an aqueous electrolyte solution containing proteins, lipids, and hyaluronic acid. It is expected that most of the important constituents in lubricant can play protecting roles with low friction, but some active constituents may act as adverse adhesive role with high friction. First, the phenomena with proteins are described.

As shown in Fig. 5.5 for simulator testing of knee prosthesis with PVA hydrogel layer, friction is reduced if the protein adsorbed film is effective in protection of rubbing surface. However, high friction and high wear can occur if the physical and chemical properties of adsorbed film, rubbing materials, and lubricants and the severity of the operating conditions change. On serum proteins in synovial fluid, it is indicated that the albumin as the most abundant protein acts as a boundary lubricant in joint prostheses (Heuberger et al., 2005), and the γ-globulin is effective as a boundary lubricant in natural synovial joints (Higaki & Murakami, 1996; Murakami et al., 1998). Important cases where albumin and γ-globulin were selected as representative protein constituents in synovial fluid are shown below. As estimated by the circular dichroism method and Fourier transform infrared (FT-IR) spectroscopy, albumin contains a large amount of α-helix structure (60%–67%) (Sethuraman et al., 2004; Wittemann & Ballauff, 2004), and γ-globulin contains a large amount of β-sheet structure (76%) (Vermeer et al., 1998) in the secondary structure.

In reciprocating test of FT-PVA hydrogel of high water content against itself lubricated with hyaluronate (HA, molecular weight: 9.2×10^5) solution with proteins under thin film conditions at constant load, the lubricant constituents and protein concentration gave important influence on wear property of PVA hydrogel as shown in Fig. 5.23 (Nakashima et al., 2004, 2005). It is noticed that wear of PVA increased with an increase in protein concentration in HA solutions containing single protein, but at certain combination of albumin and γ-globulin in HA solutions containing both albumin and γ-globulin, the wear was remarkably reduced. At appropriate total protein concentration of 2.1 mass% as 1/2 or 2/1 ratio of albumin and γ-globulin (A/G ratio), the wear of PVA was remarkably reduced as minimum values. In HA solution of 2.1 mass% as 1/1 AG ratio, wear was little reduced. At total protein concentration of 2.8 mass%, the destructive severe wear was observed.

It is considered that the rising of wear grade is caused by adverse adhesive action between adsorbed protein molecules on both rubbing hydrogel surfaces, and the reducing of wear grade is brought due to the protection by protein adsorbed films with low friction. Therefore, to identify the roles of the proteins, further reciprocating tests for the sliding pair of a stationary elliptical PVA specimen and a reciprocating glass plate as hydrophilic surface were performed at severe thin film condition at load of 2.94 N, sliding speed of 20 mm/s, and sliding distance of 3 km. After testing, the adsorbed films on glass plate were observed with fluorescent method (Nakashima et al., 2005). Albumin and γ-globulin were dyed by rhodamine B isothiocyanate (RITC) and fluorescein (FITC) respectively, before being mixed into solution.

The fluorescent images of adsorbed films on glass plate after testing demonstrated that the optimum adsorbed film as layered structure was formed (Fig. 5.24a) in HA solution with 0.7 mass% albumin and 1.4 mass% γ-globulin (1/2 as A/G ratio), in which low wear and low friction were attained. In HA solution with 1.4 mass% albumin and

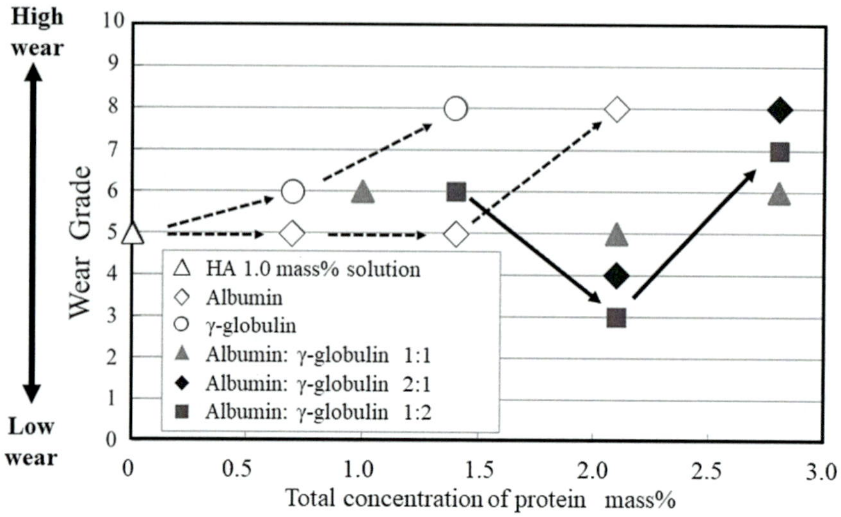

Figure 5.23 Influence of lubricant constituents and protein concentration on wear grade for PVA hydrogel against itself in reciprocating test. The wear grade is defined as the summed values of wear severity classified as 5 grades for the upper and lower rubbing surfaces. The transitions of wear grade with an increase in protein concentration are indicated by the *arrows with dotted line* for solution of single protein and with *solid line* for solution of dual proteins.
Modified from Nakashima, K., Sawae, Y., & Murakami, T. (2005). Study on wear reduction mechanisms of artificial cartilage by synergistic protein boundary film formation. *JSME International Journal Series C*, *48*(4), 555–561. https://doi.org/10.1299/jsmec.48.555.

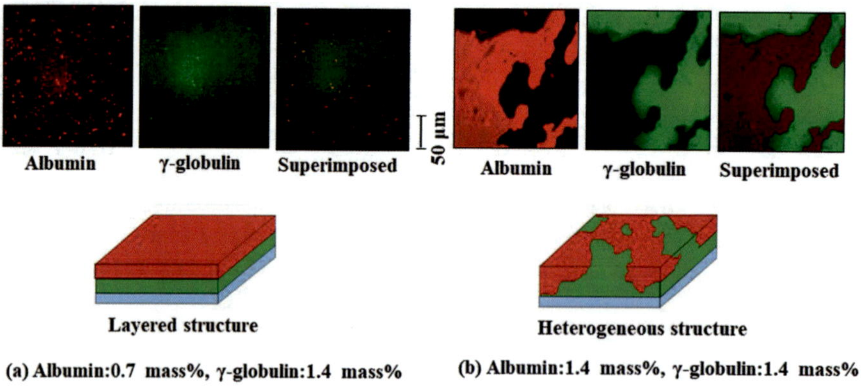

Figure 5.24 Fluorescent images of proteins for middle layer of adsorbed films in different lubricants. (a) In HA solution with 0.7 mass% albumin and 1.4 mass% γ-globulin, the adsorbed film as layered structure is formed. (b) In HA solution with 1.4 mass% albumin and 1.4 mass% γ-globulin, the adsorbed film as heterogeneous structure is formed.
Modified from Nakashima, K., Sawae, Y., & Murakami, T. (2005). Study on wear reduction mechanisms of artificial cartilage by synergistic protein boundary film formation. *JSME International Journal Series C*, *48*(4), 555–561. https://doi.org/10.1299/jsmec.48.555.

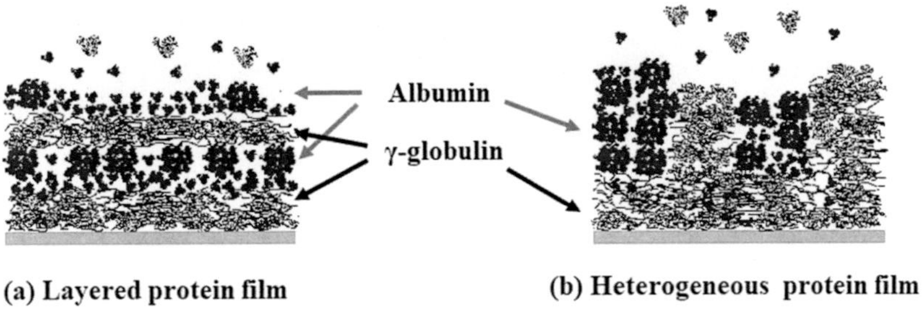

Figure 5.25 Adsorbed protein models as layered and heterogeneous structures. (a) Layered protein film composed of albumin-rich layer and γ-globulin-rich layer has an effect of reducing wear. (b) Heterogeneous protein film increases wear.
Modified from Nakashima, K., Sawae, Y., & Murakami, T. (2005). Study on wear reduction mechanisms of artificial cartilage by synergistic protein boundary film formation. *JSME International Journal Series C*, *48*(4), 555–561. https://doi.org/10.1299/jsmec.48.555.

0.7 mass% γ-globulin (2/1 as A/G ratio), similar layered structure was formed. In contrast, the adsorbed film as heterogeneous structure was formed (Fig. 5.24b) in HA solution with 1.4 mass% albumin and 1.4 mass% γ-globulin, in which severe wear occurred.

To sustain both low friction and low wear, it is necessary to form synergistic adsorbed film structure composed of albumin and γ-globulin. Fig. 5.25 shows a model of adsorbed protein film based on experimental results. Albumin is composed of mainly α-helix structure, and γ-globulin is composed of mainly β-sheet structure. Albumin containing the α-helix structure forms low shear layer because the α-helix structure is easily released from surfaces and low cohesive strength, but the γ-globulin containing β-sheet structure forms uniform adsorption layer (Nakashima et al., 2007a,b). The adsorption of proteins changes depending on surface wettability, and γ-globulin shows higher adsorption than albumin on both hydrophilic and hydrophobic surfaces (Sigal et al., 1998).

Thus, it is suggested in layered adsorbed film in Fig. 5.25a that albumin constructs low shear layer and γ-globulin layer with strong adsorption ability protects PVA hydrogel from wear. If protein adsorbed film is rubbed, the slip is likely to occur at the boundary of albumin rich layer and γ-globulin layer or within albumin-rich layer. In contrast, the heterogeneous adsorbed film (Fig. 5.25b) cannot be effective in lowering friction due to preventing the slip of albumin itself.

Subsequently, the effect of adsorbed film formation during reciprocation on wear reduction was confirmed by in situ visualization for reciprocating apparatus constructed on the stage in inverted fluorescent microscope (Yarimitsu et al., 2007). A sliding pair of a spherical reciprocating upper specimen of FT-PVA hydrogel as 7 mm radius with 2 mm thickness, and a flat stationary lower specimen of cover glass was tested in boundary lubrication regime at load of 0.736 N (average pressure of 0.104 MPa), sliding speed of 0.2 mm/s and stroke of 4.0 mm. Lubricants are single or binary protein (albumin, γ-globulin) saline solution. The theoretical film thickness (Hamrock & Dowson, 1978) is 0.025 μm. The fluorescent images for protein

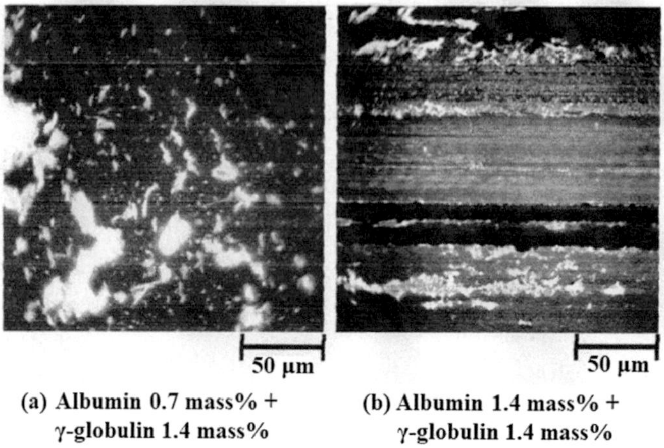

Figure 5.26 Fluorescent images of albumin boundary film on cover glass and PVA hydrogel surface at 4.0 m sliding. (a) Smooth boundary films are observed in lubricant of 0.7 mass% albumin and 1.4 mass% γ-globulin. Protein aggregates between rubbing surfaces are also observed. (b) The local exfoliation of protein boundary films is observed in lubricant of 1.4 mass% albumin and 1.4 mass% γ-globulin. Sliding direction is the lateral direction. Modified from Yarimitsu, S., Nakashima, K., Sawae, Y., & Murakami, T. (2007). Study on the mechanisms of wear reduction of artificial cartilage through in situ observation on forming protein boundary film. *Tribology Online*, 2(4), 114–119. https://doi.org/10.2474/trol.2.114.

molecules stained with RITC or FITC were recorded as movies, which include the images for protein adsorbed on glass plate, protein aggregates between rubbing surfaces, and protein adsorbed on PVA surface. The representative fluorescent images for albumin are shown in Fig. 5.26, although the fluorescent images for γ-globulin were recorded. At 0.7 mass% albumin and 1.4 mass% γ-globulin with layered adsorbed film, smooth boundary films were observed and hardly peeled off after 4.0 m sliding. In contrast, at 1.4 mass% albumin and 1.4 mass% γ-globulin with heterogeneous adsorbed film, the local exfoliation of protein boundary films was observed after 4.0 m sliding. The formation of stable adsorbed films with optimum structure is an important factor to establish longer durability of artificial cartilage.

In this test, the rolling motion of cylindrical or elliptical protein aggregates between rubbing surfaces was observed. This rolling motion may be effective in lowering of friction as proposed by Chikama (1985).

Next, to examine the composition of the bottom layer of protein boundary film, the behaviors of protein adsorption at the bottom layer were observed by using total internal reflection fluorescence microscopy (TIRFM), where only fluorescently labeled molecules within evanescent field of about 200 nm from glass-water interface are excited (Yarimitsu et al., 2009). In the reciprocating test at similar condition to the above reciprocating test (Yarimitsu et al., 2007), number of both adsorbed albumin and γ-globulin (observed for fluorescently labeled immune globulin G) increased with sliding distance in lubricant of 0.7 mass% albumin and 1.4 mass% γ-globulin with smooth stable boundary films. In the lubricant with appropriate composition an increase of γ-globulin with rubbing process was remarkable, and it was found that the formation of bottom layer of protein boundary film, which is rich in γ-globulin, is a key factor to form a stable boundary film. In contrast, number of adsorbed albumin decreased with sliding in lubricant of 1.4 mass% albumin and 1.4 mass% γ-globulin with local exfoliation of protein boundary film.

In experiment for lubricants added with hyaluronic acid (HA) as theoretical film thickness of 0.207 μm, which is still in boundary lubrication regime, adsorption of HA on the rubbed surface was enhanced by coexistence with γ-globulin and rubbing action. It was indicated that gel-like film composed of HA was formed, and it contributes to lowering friction.

In previous cases, the adsorption model as native proteins on hydrophilic surface with hydroxyl group was described in assuming mild condition. However, protein conformation can be changed under severe rubbing conditions or by a change in chemical environment, pressure, or temperature. Heuberger et al. (Heuberger et al., 2005) suggested the situation for conformational change (unfolding) of albumin with high temperature in the case of the artificial joints composed of hard-on-UHMWPE in vivo. For example, it was confirmed that albumin is denatured by heating and the secondary structure is changed from some of the α-helix to random coil (Nakashima et al., 2007b). In denatured γ-globulin, some of the β-sheet was changed to random coil. It was suggested that the denatured albumin adsorbed on hydrophilic PVA hydrogel and glass plate surfaces as a bottom layer uniformly and strongly compared with native albumin and reduced friction in repeated tests changing lubricants containing denatured proteins as albumin → γ-globulin → albumin → γ-globulin (Nakashima et al., 2007b).

As discussed above, tribological performance of boundary film should be evaluated on the basis of appropriate elucidation of structures and properties of boundary films including various changes in rubbing process.

For PVA hydrogel used in synovial fluid containing various constituents, the influence of addition of single constituents in lubricant on tribological behaviors should be clarified. Frictional behaviors in repeated reciprocating tests including 5 min unloading of 9.8 N after 36 m sliding and restart are shown in Fig. 5.27, where the effect of recovering of hydration, adsorbed film formation, and deformation can be compared (Murakami et al., 2014). It is confirmed that the addition of protein affects friction adversely, but the addition of HA or DPPC contributes to remarkable reduction in friction.

Next, the influence of combination of synovia constituents on friction and wear was examined. In Fig. 5.28, the representative results are shown. It is noticed that the combination of HA and DPPC with and without protein can maintain low friction level. In comparison of wear, HA solution containing 0.01 mass% DPPC, 1.4 mass% albumin, and 0.7 mass% γ-globulin (lubricant F) showed minimum wear, which is similar to the case for articular cartilage. This indicates that the reduction in both friction and wear of FT-PVA hydrogel is possible with the aid of synovial fluid in human joint environment.

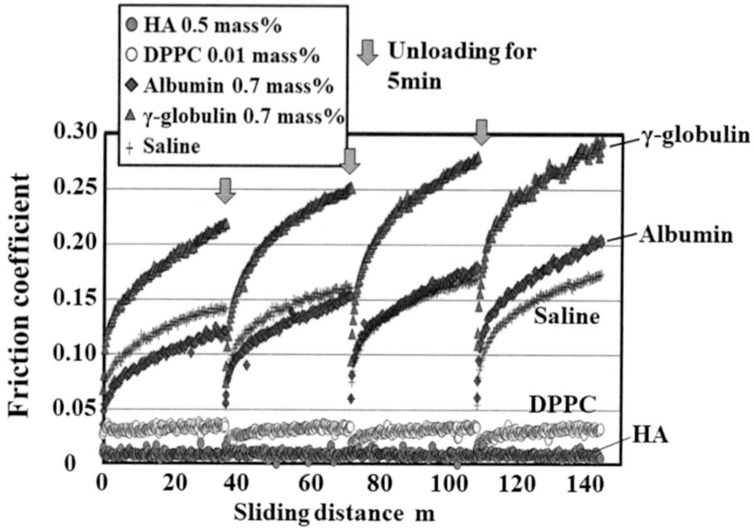

Figure 5.27 Influence of each synovia constituent on changes in friction of FT-PVA gel against glass plate. The addition of protein affects friction adversely but the addition of HA or DPPC maintains low friction level.
Modified from Murakami, T., Yarimitsu, S., Nakashima, K., Yamaguchi, T., Sawae, Y., Sakai, N., & Suzuki, A. (2014). Superior lubricity in articular cartilage and artificial hydrogel cartilage. *Proceedings of the Institution of Mechanical Engineers, Part J: Journal of Engineering Tribology*, *228*(10), 1099–1111. https://doi.org/10.1177/1350650114530273.

5.5 Influence of hydrogel wear debris

To establish the longer durability of total joint replacements composed of UHMWPE and metal or ceramic component in clinical application, it is required to prevent the aseptic loosening due to wear debris-induced osteolysis as the active resorption of bone matrix by osteoclasts associated with the wear particles. The osteolysis is induced by synthesizing various proinflammatory cytokines such as interleukin-1β (IL-1β), tumor necrosis factor-α (TNF-α), and prostaglandin E2, from immunocytes, i.e., macrophages and leukocytes (Catelas et al., 2011; Gallo et al., 2013, 2014; Ingham & Fisher, 2005; Willert & Semlitsch, 1977). The synthesized cytokines from macrophages stimulate the surrounding osteoclasts and activate bone resorption, which is induced because the activated osteoclast pumps a high concentration of proton between bone and osteoclast. Thus, unbalance of a bone remodeling occurs and induces osteolysis and aseptic loosening.

However, the influences of artificial hydrogel cartilage wear particles on the cytokine synthesis, and the osteolysis have not been elucidated. Therefore, Omata et al. (2015) evaluated the influence of PVA hydrogel wear particles on the immune response of THP-1-derived macrophages by assaying two cytokines IL-1β and TNF-α.

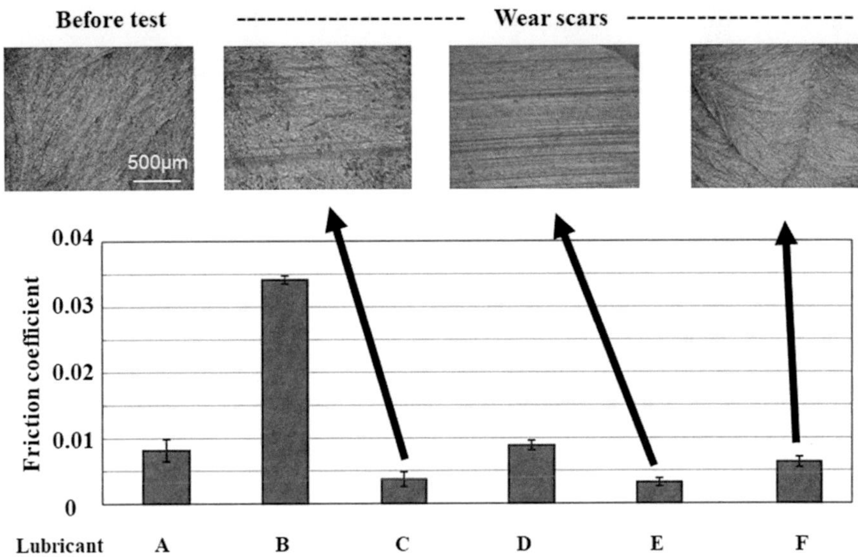

Figure 5.28 Influence of lubricant on average friction coefficient after each 36 m sliding and wear scars for FT-PVA hydrogel. Lubricant A: HA 0.5 mass%, Lubricant B: DPPC 0.01 mass%, Lubricant C: HA 0.5 mass% + DPPC 0.01 mass%, Lubricant D: HA 0.5 mass% + DPPC 0.01 mass% + Albumin 0.7 mass%, Lubricant E: HA 0.5 mass% + DPPC 0.01 mass% + γ-globulin 0.7 mass%, Lubricant F: HA 0.5 mass% + DPPC 0.01 mass% + Albumin 1.4 mass% + γ-globulin 0.7 mass%. Friction is minimum in lubricant E but wear is minimum in lubricant F. Error bars indicate standard deviation.
Modified from Murakami, T., Yarimitsu, S., Nakashima, K., Yamaguchi, T., Sawae, Y., Sakai, N., & Suzuki, A. (2014). Superior lubricity in articular cartilage and artificial hydrogel cartilage. *Proceedings of the Institution of Mechanical Engineers, Part J: Journal of Engineering Tribology*, *228*(10), 1099−1111. https://doi.org/10.1177/1350650114530273.

In this study, wear particles of PVA hydrogels as the repeated freeze-thawing (FT) and the cast-drying (CD) gels were collected, which were obtained in uniaxial ball (Co-Cr−Mo alloy, 26 mm diameter)-on-plate(PVA hydrogel, 2 mm thickness) reciprocating test at stroke of 50 mm in distilled water used for soaking of gel specimens, at 37°C. The maximum contact pressure at 4.9 N loading was about 0.1 MPa, and the sliding distance was 3.0 km. To evaluate total wear amount of PVA hydrogel, the concentration of PVA molecules in lubricant was assessed by a total organic carbon (TOC) analysis system (Suzuki et al., 2013) because of difficulty in gravimetric method. Total amounts of wear particles for FT and CD gels were 5.9 ± 0.2 mg ($N = 4$) and 0.93 ± 0.07 mg ($N = 4$), respectively. CD gel exhibited lower wear than FT gel.

In the particle size measurement system that utilizes dynamic light scattering, it was shown that wear particles 10−50 nm in diameter were dominantly dispersed as particle number distribution in both wear tests. In particle volume distribution, wear particles in the range from 5 to 10 µm were similar to those in the range of 100−600 nm. There

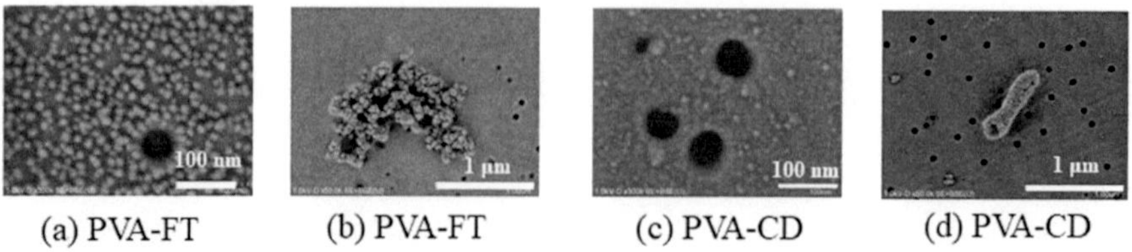

Figure 5.29 SEM images of PVA wear particles. Nano- and micron-sized wear debris for PVA-FT and CD gels are shown.
Modified from Omata, S., Sawae, Y., & Murakami, T. (2015). Effect of poly(vinyl alcohol) (PVA) wear particles generated in water lubricant on immune response of macrophage. *Biosurface and Biotribology*, *1*(1), 71–79. https://doi.org/10.1016/j.bsbt.2015.02.003.

were few particles of approximately 1 μm in size, which are known to induce the macrophage immune response (Green et al., 1998). There were negligible differences in the distributions between FT and CD gels. Scanning electron microscope (SEM) images of nano- and micron-sized wear particles are shown in Fig. 5.29. Most micron-sized particles appeared to be aggregations of nano-sized PVA particles of approximately 20 nm in diameter.

Next, the influences of PVA wear particles on the immune response of THP-1-derived macrophages were examined by assaying two cytokines, IL-1β, and TNF-α. To evaluate the effect of PVA wear particles of PVA-FT and PVA-CD gels, not only the measurement of cytokines released into the culture medium using an enzyme-linked immune sorbent assay (ELIZA) but also the observation of fluorescently immunostained cytokines remaining in their cells with confocal laser scanning microscopy (CLSM). Biochemical and immunological evaluations revealed no cytotoxic effects of wear particles on macrophages.

In Fig. 5.30, the cytokine syntheses of IL-1β and TNF-α are evaluated by applying each PVA wear particles of PVA-FT and PVA-CD gels to macrophages for a 24 h culture period. Here, two control media were prepared; the basal medium was the negative control (NegC), and the culture medium with 1 μg/mL of lipopolysaccharide (LPS)

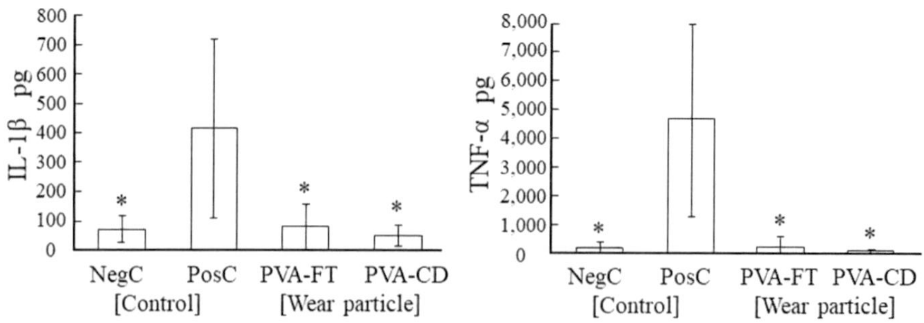

Figure 5.30 Effects of PVA wear particles on cytokine syntheses of IL-1β and TNF-α NegC and PosC indicate negative control and positive control groups, respectively. Error bar means standard deviation. Asterisks (*) indicate statistically significant respective culture groups compared to PosC, *: $d > 1$, $P < .01$, $(1-\beta) > 0.8$.
Modified from Omata, S., Sawae, Y., & Murakami, T. (2015). Effect of poly(vinyl alcohol) (PVA) wear particles generated in water lubricant on immune response of macrophage. *Biosurface and Biotribology*, *1*(1), 71–79. https://doi.org/10.1016/j.bsbt.2015.02.003.

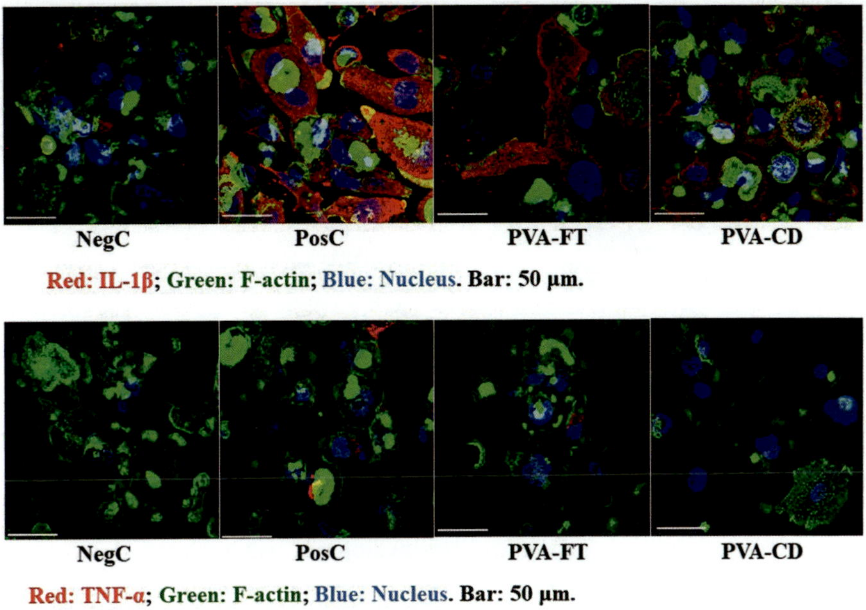

Figure 5.31 Immunohistochemical observation of the distribution of cytokines, IL-1β and TNF-α in macrophage. NegC and PosC indicate negative control and positive control groups, respectively.
Modified from Omata, S., Sawae, Y., & Murakami, T. (2015). Effect of poly(vinyl alcohol) (PVA) wear particles generated in water lubricant on immune response of macrophage. *Biosurface and Biotribology*, *1*(1), 71–79. https://doi.org/10.1016/j.bsbt.2015.02.003.

was the positive control (PosC). LPS was an endotoxin for inducing a powerful immune response. It is noticed that exposing macrophage to wear particles in both particle stimulated groups did not change cytokine synthesis compared to positive control. The cytokine distribution in macrophage is shown in Fig. 5.31. In PosC, both IL-1β and TNF-α were synthesized compared to NegC. On the contrary, groups exposed to PVA wear particles exhibited low synthesis of each cytokine compared to the positive control (PosC).

As described above, cytokine syntheses of IL-1β and TNF-α in both wear particle-stimulated groups for PVA-FT and PVA-CD gels were significantly lower than that of the lipopolysaccharide-stimulated positive control. Therefore, it is suggested that PVA hydrogel wear particles do not affect the macrophage immune response.

The information on the physical and chemical properties and morphology of wear debris of other hydrogels is insufficient and the evaluation of effect of wear particles on cytokine synthesis is required before clinical application.

5.6 Bionic design with biomimetic artificial cartilage

Natural articular cartilage with superior lubricity and load-carrying capacity is natural hydrogel, which contains plenty water in extracellular matrix composed of

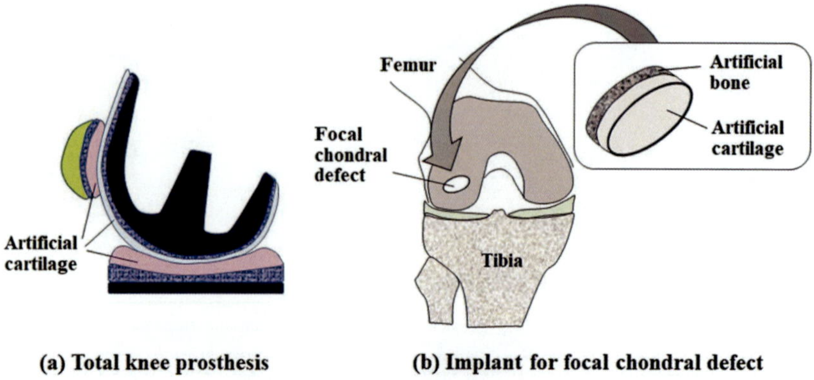

Figure 5.32 Application of compliant artificial cartilage in knee joints. (a) Total knee prosthesis with artificial hydrogel cartilage and (b) implant for focal chondral defect with artificial hydrogel cartilage.

proteoglycan reinforced by collagen network with the lubricious surface layer. Therefore, various artificial hydrogels as complex hydrophilic polymer network have been developed to mimic natural articular cartilage. The artificial biphasic hydrogel cartilage materials are potential candidates, although the clinical applications of artificial hydrogel cartilage have been restricted within some limited fields at the present stage.

As clinical replacements with artificial cartilage, for example, acetabular cup and/or femoral head in hip joints, femoral condyle, tibial plateau, patella and/or meniscus etc. in knee joints (Fig. 5.32a) are applicable as total joint replacement or hemiarthroplasty. Another application is the implanting for focal chondral defect (Fig. 5.32b). In clinical application of artificial cartilage, not only tribological property but also the fixation stability are important. The application of composite bearing/device supported with stiffer biomaterials or artificial bone is expected. Oka et al. (2000) successfully applied the composite osteochondral device (COD) composed of PVA hydrogel and titanium fiber mesh for hemiarthroplasty. Yarimitsu et al. (2015) confirmed the stable fixation by bone ingrowth into through-hole in PVA hydrogel.

The usual biocompatible compliant materials have not so high strength, although some new materials with higher strength have been developed with special preparation or treatment, which sometimes may give the biological reaction during longer clinical usage. The development of biocompatible compliant artificial cartilage with high strength and superior lubricity has been progressed, but it is concurrently important to reduce the operating severity for artificial cartilage by exploring of optimum geometric and material design. The improvement of geometrical congruity with appropriate layer thickness is capable to reduce the contact stress and friction shear stress with enhancement of fluid film formation or lubricating abilities, which induce longer durability. In contrast, excessive geometrical congruity can induce the starved lubrication due to the prevention of the entrainment of lubricants with larger deformation. For hydrogel materials with biphasic property, the translation of loading conjunction area specified by geometric design and motion modes can control the sustaining efficient interstitial pressurization for various daily activities. Effective hydration lubrication and adsorbed film formation are beneficial.

In thin film condition at slow speed even at contact pressure as several MPa, it was proposed by Kurokawa et al (2005) that the elastic friction mechanism increases friction level with maximum value μ_{max} at specified speed according to elastic deformation of adsorbing polymer chain in an adsorption-repulsion model for sliding of polymer gel on a solid as shown in Fig. 5.33. When the sliding speed is sufficiently higher than specified speed, the polymer of gel surface has no enough time to form an adsorbing site, so the hydrodynamic lubrication becomes effective. As an effective method to suppress friction level for slow speed, the polymer brushes on gel surfaces with repulsive interaction with the sliding substrate can reduce surface friction (Gong et al., 2001).

The improving methods for properties of PVA hydrogel are to additionally control the micro- and nano-scopic structures, to reinforce the structure with fibrous element or other method and thus to enhance biphasic properties for efficient lubrication and load-carrying property. The PVA FT gel reinforced by PVA fiber showed the improvement of friction in cylindrical knee prosthesis during walking (Murakami, 2012; Sakai, Hosoda, et al., 2012; Sakai et al., 2019), probably due to increasing fluid load support, as discussed at 5.3.3. In PVA hybrid (CD on FT) gel composed of CD gel with low permeability and FT gel with high permeability, the layered structure can be optimized by preparation conditions and can improve the interstitial fluid pressurization (Figs. 5.16–5.19).

For further improvement of friction and wear properties at high contact pressure, Yarimitsu et al. (2018) applied the hybridization method compositing cross-linking

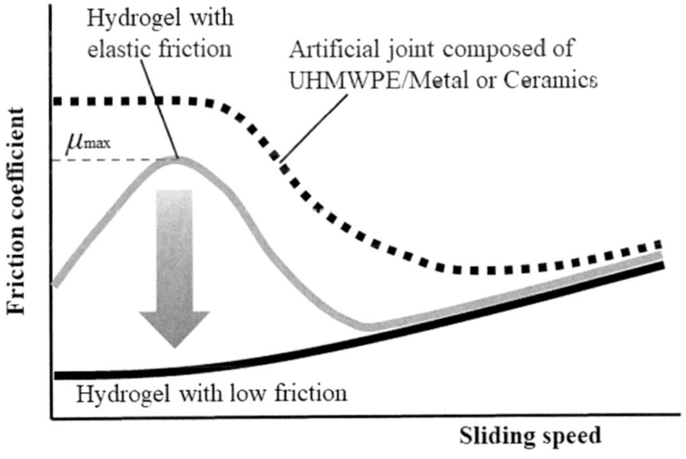

Figure 5.33 Frictional behaviors of hydrogels and UHMWPE with sliding speed. High elastic friction for hydrogel at slow speed can be reduced with a water layer at the interface when the interfacial interaction is repulsive between gel and the opposing substrate.
Modified from Murakami, T., Yarimitsu, S., Nakashima, K., Sakai, N., Yamaguchi, T., Sawae, Y., & Suzuki, A. (2015). Biphasic and boundary lubrication mechanisms in artificial hydrogel cartilage: A review. *Proceedings of the Institution of Mechanical Engineers, Part H: Journal of Engineering in Medicine*, *229*(12), 864–878. https://doi.org/10.1177/0954411915611160.

network of FT and CD gels in single layer gel, i.e., the composite-type PVA hybrid gel. PVA FT gel was firstly prepared by single FT treatment, and then the sample was dried at specified temperature and humidity. After drying, the sample was treated by single FT process. Finally, after drying at lower temperature and humidity, the sample was swollen in pure water, and the composite hybrid gel was obtained. This composite PVA hybrid gel exhibited lower friction and lower wear than the laminated CD on FT hybrid gel in reciprocating test in saline at higher contact pressure.

The extent of water content in hydrogel usually controls biphasic properties. The decrease in water content of hydrogel tends to increase mechanical strength and stiffness but has janus-faced result which reduces wear (Oka et al., 2000; Suciu et al., 2004) or conversely increases wear (Murakami et al., 2000). In the latter case, the biphasic lubrication becomes ineffective in hydrogel with low water content.

On the effectiveness of boundary lubrication at slow speed, Kosukegawa et al. (2015) pointed out that the presence of plasma proteins into lubricant largely reduced maximum friction μ_{max} (shown in Fig. 5.33) for sliding pair of stainless steel ball on PVA hydrogel disk. It is noted that serum and albumin single solution showed lowest friction compared with that in IgG single solution. They assume that this difference is attributed to different conformation of protein domains adsorbed on PVA hydrogel, on which IgG may be denatured, but albumin on hydrophilic PVA hydrogel may be kept in natural conformation. The changes in conformation of serum proteins exert influence on adsorption as pointed out by Nakashima et al. (2007b). Common phenomena were observed in related studies, where DPPC and albumin are likely to act as low shearing layer, and γ-globulin acts as the protective film as strongly adsorbed molecules on PVA surface. HA has ability to thicken the lubricating fluid film and form some lubricating gel-like layer. It is noteworthy that the coexistence of DPPC, albumin, γ-globulin, and HA at appropriate concentration is expected to bring about very low friction and minimal wear of PVA hydrogels (Murakami et al., 2014).

Another merit of PVA hydrogel is superior biocompatibility. The PVA hydrogels based on hydrogen bonding network as physical bonding prepared without any additives or irradiative stimulus have superiority in safety in human body. Oka et al. (2000) showed good biocompatibility of PVA hydrogel in implantation of plug specimen in rabbit patellar groove and small particles in wister rat knee joint with histological study. Omata et al. (Omata et al., 2015) indicated that the wear debris of PVA FT and CD gels generated in reciprocating ball-on-plate test does not induce cytotoxity and affect immune response of macrophage in evaluation of cytokines of interleukin-1β (IL-1β) and tumor necrosis factor-α (TNF-α). These facts indicate beneficial property for clinical application. For physically cross-linked PVA gels having hydrogen bonds without using a cross-linking agent, the gamma ray sterilization for clinical applications usually causes additional chemical cross-linking and changes physical properties of gels. To evaluate the effect of gamma ray irradiation on friction and wear properties, the laminated CD on FT hybrid gels were irradiated using gamma rays at a different dose rate and irradiance (Sasaki et al., 2018). At the optimized irradiation conditions of 10 kGy at 5 kGy/h for gamma irradiated gels, the excellent friction characteristics and minimal wear at 37°C were retained. Higher gamma irradiation increased friction and wear.

So far, PVA hydrogels have been successfully applied for cartilage replacement as repair implant for the focal chondral defects in human clinical studies (Baker et al., 2012). It is indicated in follow-up periods of up to 2 years that isolated implants surrounded by high quality bone, a flush presentation, and about 10% radial compression improve outcome in vivo. PVA gave no apparent damage on opposing cartilage, but misplacement of these implants was the major reason for dislocation and loosening. Both the repair device for focal chondral defects and the application as rubbing surface layer in total joint replacements or hemi-arthroplasty are considered as possible clinical applications of PVA hydrogels (Fig. 5.32).

Another combination is gel-to-gel rubbing pair (Yarimitsu et al., 2013), where the interstitial fluid pressurization on moving hydrogel can contribute to sustain significant fluid load support after rehydration in migrating part under low contact pressure. However, under thin film severe condition where boundary lubrication plays main role for larger contact areas with intense interaction, both physical and chemical properties of lubricant and hydrogel appear to become much more important to maintain good tribological behavior than for the combination of hydrogel against hard mating materials. The synergistic function of the boundary lubrication by adsorbed films with phospholipids/proteins/HA and the biphasic lubrication becomes important to maintain low friction and low wear.

For PVA gel-articular cartilage combination, PVA hybrid gel showed superior tribological behavior (Yarimitsu, Sasaki, & Murakami, 2016).

For the extensive development of clinical application of biocompatible PVA hydrogel, durability with appropriate mechanical and tribological properties, synergistic interaction with lubricants, ease of production, formability, appropriate fixation methods with stability, optimum geometric design with surface texture, and less damage to the opposing cartilage or rubbing surface are required (Katta et al., 2008; Kobayashi & Hyu, 2010). In particular, the establishment of appropriate design in consideration of a multimode lubrication mechanism is required not only for low friction but also for zero-wear and high fatigue resistance in the human body. The optimization of tribological properties of hybrid gel and other candidate hydrogels should be actualized on the basis of discussion described above. The effective utilization of biphasic and boundary lubrication mechanisms in close linkage with fluid film and hydration lubrication mechanisms is expected to reduce friction and protect surface of hydrogels as not only physically crosslinked PVA but also modified PVA (Bodugoz-Senturk et al., 2009; Sardinha et al., 2013), chemically crosslinked DN hydrogels (Arakaki et al., 2010; Yasuda et al., 2005), silk protein hydrogel for cartilage scaffolds (Parkes et al., 2015) and other related hydrogels.

References

Arakaki, K., Kitamura, N., Fujiki, H., Kurokawa, T., Iwamoto, M., Ueno, M., Kanaya, F., Osada, Y., Gong, J. P., & Yasuda, K. (2010). Artificial cartilage made from a novel double-network hydrogel: In vivo effects on the normal cartilage and ex vivo evaluation of the friction property. *Journal of Biomedical Materials Research - Part A, 93*(3), 1160–1168. https://doi.org/10.1002/jbm.a.32613

Ateshian, G. A. (2009). The role of interstitial fluid pressurization in articular cartilage lubrication. *Journal of Biomechanics, 42*(9), 1163–1176. https://doi.org/10.1016/j.jbiomech.2009.04.040

Auger, D. D., Dowson, D., & Fisher, J. (1995a). Cushion form bearings for total knee joint replacement part 1: Design, friction and lubrication. *Proceedings of the Institution of Mechanical Engineers, Part H: Journal of Engineering in Medicine, 209*(2), 73–81. https://doi.org/10.1243/PIME_PROC_1995_209_323_02

Auger, D. D., Dowson, D., & Fisher, J. (1995b). Cushion form bearings for total knee joint replacement part 2: Wear and durability. *Proceedings of the Institution of Mechanical Engineers, Part H: Journal of Engineering in Medicine, 209*(2), 83–91. https://doi.org/10.1243/PIME_PROC_1995_209_324_02

Auger, D. D., Dowson, D., Fisher, J., & Jin, Z. M. (1993). Friction and lubrication in cushion form bearings for artificial hip joints. *Proceedings of the Institution of Mechanical Engineers, Part H: Journal of Engineering in Medicine, 207*(1), 25–33. https://doi.org/10.1243/PIME_PROC_1993_207_265_02

Baker, M. I., Walsh, S. P., Schwartz, Z., & Boyan, B. D. (2012). A review of polyvinyl alcohol and its uses in cartilage and orthopedic applications. *Journal of Biomedical Materials Research - Part B Applied Biomaterials, 100*(5), 1451–1457. https://doi.org/10.1002/jbm.b.32694

Bodugoz-Senturk, H., Macias, C. E., Kung, J. H., & Muratoglu, O. K. (2009). Poly(vinyl alcohol)-acrylamide hydrogels as load-bearing cartilage substitute. *Biomaterials, 30*(4), 589–596. https://doi.org/10.1016/j.biomaterials.2008.10.010

Catelas, I., Wimmer, M. A., & Utzschneider, S. (2011). Polyethylene and metal wear particles: Characteristics and biological effects. *Seminars in Immunopathology, 33*(3), 257–271. https://doi.org/10.1007/s00281-011-0242-3

Chikama, H. (1985). The role of the protein and the hyaluronic acid in the synovial fluid in animal joint lubrication (In Japanese). *Journal of the Japanese Orthopaedic Association, 59*(5), 559–572.

Corkhill, P. H., Trevett, A. S., & Tighe, B. J. (1990). The potential of hydrogels as synthetic articular cartilage. *Proceedings of the Institution of Mechanical Engineers, Part H: Journal of Engineering in Medicine, 204*(3), 147–155. https://doi.org/10.1243/PIME_PROC_1990_204_249_02

Dowson, D. (1966–67). Modes of lubrication in human joints. Proceedings of the Institution of Mechanical Engineers, 181(Pt3J) 45–54.

Dowson, D. (1989). Are our joint replacement materials adequate? *Proceedings of the Institution of Mechanical Engineers, Part H: Journal of Engineering in Medicine, 203*, 1–5.

Dowson, D., Fisher, J., Jin, Z. M., Auger, D. D., & Jobbins, B. (1991). Design considerations for cushion form bearings in artificial hip joints. *Proceedings of the Institution of Mechanical Engineers, Part H: Journal of Engineering in Medicine, 205*(2), 59–68. https://doi.org/10.1243/PIME_PROC_1991_205_269_02

Dowson, D., & Jin, Z.-M. (1986). Micro-elastohydrodynamic lubrication of synovial joints. *Engineering in Medicine, 15*(2), 63–65. https://doi.org/10.1243/EMED_JOUR_1986_015_019_02

Dowson, D., & Yao, J. Q. (1994). Elastohydrodynamic lubrication of soft-layered solids at elliptical contacts: Part 2: Film thickness analysis. *Proceedings of the Institution of Mechanical Engineers, Part J: Journal of Engineering Tribology, 208*(1), 43–52. https://doi.org/10.1243/PIME_PROC_1994_208_348_02

Gallo, J., Goodman, S. B., Konttinen, Y. T., Wimmer, M. A., & Holinka, M. (2013). Osteolysis around total knee arthroplasty: A review of pathogenetic mechanisms. *Acta Biomaterialia, 9*(9), 8046–8058. https://doi.org/10.1016/j.actbio.2013.05.005

Gallo, J., Vaculova, J., Goodman, S. B., Konttinen, Y. T., & Thyssen, J. P. (2014). Contributions of human tissue analysis to understanding the mechanisms of loosening and osteolysis in total hip replacement. *Acta Biomaterialia, 10*(6), 2354–2366. https://doi.org/10.1016/j.actbio.2014.02.003

Gladstone, J. R., & Medley, J. B. (1990). Paper X (ii) Comparison of theoretical and experimental values for friction of lubricated elastomeric surface layers under transient conditions. *Tribology Series, 17*(C), 241–250. https://doi.org/10.1016/S0167-8922(08)70262-1

Gong, J. P., Katsuyama, Y., Kurokawa, T., & Osada, Y. (2003). Double-network hydrogels with extremely high mechanical strength. *Advanced Materials, 15*(14), 1155–1158. https://doi.org/10.1002/adma.200304907

Gong, J. P., Kurokawa, T., Narita, T., Kagata, G., Osada, Y., Nishimura, G., & Kinjo, M. (2001). Synthesis of hydrogels with extremely low surface friction. *Journal of the American Chemical Society, 123*(23), 5582–5583. https://doi.org/10.1021/ja003794q

Green, T. R., Fisher, J., Stone, M., Wroblewski, B. M., & Ingham, E. (1998). Polyethylene particles of a "critical size" are necessary for the induction of cytokines by macrophages in vitro. *Biomaterials, 19*(24), 2297–2302. https://doi.org/10.1016/S0142-9612(98)00140-9

Hamrock, B. J., & Dowson, D. (1978). Elastohydrodynamic lubrication of elliptical contacts for materials of low elastic modulus. Part 1: Fully flooded conjunction. *Transactions of the ASME. Journal of Lubrication Technology, 100*(1), 236–245.

Heuberger, M. P., Widmer, M. R., Zobeley, E., Glockshuber, R., & Spencer, N. D. (2005). Protein-mediated boundary lubrication in arthroplasty. *Biomaterials, 26*(10), 1165–1173. https://doi.org/10.1016/j.biomaterials.2004.05.020

Higaki, H., & Murakami, T. (1996). Role of constituents in synovial fluid and surface layer of articular cartilage in joint lubrication (part 2): The boundary lubricating ability of proteins. *Japanese Journal of Tribology, 40*(7), 691–700.

Higginson, G. R. (1978). Squeeze films between compliant solids. *Wear, 46*(2), 387–395. https://doi.org/10.1016/0043-1648(78)90042-X

Ingham, E., & Fisher, J. (2005). The role of macrophages in osteolysis of total joint replacement. *Biomaterials, 26*(11), 1271–1286. https://doi.org/10.1016/j.biomaterials.2004.04.035

Jalali-Vahid, D., Jagatia, M., Jin, Z. M., & Dowson, D. (2001). Prediction of lubricating film thickness in UHMWPE hip joint replacements. *Journal of Biomechanics, 34*(2), 261–266. https://doi.org/10.1016/S0021-9290(00)00181-0

Jones, E., Scholes, S. C., Unsworth, A., & Burgess, I. C. (2008). Compliant-layer tibial bearing inserts: Friction testing of different materials and designs for a new generation of prostheses that mimic the natural joint. *Proceedings of the Institution of Mechanical Engineers, Part H: Journal of Engineering in Medicine, 222*(8), 1197–1208. https://doi.org/10.1243/09544119JEIM442

Katta, J., Jin, Z., Ingham, E., & Fisher, J. (2008). Biotribology of articular cartilage-A review of the recent advances. *Medical Engineering and Physics, 30*(10), 1349–1363. https://doi.org/10.1016/j.medengphy.2008.09.004

Kobayashi, M., & Hyu, H. S. (2010). Development and evaluation of polyvinyl alcohol-hydrogels as an artificial atrticular cartilage for orthopedic implants. *Materials, 3*(4), 2753–2771. https://doi.org/10.3390/ma3042753

Kosukegawa, H., Fridrici, V., Laurenceau, E., Kapsa, P., & Ohta, M. (2015). Friction of 316L stainless steel on soft-tissue-like poly(vinyl alcohol) hydrogel in physiological liquid. *Tribology International, 82*, 407–414. https://doi.org/10.1016/j.triboint.2014.05.029

Kurokawa, T., Tominaga, T., Katsuyama, Y., Kuwabara, R., Furukawa, H., Osada, Y., & Gong, J. P. (2005). Elastic-hydrodynamic transition of gel friction. *Langmuir, 21*(19), 8643–8648. https://doi.org/10.1021/la050635h

Kurtz, Steven M., Gawel, H. A., & Patel, J. D. (2011). History and systematic review of wear and osteolysis outcomes for first-generation highly crosslinked polyethylene. *Clinical Orthopaedics & Related Research, 469*(8), 2262–2277. https://doi.org/10.1007/s11999-011-1872-4

Kurtz, S. M., Siskey, R., & Shah, P. (2008). Validation of microCT of in vivo volumetric wear for retrieved PCU acetabular components. In *Proc. 8th world biomaterials congress.*

Kurtz, S. M., Siskey, R., Wippermann, B., Siebert, W., & Mai, S. (2008). Wear mechanisms of human retrieved polycarbonate urethane acetabular components. In *8th world biomaterials congress 2008* (Vol. 2, p. 618).

Luo, Y., McCann, L., Ingham, E., Jin, Z.-M., Ge, S., & Fisher, J. (2010). Polyurethane as a potential knee hemiarthroplasty biomaterial: An in-vitro simulation of its tribological performance. *Proceedings of the Institution of Mechanical Engineers, Part H: Journal of Engineering in Medicine, 224*(3), 415–425. https://doi.org/10.1243/09544119jeim657

Medley, J. B., & Dowson, D. (2008). Lubrication of elastic-isoviscous line contacts subject to cyclic time-varying loads and entrainment velocities. *A S L E Transactions, 27*(3), 243–251. https://doi.org/10.1080/05698198408981567

Medley, J. B., Dowson, D., & Wright, V. (1984). Transient elastohydrodynamic lubrication models for the human ankle joint. *Engineering in Medicine, 13*(3), 137–151. https://doi.org/10.1243/EMED_JOUR_1984_013_035_02

Mihalko, W. M., Wimmer, M. A., Pacione, C. A., Laurent, M. P., Murphy, R. F., & Rider, C. (2014). How have alternative bearings and modularity affected revision rates in total hip arthroplasty? *Clinical Orthopaedics and Related Research, 472*(12), 3747–3758. https://doi.org/10.1007/s11999-014-3816-2

Moro, T., Takatori, Y., Ishihara, K., Konno, T., Takigawa, Y., Matsushita, T., Chung, U. I. L., Nakamura, K., & Kawaguchi, H. (2004). Surface grafting of artificial joints with a biocompatible polymer for preventing periprosthetic osteolysis. *Nature Materials, 3*(11), 829–836. https://doi.org/10.1038/nmat1233

Moro, T., Takatori, Y., Tanaka, S., Ishihara, K., Oda, H., Kim, Y. T., Umeyama, T., Fukatani, E., Ito, H., Kyomoto, M., Oshima, H., Tanaka, T., Kawaguchi, H., & Nakamura, K. (2017). Clinical safety and wear resistance of the phospholipid polymer-grafted highly cross-linked polyethylene liner. *Journal of Orthopaedic Research, 35*(9), 2007–2016. https://doi.org/10.1002/jor.23473

Mow, V., C., Hou, J., S., Owens, J., M., & Ratcliffe, A. (1990). Biphasic and quasilinear viscoelastic theories for hydrated soft tissues. In Ratcliffe A., Woo S., L.-Y., & Mow V., C. (Eds.), *Biomechanics of diarthroidal joint* (Vol. 1, pp. 215–260). Springer, New York, NY.

Mow, V. C., Kuei, S. C., Lai, W. M., & Armstrong, C. G. (1980). Biphasic creep and stress relaxation of articular cartilage in compression: Theory and experiments. *Journal of Biomechanical Engineering, 102*(1), 73–84. https://doi.org/10.1115/1.3138202

Murakami, T. (1990). The lubrication in natural synovial joints and joint prostheses. *JSME International Journal. Ser. 3, 33*(4), 465–474. https://doi.org/10.1299/jsmec1988.33.465

Murakami, T. (2012). Importance of adaptive multimode lubrication mechanism in natural and artificial joints. *Proceedings of the Institution of Mechanical Engineers, Part J: Journal of Engineering Tribology, 226*(10), 827–837. https://doi.org/10.1177/1350650112451377

Murakami, T., Higaki, H., Sawae, Y., Ohtsuki, N., Moriyama, S., & Nakanishi, Y. (1998). Adaptive multimode lubrication in natural synovial joints and artificial joints. *Proceedings of the Institution of Mechanical Engineers, Part H: Journal of Engineering in Medicine, 212*(1), 23–35. https://doi.org/10.1243/0954411981533791

Murakami, T., & Ohtsuki, N. (1987). Lubricating film formation in knee prostheses under walking conditions. *Tribology Series, 11*, 387–392.

Murakami, T., Ohtsuki, N., & Higaki, H. (1993). The adaptive multimode lubrication in knee prostheses with compliant layer during walking motion. *Tribology Series, 25*(C), 673–682. https://doi.org/10.1016/S0167-8922(08)70420-6

Murakami, T., Sakai, N., Yamaguchi, T., Yarimitsu, S., Nakashima, K., Sawae, Y., & Suzuki, A. (2015). Evaluation of a superior lubrication mechanism with biphasic hydrogels for artificial cartilage. *Tribology International, 89*, 19–26. https://doi.org/10.1016/j.triboint.2014.12.013

Murakami, T., Sawae, Y., Higaki, H., Ohtsuki, N., & Moriyama, S. (1997). The adaptive multimode lubrication in knee prostheses with artificial cartilage during walking. *Tribology Series, 32*, 371–382. https://doi.org/10.1016/s0167-8922(08)70466-8

Murakami, T., Sawae, Y., Nakashima, K., & Fisher, J. (2000). Tribological behaviour of artificial cartilage in thin film lubrication. *Tribology Series, 38*, 317–327. https://doi.org/10.1016/s0167-8922(00)80136-4

Murakami, T., Yarimitsu, S., Nakashima, K., Sakai, N., Yamaguchi, T., Sawae, Y., & Suzuki, A. (2015). Biphasic and boundary lubrication mechanisms in artificial hydrogel cartilage: A review. *Proceedings of the Institution of Mechanical Engineers, Part H: Journal of Engineering in Medicine, 229*(12), 864–878. https://doi.org/10.1177/0954411915611160

Murakami, T., Yarimitsu, S., Nakashima, K., Yamaguchi, T., Sawae, Y., Sakai, N., & Suzuki, A. (2014). Superior lubricity in articular cartilage and artificial hydrogel cartilage. *Proceedings of the Institution of Mechanical Engineers, Part J: Journal of Engineering Tribology, 228*(10), 1099–1111. https://doi.org/10.1177/1350650114530273

Murakami, T., Yarimitsu, S., Sakai, N., Nakashima, K., Yamaguchi, T., & Sawae, Y. (2017). Importance of adaptive multimode lubrication mechanism in natural synovial joints. *Tribology International, 113*, 306–315. https://doi.org/10.1016/j.triboint.2016.12.052

Murakami, T., Yarimitsu, S., Sakai, N., Nakashima, K., Yamaguchi, T., Sawae, Y., & Suzuki, A. (2017). Superior lubrication mechanism in poly(vinyl alcohol) hybrid gel as artificial cartilage. *Proceedings of the Institution of Mechanical Engineers, Part J: Journal of Engineering Tribology, 231*(9), 1160–1170. https://doi.org/10.1177/1350650117712881

Nakashima, K., Murakami, T., & Sawae, Y. (2001). Evaluation of tribological properties of polyvinyl alcohol hydrogel as artificial cartilage. In JAST (Ed.), *Proc. ITC Nagasaki 2000* (Vol. 2, pp. 1537–1540).

Nakashima, K., Murakami, T., & Sawae, Y. (2004). Evaluation of wear property of PVA hydrogel as artifical cartilage and effect of protein film on wear-resistant property. *Transactions of the Japan Society of Mechanical Engineers, Part C (In Japanese), 70*(9), 2780–2787. https://doi.org/10.1299/kikaic.70.2780

Nakashima, Kazuhiro, Sawae, Y., & Murakami, T. (2005). Study on wear reduction mechanisms of artificial cartilage by synergistic protein boundary film formation. *JSME International Journal Series C, 48*(4), 555–561. https://doi.org/10.1299/jsmec.48.555

Nakashima, K., Sawae, Y., & Murakami, T. (2007a). Effect of conformational changes and differences of proteins on frictional properties of poly(vinyl alcohol) hydrogel. *Tribology International, 40*(10–12), 1423–1427. https://doi.org/10.1016/j.triboint.2007.02.010

Nakashima, K., Sawae, Y., & Murakami, T. (2007b). Influence of protein conformation on frictional properties of poly (vinyl alcohol) hydrogel for artificial cartilage. *Tribology Letters, 26*(2), 145–151. https://doi.org/10.1007/s11249-006-9185-6

Nambu, M. (1982). *Japanese patent kokai.*

Oka, M., Ushio, K., Kumar, P., Ikeuchi, K., Hyon, S. H., Nakamura, T., & Fujita, H. (2000). Development of artificial articular cartilage. *Proceedings of the Institution of Mechanical Engineers, Part H: Journal of Engineering in Medicine, 214*(1), 59–68. https://doi.org/10.1243/0954411001535246

Omata, S., Sawae, Y., & Murakami, T. (2015). Effect of poly(vinyl alcohol) (PVA) wear particles generated in water lubricant on immune response of macrophage. *Biosurface and Biotribology, 1*(1), 71–79. https://doi.org/10.1016/j.bsbt.2015.02.003

Oonishi, H., Kadoya, Y., & Masuda, S. (2001). Gamma-irradiated cross-linked polyethylene in total hip replacements - analysis of retrieved sockets after long-term implantation. *Journal of Biomedical Materials Research, 58*(2), 167–171. https://doi.org/10.1002/1097-4636(2001)58:2<167::AID-JBM1003>3.0.CO;2-A

Oral, E., & Muratoglu, O. K. (2011). Vitamin E diffused, highly crosslinked UHMWPE: A review. *International Orthopaedics, 35*(2), 215–223. https://doi.org/10.1007/s00264-010-1161-y

Otsuka, E., Sugiyama, M., & Suzuki, A. (2010). Network microstructure of PVA cast gels observed by SAXS measurements. *Journal of Physics: Conference Series, 247*, 012043. https://doi.org/10.1088/1742-6596/247/1/012043

Otsuka, E., & Suzuki, A. (2009a). A simple method to obtain a swollen PVA gel crosslinked by hydrogen bonds. *Journal of Applied Polymer Science, 114*(1), 10–16. https://doi.org/10.1002/app.30546

Otsuka, E., & Suzuki, A. (2009b). Swelling properties of physically cross-linked PVA gels prepared by a cast-drying method. *Progress in Colloid and Polymer Science, 136*, 121–126. https://doi.org/10.1007/2882_2009_17

Parkes, M., Myant, C., Dini, D., & Cann, P. (2015). Tribology-optimised silk protein hydrogels for articular cartilage repair. *Tribology International, 89*, 9–18. https://doi.org/10.1016/j.triboint.2014.11.024

Sakai, N., Hagihara, Y., Furusawa, T., Hosoda, N., Sawae, Y., & Murakami, T. (2012). Analysis of biphasic lubrication of articular cartilage loaded by cylindrical indenter. *Tribology International, 46*(1), 225–236. https://doi.org/10.1016/j.triboint.2011.03.016

Sakai, N., Hashimoto, C., Yarimitsu, S., Sawae, Y., Komori, M., & Murakami, T. (2016). A functional effect of the superficial mechanical properties of articular cartilage as a load bearing system in a sliding condition. *Biosurface and Biotribology, 2*(1), 26–39. https://doi.org/10.1016/j.bsbt.2016.02.004

Sakai, N., Hosoda, N., Hagihara, Y., Sawae, Y., & Murakami, T. (2012). Analyses of functional mechanism of articular cartilage. *Biomechanisms (In Japanese), 21*, 251–263. https://doi.org/10.3951/biomechanisms.21.251

Sakai, N., Yarimitsu, S., Sawae, Y., Komori, M., & Murakami, T. (2018). Transitional behaviour between biphasic lubrication and soft elastohydrodynamic lubrication of poly(vinyl alcohol) hydrogel using microelectromechanical system pressure sensor. *Biosurface and Biotribology, 4*(1), 24–33. https://doi.org/10.1049/bsbt.2018.0001

Sakai, N., Yarimitsu, S., Sawae, Y., Komori, M., & Murakami, T. (2019). Biomimetic artificial cartilage: Fibre-reinforcement of PVA hydrogel to promote biphasic lubrication mechanism. *Biosurface and Biotribology, 5*(1), 13–19. https://doi.org/10.1049/bsbt.2018.0031

Sardinha, V. M., Lima, L. L., Belangero, W. D., Zavaglia, C. A., Bavaresco, V. P., & Gomes, J. R. (2013). Tribological characterization of polyvinyl alcohol hydrogel as substitute of articular cartilage. *Wear, 301*(1–2), 218–225. https://doi.org/10.1016/j.wear.2012.11.054

Sasada, T. (1988). Biomechanics and biomaterials – friction behaviour of an artificial articular cartilage. *Trans. 3rd World Biomaterials Congress*, 11.

Sasada, T., Tsukamoto, Y., & Mabuchi, K. (1988). *Biotribology - Friction and lubrication of synovial joints - (In Japanese)*. Sangyo Tosyo.

Sasaki, S., Omata, S., Murakami, T., Nagasawa, N., Taguchi, M., & Suzuki, A. (2018). Effect of gamma ray irradiation on friction property of poly(vinyl alcohol) cast-drying on freeze-thawed hybrid gel. *Gels, 4*(2), 30. https://doi.org/10.3390/gels4020030

Scholes, S. C., Burgess, I. C., Marsden, H. R., Unsworth, A., Jones, E., & Smith, N. (2006). Compliant layer acetabular cups: Friction testing of a range of materials and designs for a new generation of prosthesis that mimics the natural joint. *Proceedings of the Institution of Mechanical Engineers, Part H: Journal of Engineering in Medicine, 220*(5), 583–596. https://doi.org/10.1243/09544119H06404

Scholes, S. C., Unsworth, A., Blamey, J. M., Burges, I. C., Jones, E., & Smith, N. (2005). Design aspects of compliant, soft layer bearings for an experimental hip prosthesis. *Proceedings of the Institution of Mechanical Engineers, Part H: Journal of Engineering in Medicine, 219*(2), 79–87. https://doi.org/10.1243/095441105x9318

Sethuraman, A., Han, M., Kane, R. S., & Belfort, G. (2004). Effect of surface wettability on the adhesion of proteins. *Langmuir, 20*(18), 7779–7788. https://doi.org/10.1021/la049454q

Sigal, G. B., Mrksich, M., & Whitesides, G. M. (1998). Effect of surface wettability on the adsorption of proteins and detergents. *Journal of the American Chemical Society, 120*(14), 3464–3473. https://doi.org/10.1021/ja970819l

Stewart, T., Jin, Z. M., & Fisher, J. (1997). Friction of composite cushion bearings for total knee joint replacements under adverse lubrication conditions. *Proceedings of the Institution of Mechanical Engineers, Part H: Journal of Engineering in Medicine, 211*(6), 451–465. https://doi.org/10.1243/0954411981534574

Stoimenov, B., Fridrici, V., Kapsa, P., Kosukegawa, H., & Ohta, M. (2013). Bioengineering materials and conditions for obtaining low friction with PVA hydrogels. *Tribology Online, 8*(1), 140–152. https://doi.org/10.2474/trol.8.140

Suciu, A. N., Iwatsubo, T., Matsuda, M., & Nshino, T. (2004). A study upon durability of the artificial knee joint with PVA hydrogel cartilage. *JSME International Journal Series C, 47*(1), 199–208. https://doi.org/10.1299/jsmec.47.199

Suzuki, A., & Murakami, T. (2016). High-strength poly(vinyl-alcohol) hydrogels for artificial cartilage. In Ohshima H. (Ed.), *Encyclopedia of biocolloid and biointerface science* (Vol. 1, pp. 269–277). New York: John Wiley & Sons.

Suzuki, A., & Sasaki, S. (2015). Swelling and mechanical properties of physically crosslinked poly(vinyl alcohol) hydrogels. *Proceedings of the Institution of Mechanical Engineers, Part H: Journal of Engineering in Medicine, 229*(12), 828–844. https://doi.org/10.1177/0954411915615469

Suzuki, A., Sasaki, S., Sasaki, S., Noh, T., Nakashima, K., Yarimitsu, S., & Murakami, T. (2013). Elution and wear of PVA hydrogels by reciprocating friction. In *Extended abstract for world tribology congress 2013*.

Tomita, N., Kitakura, T., Onmori, N., Ikada, Y., & Aoyama, E. (1999). Prevention of fatigue cracks in ultrahigh molecular weight polyethylene joint components by the addition of vitamin E. *Journal of Biomedical Materials Research, 48*(4), 474–478. https://doi.org/10.1002/(SICI)1097-4636(1999)48:4<474::AID-JBM11>3.0.CO;2-T

Unsworth, A., Dowson, D., & Wright, V. (1975). Some new evidence on human joint lubrication. *Annals of the Rheumatic Diseases, 34*(4), 277–285. https://doi.org/10.1136/ard.34.4.277

Unsworth, A., Pearcy, M. J., White, E. F. T., & White, G. (1987). Soft layer lubrication of artificial hip joints. In *Proc. IMechE int. conf. on tribology - friction, lubrication and wear, FIFTY years on, (LONDON, U.K.: July. 1-3, 1987)* (Vol. II, pp. 715–724).

Vermeer, A. W. P., Bremer, M. G. E. G., & Norde, W. (1998). Structural changes of IgG induced by heat treatment and by adsorption onto a hydrophobic Teflon surface studied by circular dichroism spectroscopy. *Biochimica et Biophysica Acta - General Subjects, 1425*(1), 1–12. https://doi.org/10.1016/S0304-4165(98)00048-8

Willert, H.-., & Semlitsch, M. (1977). Reactions of the articular capsule to wear products of artificial joint prostheses. *Journal of Biomedical Materials Research, 11*(2), 157–164. https://doi.org/10.1002/jbm.820110202

Wittemann, A., & Ballauff, M. (2004). Secondary structure analysis of proteins embedded in spherical polyelectrolyte brushes by FT-IR spectroscopy. *Analytical Chemistry, 76*(10), 2813–2819. https://doi.org/10.1021/ac0354692

Yamaguchi, T., & Murakami, T. (2012). Surface friction and bulk transport properties in hydrogels. *Preprint JAST Conference*, 199–200.

Yarimitsu, S., Murakami, T., & Suzuki, A. (2018). Development of Poly(vinyl alcohol) hybrid gel as artificial cartilage material with low friction and wear. In *Abstract S6.4 of 4th International conference of BioTribology*.

Yarimitsu, S., Nakashima, K., Murakami, T., Okazaki, K., Matsuda, S., & Iwamoto, Y. (2015). Evaluation of in vivo fixation and stabilization of hydrogel artificial cartilage (In Japanese). *Japanese Journal of Clinical Biomechanics, 36*, 201–205.

Yarimitsu, S., Nakashima, K., Sawae, Y., & Murakami, T. (2007). Study on the mechanisms of wear reduction of artificial cartilage through in situ observation on forming protein boundary film. *Tribology Online, 2*(4), 114–119. https://doi.org/10.2474/trol.2.114

Yarimitsu, S., Nakashima, K., Sawae, Y., & Murakami, T. (2009). Influences of lubricant composition on forming boundary film composed of synovia constituents. *Tribology International, 42*(11–12), 1615–1623. https://doi.org/10.1016/j.triboint.2008.11.005

Yarimitsu, S., Nakashima, K., Sawae, Y., & Murakami, T. (2013). Influence of phospholipid and protein constituents on tribological properties of artificial hydrogel cartilage material. *Journal of Biomechanical Science and Engineering, 8*(3), 257–267. https://doi.org/10.1299/jbse.8.257

Yarimitsu, S., Sasaki, S., & Murakami, T. (2016). Evaluation of lubrication property of hydrogel artificial cartilage rubbed against natural articular cartilage (In Japanese). *Japanese Journal of Clinical Biomechanics, 37*, 225–229.

Yarimitsu, S., Yoshida, A., Sasaki, S., Murakami, T., & Suzuki, A. (2016). Evaluation of lubrication property of poly(vinyl alcohol) hybrid gel for artificial articular cartilage. *Tribology Online, 11*(2), 360–365. https://doi.org/10.2474/trol.11.360

Yasuda, K., Gong, J. P., Katsuyama, Y., Nakayama, A., Tanabe, Y., Kondo, E., Ueno, M., & Osada, Y. (2005). Biomechanical properties of high-toughness double network hydrogels. *Biomaterials, 26*(21), 4468–4475. https://doi.org/10.1016/j.biomaterials.2004.11.021

Zywiel, M. G., Sayeed, S. A., Johnson, A. J., Schmalzried, T. P., & Mont, M. A. (2011). Survival of hard-on-hard bearings in total hip arthroplasty: A systematic review. *Clinical orthopaedics and related research, 469*(6), 1536–1546. http://doi.org/10.1007/s11999-010-1658-0.

Biotribology of regenerated cartilage

6

6.1 Medical treatments to repair focal chondral defects

Articular cartilage plays very important roles in load-support and smooth movement with very low friction and low wear property as a lubricious rubbing surface in synovial joints. Healthy cartilage sustains its excellent tribological property for a whole life, but cartilage tissue can be deteriorated or damaged due to various diseases with aging, inflammatory activation, and excessive stresses and injuries by high loading. As mentioned in Chapter 3, the articular cartilage is compliant and hydrated tissue with high water content of 70%−80% as biphasic material which is composed of the chondrocytes and the extracellular matrix of proteoglycan reinforced by collagen fibers (predominately type II). It has minimal self-healing ability due to the lack of blood supply as avascular nature, lack of nerve supply, and the sparse population of immobile chondrocytes. Here, chondrocytes which are surrounded by the extracellular matrix cannot migrate to the site of injury.

So far, several kinds of medical treatments to repair focal defects of articular cartilage have been applied in clinical fields (Davies & Kuiper, 2019). As a medical treatment without cartilage transfer, in microfracture surgery, drilling is performed at the injury site by creating tiny fractures in the underlying bone to encourage cell migration to undertake natural repair mechanism. In this section, the essential points in the following three kinds of medical treatments are described.

6.1.1 Osteochondral autograft transplant (mosaicplasty)

Osteochondral autograft transplant (mosaicplasty) in knee joint involves filling the focal chondral defect in the weight bearing portion by harvesting bone and intact articular cartilage as tubes from a less weight bearing portion of the knee of the same patient (Fig. 6.1). After osteochondral autografting at the lesion area, the repaired cartilage has load-carrying ability, but this technique is limited for harvesting intact cartilage and risk in residual damage in harvested parts. As healthy osteochondral plugs, usually cylindrical shapes were adopted and the gap space left between the cylindrical plugs. Fibrocartilage filling of the gap space between cylindrical plugs and insufficient donor cartilage reserve are among the disadvantage of classical mosaicplasty. Erol and Karakoyun (2016) indicated that the application of hexagonally shaped osteochondral plugs was effective in honeycomb pattern to reduce the gap space. It is expected that the hexagonal geometry of plugs has advantages including surface congruity and stability.

In contrast, for large chondral and osteochondral defects, the osteochondral allograft transfer has been applied as transfer of a cartilage allograft sourced from a

Biotribology of Natural and Artificial Joints. https://doi.org/10.1016/B978-0-12-823669-7.00005-3

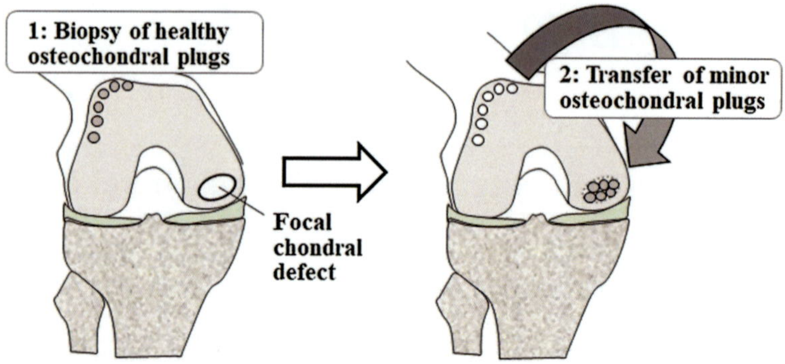

Figure 6.1 Osteochondral autograft transplant (mosaicplasty). Osteochondral grafts harvested from the nonweight bearing areas of knee are transferred to the focal chondral defect lesion.

cadaver or tissue bank into the patient at the site of injury. This treatment has no risk of donor site morbidity, but has allogenic tissue problems with graft availability, cell viability, disease and immune responses.

6.1.2 Autologous chondrocyte implantation (ACI)

Autologous chondrocyte implantation (ACI) was introduced for humans by Brittberg et al. (1994) as the first attempt at a cell-based biological therapy for the treatment of deep focal cartilage defects (Davies & Kuiper, 2019). As shown in Fig. 6.2 for ACI, cartilage samples are harvested from the lesser weight bearing area of knee joint by arthroscopy. The cartilage biopsy sample is enzymatically digested in order to release

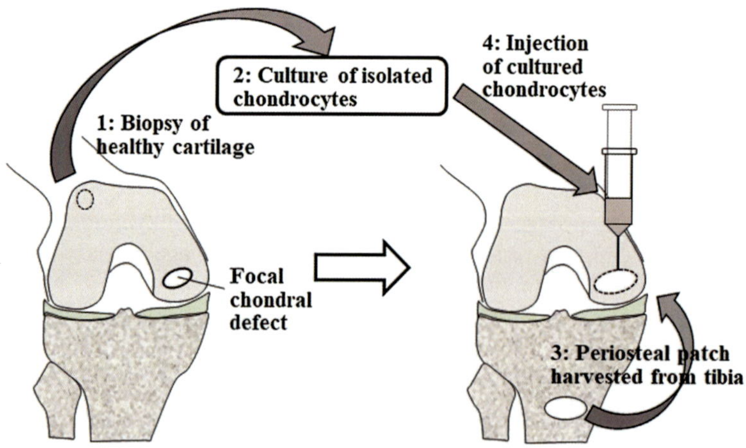

Figure 6.2 Autologous chondrocyte implantation. The cultured chondrocytes in suspension are injected under the periosteal patch into the defect.

the autologous chondrocytes which are then cultured for several weeks. At the second stage of the surgical procedure, the lesion is initially debrided before a periosteal patch is harvested from the tibia. Then, the periosteal patch is sutured over the lesion site and the edges are sealed, leaving one interface free. The cultured chondrocytes in suspension are injected under the periosteal patch into the defect and the final patch-lesion interface is sealed. ACI has been a worldwide, well-established surgical technique. In this method using monolayer culture, there is difficulty in the maintenance of chondrocyte phenotype during a prolonged culture. In addition, this method has the risk of chondrocyte leakage from the graft site during load bearing. In contrast, in the third generation ACI which consists of suspending expanded chondrocytes in a hydrated scaffold, a native or synthetic periosteal patch was removed and better control of cell distribution was attained through the defect.

6.1.3 Implantation of regenerated cartilage tissues derived from chondrocytes or chondrocytes differentiated from related stem cells

As described above, the autologous chondrocyte implantation (ACI) using the monolayer culture method for chondrocyte transplantation has issues to be solved for the maintenance of chondrocyte phenotype during monolayer culture and the risk of chondrocyte leakage from the graft site. In a monolayer culture, the chondrocytes tend to assume a fibroblastic dedifferentiated morphology and synthesized predominantly type I collagen. In contrast, the culture in 3-dimensional systems using various scaffolds such as collagen, alginate, and agarose gels has superior maintenance of the chondrocyte phenotype.

For clinical application, appropriate strategy for immunogenicity and safety should be found. Uchio et al. (2000) evaluated human chondrocyte proliferation and matrix synthesis cultured in atelocollagen gel. It was shown in this study that chondrocytes embedded in the atelocollagen gel gradually proliferated and produced chondroitin 6-sulfate, maintaining the chondrocyte phenotype for up to 4 weeks. Ochi et al. (2002) indicated that transplanting chondrocytes in atelocollagen gel for focal chon dral defect can promote restoration of the articular cartilage of the knee in clinical application for patients with trauma and osteochondritis dissecans. The transplanting chondrocytes retained phenotype, reduced leakage, and had more homogeneous distribution of cells in the defect. Transplanting process for autologous cultured cartilage (cultured chondrocyte-atelocollagen gel construct) is shown in Fig. 6.3.

From the viewpoint of tissue engineering, three factors of "cell sources," "scaffolds," and "culture conditions (environment) including signaling factors" appear to play main roles to improve the function of regenerated cartilage (Davies & Kuiper, 2019; Lee et al., 2005). Cell sources are selected from native chondrocytes, mesenchymal stem cells, human embryonic stem (ES) cells, and induced pluripotent stem (iPS) cells (Tsumaki et al., 2015).

In this chapter some examples for regenerated cartilage tissues are shown.

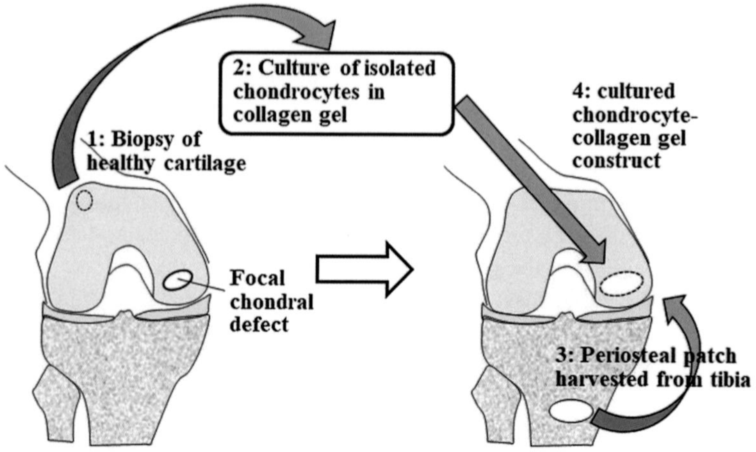

Figure 6.3 Implantation of regenerated cartilage tissues. The cultured chondrocyte-atelocollagen gel construct is transplanted in the chondral defect.

6.2 The influence of mechanical stimulation on regenerated cartilage tissue

The chondrocytes in articular cartilage play important roles in metabolism of cartilage tissue under various mechanical stimulations such as compressive load and shearing. In development of tissue engineered articular cartilage, mechanical stimuli such as dynamic compression (Lee et al., 2000; Sah et al., 1989), tissue shear (Jin et al., 2001), and hydrostatic pressure (Mizuno et al., 2002) have been applied. For example, it is expected that the applying a dynamic compressive deformation to the tissue model would not only improve diffusion of culture medium into the tissue and accelerate the supply of nutrients but also activate cell-signaling via mechanosensor and mechanotransduction pathways. It is well known that loading mechanical stimuli to regenerated-cartilage tissue and chondrocytes is distinctly important for both enhancing synthesis of ECM and inducing chondrogenesis differentiation of stem cells (Grad et al., 2011).

6.2.1 *Effects of dynamic compression loading*

Omata et al. (2012) examined the influence of both mechanical stimulation and vitamin C (VC) in culture medium on development of the extracellular matrix (ECM) and improvement in mechanical properties of a chondrocyte-agarose construct in a regenerated cartilage model of hyaline cartilage. In this study, the primary bovine chondrocytes were isolated from metacarpo-phalangeal joints of steers. The cylindrical chondrocyte-agarose constructs (diameter of 4 mm and height of 2 mm) were used in culture conditions at free-swelling or at uniaxial compressive strain, and two types of VC in culture medium to control collagen synthesis, ascorbic acid (AsA) as an acidic form and ascorbic acid 2-phosphate (A2P) as a nonacidic form, for suppressing the cytotoxicity and effective development of the mechanical characteristics of the tissue, were given in a purpose-built bioreactor (Fig. 6.4). Here, uniaxial unconfined compression up to a maximum amplitude of 15% was applied in a triangular waveform at a frequency of 1 Hz for 6 h and subsequently offloaded with the plate resting for each construct over the subsequent 18 h. Control constructs were cultured with both upper and lower plates for diffusion through the sides alone. Individual constructs were evaluated after culture periods of 22 days.

To evaluate the influence of the mechanical stimulation on the cultured construct, the tangent moduli of each group after 22 days of culture is shown in Fig. 6.5. Comparison within the four free-swelling groups only indicated significant differences between the without VC and the AsA (2.2 pmol/10^9 cells) groups, the latter revealing an increased tangent modulus. By applying cyclic compression to the construct with addition of VC to the culture medium, there was not a clear difference between free-swelling and compression groups for AsA. In contrast, both tangent moduli of the two compression groups were higher than those of several free-swelling samples in each A2P dose group (A2P(6.4) and A2P(32)). Each tangent modulus of the control groups with A2P was ranked against the free-swelling and compression groups. It is not evident that there were differences between each pair of the three compression groups (AsA(2.2), A2P(6.4) and A2P(32)).

On the effects of VC when added to the medium in free-swelling culture condition, the chondrocyte-agarose construct for AsA with uniform distribution of ECM showed significantly higher tangent modulus than the construct without VC, in which there

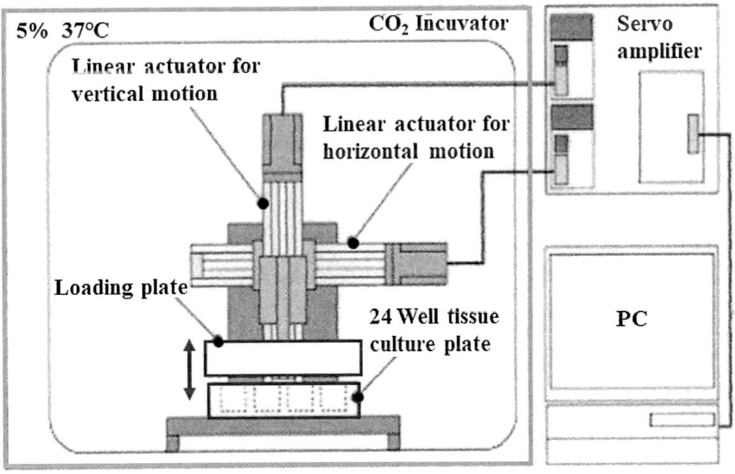

Figure 6.4 Mechanical loading system mounted within tissue culture incubator. Loading plate with 22 plungers are coupled with a 24 culture plate.
Modified from Omata, S., Sonokawa, S., Sawae, Y. & Murakami, T. (2012). Effect of both vitamin C and mechanical stimulation on improving the mechanical characteristics of regenerated cartilage. *Biochemical and Biophysical Research Communications*, *424*(4), 724–729. https://doi.org/10.1016/j.bbrc.2012.07.019.

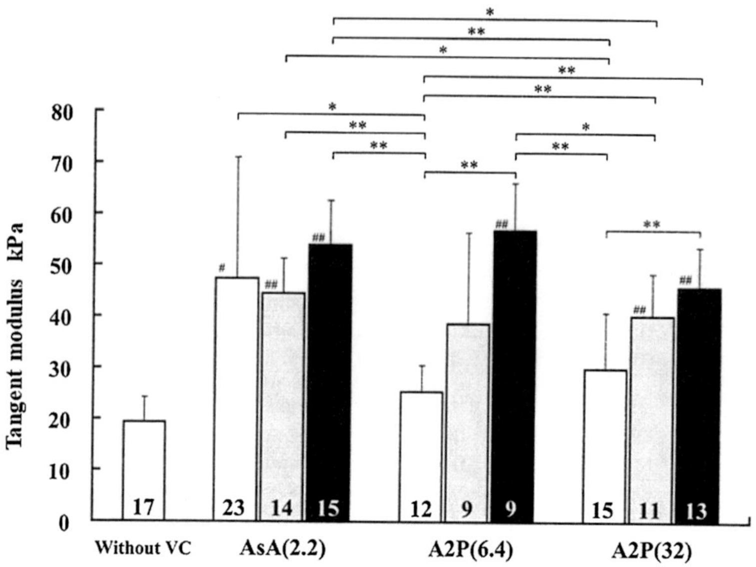

Figure 6.5 Comparison of the tangent moduli after 22 days of culture. Cultures were terminated following free-swelling (white column), control (gray), or compression (black) conditions. Each number in the columns is the cultured sample number and each sample represents an individual culture. The error bar represents SD. Sharps (# and ##) and asterisks (*and **) show respective statistical significance between control group and each group in Welch's t-test. #, *: $d > 1$, $P < .01$, $(1-\beta) > 0.8$; ##, **: $d > 2$, $P < .01$, $(1-\beta) > 0.8$. Modified from Omata, S., Sonokawa, S., Sawae, Y. & Murakami, T. (2012). Effect of both vitamin C and mechanical stimulation on improving the mechanical characteristics of regenerated cartilage. *Biochemical and Biophysical Research Communications*, *424*(4), 724–729. https://doi.org/10.1016/j.bbrc.2012.07.019.

was only a limited amount of ECM. A2P downregulated development of ECM and suppressed improvement of the tangent modulus more than AsA. In contrast, the application of mechanical stimulation as compression to the construct both increased the tangent modulus more than the free-swelling group containing A2P and enhanced the ECM network of inner tissue.

The effects of compression stimulation and addition of VC after 22 days were confirmed in immunofluorescence images for type I and type II collagen and chondroitin sulfate of ECM. As shown for A2P(32) in Fig. 6.6, it is noticed that the collagen network in compression culture had become dense compared with free-swelling construct. Furthermore, at the compression stimulation, the collagen network appeared to have interconnected among chondrocytes by type II collagen network as indicted by arrows.

Thus, mechanical stimulation and strain appear to enhance the supply of nutrients and improve the synthesis of ECM via mechanotransduction pathways of chondrocytes. Therefore, it is suggested that mechanical stimulation is necessary for homogeneous development of ECM in a cell-associated construct with a view to implantation of a large-sized articular cartilage defect.

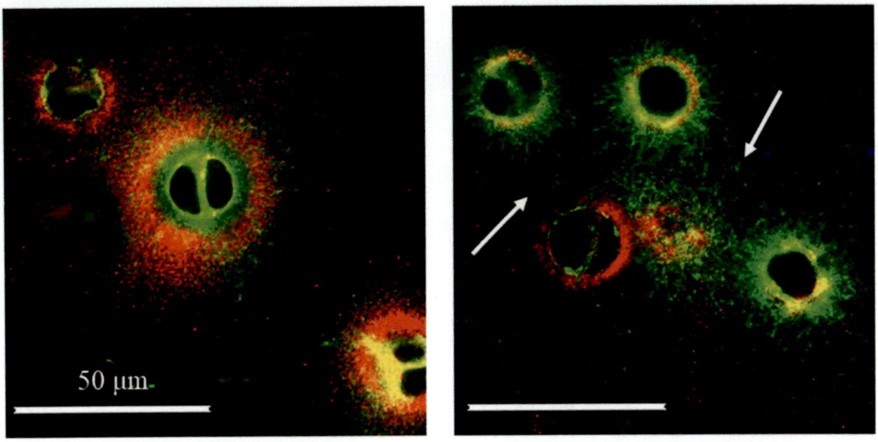

(a) Free swelling culture (b) Compression culture

Figure 6.6 High-magnified immunofluorescence images of ECMs in the chondrocyte-agarose constructs. (a) free swelling and (b) compression culture conditions with A2P(32) for 22 days. Blue: type I collagen; green: type II collagen; red: chondroitin sulfate. *Arrows* indicate the interconnect among chondrocytes by type II collagen network.
Modified from Omata, S., Sonokawa, S., Sawae, Y. & Murakami, T. (2012). Effect of both vitamin C and mechanical stimulation on improving the mechanical characteristics of regenerated cartilage. *Biochemical and Biophysical Research Communications*, *424*(4), 724–729. https://doi.org/10.1016/j.bbrc.2012.07.019.

Here, it should be kept in mind that compressive loading at human joint level affects many different physical, electrical, and biochemical phenomena at the cellular level, including deformation, fluid flow, hydrostatic pressure, streaming potentials, osmotic pressure, nutrient and ion concentration gradients, and pH changes (Lee et al., 2005). Further elucidation on the mechanism of compression stimulation is required.

6.2.2 Effects of shearing

Waldman et al. (2003) investigated whether the intermittent application of cyclic shearing forces during tissue formation would improve the tissue quality. Porous calcium polyphosphate (CPP) cylinder was prepared as the substrate. Chondrocytes were isolated from bovine metacarpal-carpal articular cartilage. Chondrocytes were seeded on the cylindrical porous CPP substrate surfaces for static 4 week culture. Intermittent shear stimulation for the cartilaginous tissue-CPP substrate construct was performed using a two-axis (compression and translational shear) mechanical tester. Cartilaginous tissues were subjected to sinusoidal shear loading at different shear strain amplitudes (2%, 6%, or 12% of the tissue height) and duration (400 or 2000 cycles per day) at a frequency of 1 Hz followed by a rest period of 48 h. After shear stimulation was applied to the tissue every other day for 1 week, the largest increases in both collagen and proteoglycan synthesis (23 ± 6% and 20 ± 6% above control, respectively) were observed with shear stimulation applied at a 2% strain amplitude at 1 Hz for 400 cycles every second day (Fig. 6.7). Thus, lower amplitudes increased matrix synthesis whereas higher amplitudes appeared to suppress matrix synthesis.

In the long-term experiment with intermittent cyclic shear strain amplitude of 2%, collagen and proteoglycan contents in extracellular matrix were increased as shown in Fig. 6.8. Furthermore, the compressive stiffness had a sixfold increase in comparison with the unstimulated controls.

The improvement of tissue properties is expected by appropriate shearing stimulus.

6.3 The influence of tribological stimulation on regenerated cartilage tissue

6.3.1 Effects of sliding motion

To elucidate the influences of biomechanical stimuli on the functions and properties of regenerated cartilage in daily activities, evaluation under conditions including complex loading with articulating motion is required. From the viewpoint of tribology, Wimmer et al. (2004) developed the pin-on-ball bioreactor to reproduce joint kinematics closely, which consists of a rotating scaffold and/or cartilage pin and orthogonally rotating alumina ball under compressive loading. Grad et al. (2005) investigated the effect of articular motion on the gene expression of superficial zone protein (SZP) (also known as lubricin, proteoglycan 4 (PRG4), or megakaryocyte stimulating factor)

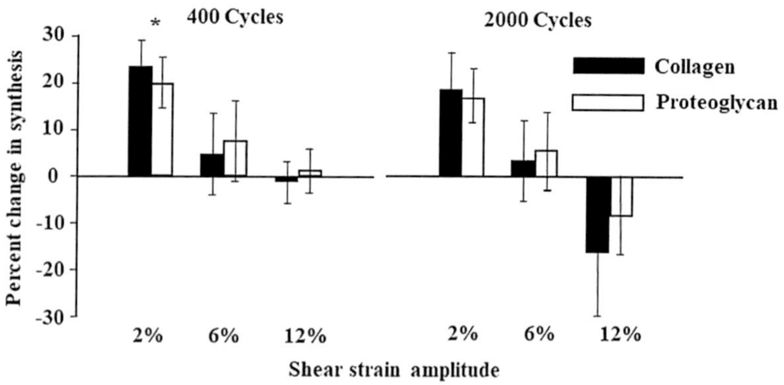

Figure 6.7 Changes in proteoglycan and collagen synthesis expressed relative to the unstimulated controls following 1 week of intermittent shear stimulation (three stimulation periods). Two separate experiments were performed, results pooled and expressed as the mean standard error of the mean (n = six to eight for all groups). (*) Denotes the condition used in the "long-term" experiments.

Modified from Waldman, S. D., Spiteri, C. G., Grynpas, M. D., Pilliar, R. M. & Kandel, R. A. (2003). Long-term intermittent shear deformation improves the quality of cartilaginous tissue formed in vitro. *Journal of Orthopedic Research*, *21*(4), 590–596. https://doi.org/10.1016/S0736-0266(03)00009-3

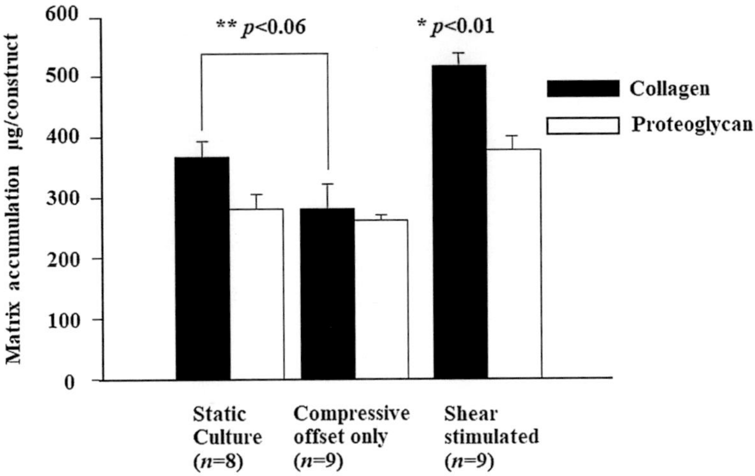

Figure 6.8 Collagen (solid bars) and proteoglycan (open bars) content of the in vitro-formed cartilaginous tissues (at 8 weeks) cultured in the presence or absence of 4 weeks of shear stimulation. As a control, cultures were also subjected to only the static compressive offset for an equivalent length of time (400 s per day). Three separate experiments were performed, results pooled, and expressed as the mean ± standard error of the mean (n = total number of tissues analyzed). (*) Significantly different than all other groups ($P < 0.01$) and (**) trend between static controls and static compressive offset ($P < 0.06$).
Modified from Waldman, S. D., Spiteri, C. G., Grynpas, M. D., Pilliar, R. M. & Kandel, R. A. (2003). Long-term intermittent shear deformation improves the quality of cartilaginous tissue formed in vitro. *Journal of Orthopedic Research*, *21*(4), 590−596. https://doi.org/10.1016/S0736-0266(03)00009-3

and hyaluronan synthases (HASs) and on the release of SZP and hyaluronan by bovine articular cartilage chondrocytes seeded onto biodegradable porous polyurethane scaffolds, using this bioreactor. It is noticed that surface motion significantly upregulated the mRNA expression of SZP and HASs. Axial compression alone had no effect on SZP and increased HAS mRNA only at high strain amplitude. SZP was immunodetected only in the media of constructs exposed to surface articular motion. The release of hyaluronan into the culture medium was significantly enhanced by surface motion.

In related study (Grad et al., 2012), the effects of sliding motion on stiffness and friction were examined based on atomic force microscopy (AFM) for the chondrocyte-polyurethane construct cultured under dynamic compression loading conditions for 1 h daily with sliding (loading group: LG2) and without sliding motion (LG1) (Fig. 6.9). Unloaded construct was used as control. Construct analysis was performed after a total culture time of 4 weeks. As shown in Fig. 6.10, friction in phosphate buffered saline (PBS) containing protease inhibitor cocktail significantly decreased in LG2 with sliding motion. In AFM indentation tests, both micrometer scale and nanometer scale stiffness values were different between three groups. Stiffness values of LG2 were significantly enhanced compared to both unloaded controls and LG1. Fig. 6.11 shows immunolabeling of cell-scaffold constructs for collagen

Figure 6.9 Bioreactor used for mechanical conditioning of cell-scaffold constructs. A ceramic ball was pressed onto the cell-seeded scaffold. In scaffolds of LG1 and LG2, the ball oscillated at 1 Hz between 10% and 20% of the scaffold height (in the center of the construct). In scaffolds of LG2, the ball additionally oscillated about an axis perpendicular to the scaffold axis at amplitude of ±25 degrees and 1 Hz.
Modified from Grad, S., Loparic, M., Peter, R., Stolz, M., Aebi, U. & Alini, M. (2012). Sliding motion modulates stiffness and friction coefficient at the surface of tissue engineered cartilage. *Osteoarthritis and Cartilage*, *20*(4), 288–295. https://doi.org/10.1016/j.joca.2011.12.010.

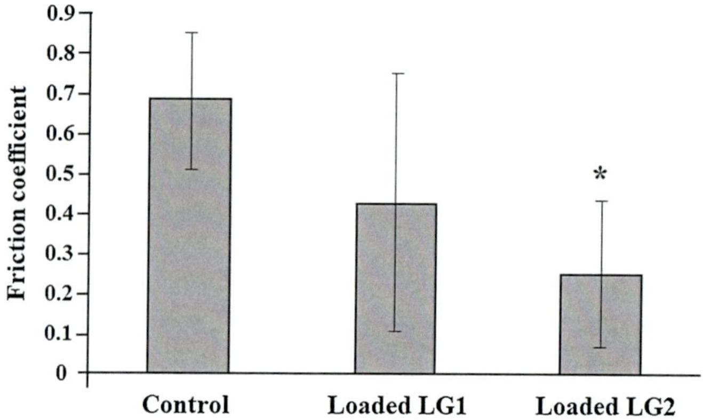

Figure 6.10 Coefficient of friction of the surface of cell-scaffold constructs measured by AFM. Constructs of LG1 were stimulated by dynamic compression only; constructs of LG2 were stimulated by dynamic compression and sliding surface motion; controls were not loaded. * = significant difference versus control ($P = .033$) (mean ± 95% confidence interval; $n = 4$).
Modified from Grad, S., Loparic, M., Peter, R., Stolz, M., Aebi, U. & Alini, M. (2012). Sliding motion modulates stiffness and friction coefficient at the surface of tissue engineered cartilage. *Osteoarthritis and Cartilage*, *20*(4), 288–295. https://doi.org/10.1016/j.joca.2011.12.010.

II, collagen I, aggrecan, and lubricin. Type II collagen staining was observed in both loaded and unloaded constructs, but in construct of LG2 the staining appeared more uniform than others. Collagen type I staining was visible as a thin layer at the surface of loaded constructs, especially from LG2. Aggrecan staining was readily observed in all constructs. Here, it is noticed that LG2 exhibited strong immunoreactivity for lubricin at the top surface. Lubricin at surface appeared to have contributed to reduction of friction in LG2 with sliding motion.

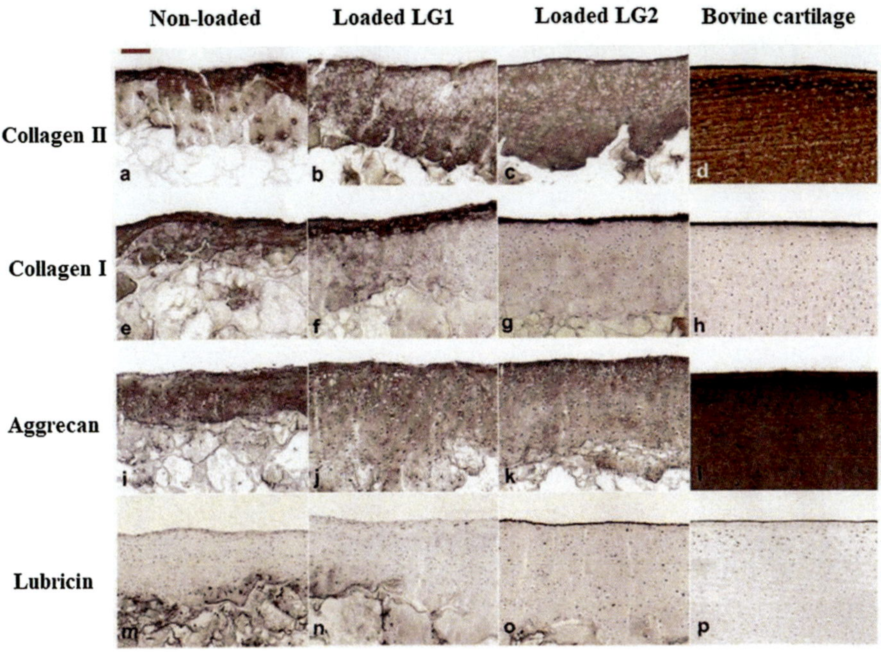

Figure 6.11 Immunolabeling of cell-scaffold constructs for collagen type II, collagen type I, aggrecan, and lubricin. Constructs of LG1 were stimulated by dynamic compression only; constructs of LG2 were stimulated by dynamic compression and sliding surface motion. Immunolabeling characteristics of bovine articular cartilage used as positive control are also shown. Scale bar = 100 μm.
Modified from Grad, S., Loparic, M., Peter, R., Stolz, M., Aebi, U. & Alini, M. (2012). Sliding motion modulates stiffness and friction coefficient at the surface of tissue engineered cartilage. *Osteoarthritis and Cartilage*, *20*(4), 288–295. https://doi.org/10.1016/j.joca.2011.12.010.

Effects of dynamic compression (10%–20%, 0.1 Hz) with oscillating rotation (± 7.5°at 0.1 Hz) of the scaffold and/or ball oscillation (± 25°at 0.1 Hz) as unidirectional and multidirectional motion patterns on gene expression of chondrocytes seeded in 3D scaffolds were investigated by Grad et al. (2006). It is noticed that the proteoglycan 4 (PRG4, SZP or lubricin) and cartilage oligomeric matrix protein (COMP) mRNA expression were markedly increased by ball oscillation. In addition, the collagen type II mRNA expression was enhanced in ball oscillation, while the aggrecan and tissue inhibitor of metalloproteinase-3 (TIMP-3) mRNA expression levels were upregulated by multidirectional articular motion (both ball and scaffold oscillation). Ball oscillation also increased the release of PRG4, COMP, and HA into the culture media. It is indicated that these applied stimuli can contribute to the maintenance of the chondrocytic phenotype. It is suggested that the mechanical effects causing cell stimulation via surface motion might be related to fluid film buildup and/or shear within the joint space. The oscillating ball drags the fluid into the joint space, thereby probably causing biophysical effects similar to those of fluid flow.

6.3.2 Frictional behaviors of regenerated cartilage different in cultivation conditions

Morita et al. (2006) examined the frictional behaviors of the regenerated cartilage as chondrocyte-fibroin sponge construct against a stainless steel plate in physiological saline solution in reciprocating apparatus. The applied load was 0.03 N, the stroke length was 20 mm, and the mean sliding velocity was 0.8 mm/s. Friction testing, histological observation, and measurements of the amount of sulfated glycosaminoglycan were carried out at 0, 3, 7, 14, and 28 days after the beginning of cultivation. Fig. 6.12 shows representative result for the transient friction coefficient throughout the sliding test in the case of the 30 s preload. It is noticed that at early stage during sliding test, friction coefficient decreased remarkably after 7 days cultivation. Friction after 28 days cultivation approached to that of porcine cartilage at initial stage but deviated to higher level with increased sliding time. Fig. 6.13 shows the histology of the regenerated cartilage. After 7 days cultivation (Fig. 6.13c), the newly formed tissue exhibited intense staining with safranin-O (- fast green for glycosaminoglycan), and the pores of fibroin sponge were filled with extracellular matrix, which was also present at the surface of fibroin sponge. The thickness of the surface layer increased with duration of cultivation (Fig. 6.13c-e). The increase in concentration of chondroitin sulfate with cultivation was confirmed. In histological photomicrograph of regenerated cartilage (28 days cultivation) after the sliding test, the newly formed surface layer containing proteoglycan remained significantly but partially removed (Fig. 6.13f), which might have increased friction accompanied with the exudation of interstitial water.

In related study on the effects of the loading history for tribological stimulation on frictional properties of regenerated cartilage as chondrocytes-fibroin constructs (Yamamoto et al., 2008), the combination of tribological stimulation (dynamic culture) and static culture showed lower friction than the combination of tribological stimulation and tribological one. On this mechanism, it was suggested that tribological movements in pre-conditioning period can lead to effective tribological functions and the post-conditioning in static culture can reinforce the effects under in-vitro cartilage regeneration.

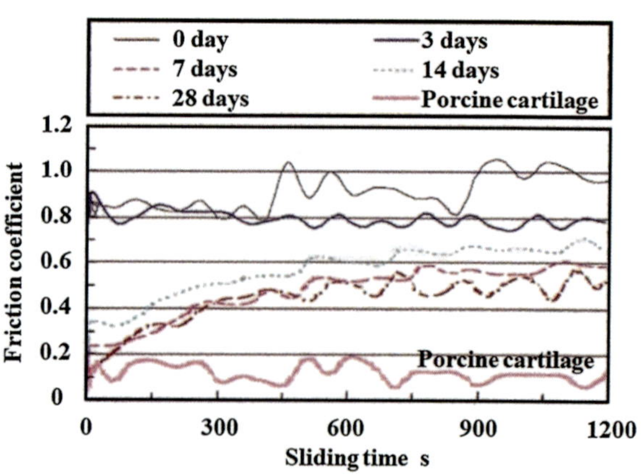

Figure 6.12 Transient friction coefficient through the test in the case of the 30 s preload. Initial friction for regenerated cartilage decreases after 7 days' cultivation but gradually increases with rubbing. Modified from Morita, Y., Tomita, N., Aoki, H., Sonobe, M., Wakitani, S., Tamada, Y., Suguro, T. & Ikeuchi, K. (2006). Frictional properties of regenerated cartilage in vitro. *Journal of Biomechanics, 39*(1), 103–109. https://doi.org/10.1016/j.jbiomech.2004.10.031.

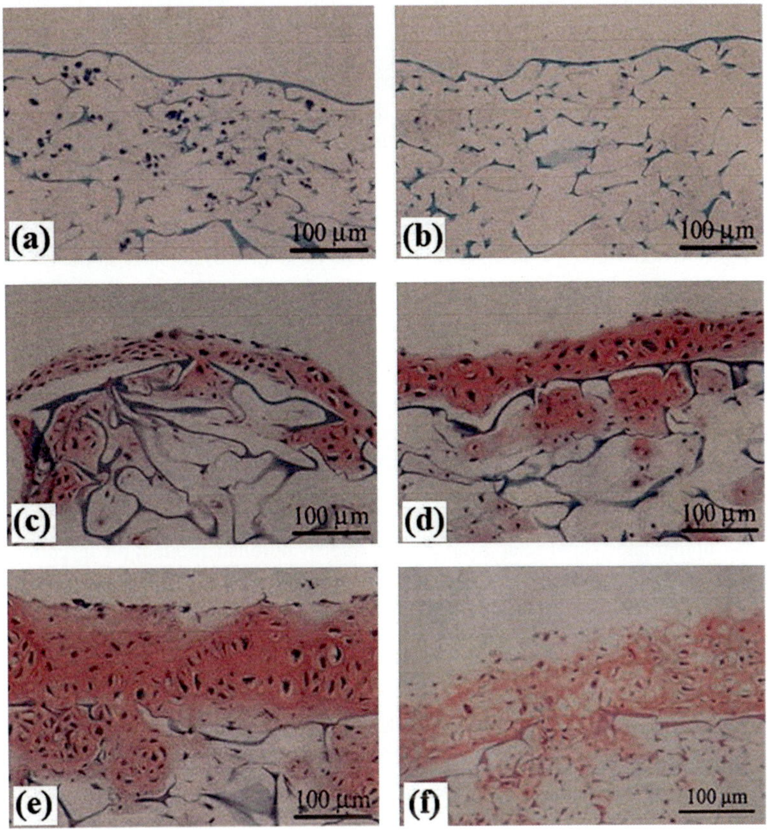

Figure 6.13 Histological photomicrographs of the regenerated cartilage (Safranin-O staining). (a) 1 day, (b) 3 days, (c) 7 days, (d) 14 days, (e) 28 days, (f) after sliding test (30 s preload) at 28 days' cultivation.
Modified from Morita, Y., Tomita, N., Aoki, H., Sonobe, M., Wakitani, S., Tamada, Y., Suguro, T. & Ikeuchi, K. (2006). Frictional properties of regenerated cartilage in vitro. *Journal of Biomechanics*, *39*(1), 103–109. https://doi.org/10.1016/j.jbiomech.2004.10.031.

In the study on cell aggregate (spheroid) culture for fibroin sponge seeded with chondrocyte aggregates (Kachi et al., 2009), the aggregate-derived cartilage tissue exhibited higher safranin O staining intensity and positive immunostaining for collagen type II. However, its friction was higher than that of regenerated cartilage derived from individual cells. Poor tribological function of aggregate-derived cartilage tissue is probably caused by its immature functional layer structure. Appropriate stimulation is expected to be effective in improving of tribological property of regenerated cartilage.

The effects of tribological stimulation on the function of regenerated cartilage have been confirmed, but further research is required to elucidate the detailed mechanism.

References

Brittberg, M., Lindahl, A., Nilsson, A., Ohlsson, C., Isaksson, O., & Peterson, L. (1994). Treatment of deep cartilage defects in the knee with autologous chondrocyte transplantation. *New England Journal of Medicine, 331*(14), 889–895. https://doi.org/10.1056/NEJM199410063311401

Davies, R. L., & Kuiper, N. J. (2019). Regenerated medicine: A review of the evolution of autologous chondrocyte implantation (ACI) therapy. *Bioengineering, 6*, 1–16.

Erol, M. F., & Karakoyun, O. (2016). A new point of view for mosaicplasty in the treatment of focal cartilage defects of knee joint: Honeycomb pattern. *SpringerPlus, 5*(1). https://doi.org/10.1186/s40064-016-2796-y

Grad, S., Eglin, D., Alini, M., & Stoddart, M. J. (2011). Physical stimulation of chondrogenic cells in vitro: A review. *Clinical Orthopaedics and Related Research, 469*(10), 2764–2772. https://doi.org/10.1007/s11999-011-1819-9

Grad, S., Gogolewski, S., Alini, M., & Wimmer, M. A. (2006). Effects of simple and complex motion patterns on gene expression of chondrocytes seeded in 3D scaffolds. *Tissue Engineering, 12*(11), 3171–3179. https://doi.org/10.1089/ten.2006.12.3171

Grad, S., Lee, C. R., Gorna, K., Gogolewski, S., Wimmer, M. A., & Alini, M. (2005). Surface motion upregulates superficial zone protein and hyaluronan production in chondrocyte-seeded three-dimensional scaffolds. *Tissue Engineering, 11*(1–2), 249–256. https://doi.org/10.1089/ten.2005.11.249

Grad, S., Loparic, M., Peter, R., Stolz, M., Aebi, U., & Alini, M. (2012). Sliding motion modulates stiffness and friction coefficient at the surface of tissue engineered cartilage. *Osteoarthritis and Cartilage, 20*(4), 288–295. https://doi.org/10.1016/j.joca.2011.12.010

Jin, M., Frank, E. H., Quinn, T. M., Hunziker, E. B., & Grodzinsky, A. J. (2001). Tissue shear deformation stimulates proteoglycan and protein biosynthesis in bovine cartilage explants. *Archives of Biochemistry and Biophysics, 395*(1), 41–48. https://doi.org/10.1006/abbi.2001.2543

Kachi, N., Tomita, N., Yamamoto, K., Takaya, R., & Tamada, Y. (2009). Tribological maturation of regenerated cartilage was inhibited by using chondrocyte aggregates. *Journal of Biomechanical Science and Engineering, 4*(2), 174–181. https://doi.org/10.1299/jbse.4.174

Lee, C., Grad, S., Wimmer, M., & Alini, M. (2005). The influence of mechanical stimuli on articular cartilage tissue engineering. In N. Ashammakhi R.L. Reis (Series Eds.) & N. Ashammakhi, & R. Reis, L. (Vol. Eds.), *The influence of mechanical stimuli on articular cartilage tissue engineering: Vol 2. Topics in Tissue Engineering* (pp. 1–32). Oulu University.

Lee, D. A., Noguchi, T., Frean, S. P., Lees, P., & Bader, D. L. (2000). The influence of mechanical loading on isolated chondrocytes seeded in agarose constructs. *Biorheology, 37*(1–2), 149–161. http://iospress.metapress.com/content/0006-355x/.

Mizuno, S., Tateishi, T., Ushida, T., & Glowacki, J. (2002). Hydrostatic fluid pressure enhances matrix synthesis and accumulation by bovine chondrocytes in three-dimensional culture. *Journal of Cellular Physiology, 193*(3), 319–327. https://doi.org/10.1002/jcp.10180

Morita, Y., Tomita, N., Aoki, H., Sonobe, M., Wakitani, S., Tamada, Y., Suguro, T., & Ikeuchi, K. (2006). Frictional properties of regenerated cartilage in vitro. *Journal of Biomechanics, 39*(1), 103–109. https://doi.org/10.1016/j.jbiomech.2004.10.031

Ochi, M., Uchio, Y., Kawasaki, K., Wakitani, S., & Iwasa, J. (2002). Transplantation of cartilage-like tissue made by tissue engineering in the treatment of cartilage defects of the

knee. *Journal of Bone and Joint Surgery - Series B, 84*(4), 571–578. https://doi.org/10.1302/0301-620X.84B4.11947
Omata, S., Sonokawa, S., Sawae, Y., & Murakami, T. (2012). Effect of both vitamin C and mechanical stimulation on improving the mechanical characteristics of regenerated cartilage. *Biochemical and Biophysical Research Communications, 424*(4), 724–729. https://doi.org/10.1016/j.bbrc.2012.07.019
Sah, R. L.-Y., Kim, Y., Doong, J.-Y. H., Grodzinsky, A. J., Plass, A. H. K., & Sandy, J. D. (1989). Biosynthetic response of cartilage explants to dynamic compression. *Journal of Orthopaedic Research, 7*(5), 619–636. https://doi.org/10.1002/jor.1100070502
Tsumaki, N., Okada, M., & Yamashita, A. (2015). iPS cell technologies and cartilage regeneration. *Bone, 70*, 48–54. https://doi.org/10.1016/j.bone.2014.07.011
Uchio, Y., Ochi, M., Matsusaki, M., Kurioka, H., & Katsube, K. (2000). Human chondrocyte proliferation and matrix synthesis culture in Atelocollagen® gel. *Journal of Biomedical Materials Research, 50*(2), 138–143. https://doi.org/10.1002/(SICI)1097-4636(200005)50:2<138::AID-JBM7>3.0.CO;2-K
Waldman, S. D., Spiteri, C. G., Grynpas, M. D., Pilliar, R. M., & Kandel, R. A. (2003). Long-term intermittent shear deformation improves the quality of cartilaginous tissue formed in vitro. *Journal of Orthopaedic Research, 21*(4), 590–596. https://doi.org/10.1016/S0736-0266(03)00009-3
Wimmer, M. A., Grad, S., Kaup, T., Hänni, M., Schneider, E., Gogolewski, S., & Alini, M. (2004). Tribology approach to the engineering and study of articular cartilage. *Tissue Engineering, 10*(9), 1436–1445. https://doi.org/10.1089/1076327042500454
Yamamoto, K., Takaya, R., Tamada, Y., & Tomita, N. (2008). Effects of tribological loading history on the expression of tribological function of regenerated cartilage. *Tribology Online, 3*(2), 148–152. https://doi.org/10.2474/trol3.148

Summarization and future directions of biotribology of human and artificial joints

7

7.1 Biotribology of human and artificial joints

A half-century ago, "Biotribology" as "those aspects of tribology concerned with biological systems" was introduced by Dowson and Wright (1973). Biotribology includes various areas on tribology in biological systems such as joint tribology, skin tribology, oral tribology, and other biological tribology. Particularly, the biotribology of human and artificial joints has been the most important topic since the proposal of biotribology (Davim, 2010; Dowson, 2012; Dumbleton, 1981; Zhou & Jin, 2015). As described in Chapter 1, synovial joints are superior natural joints/bearings, and have intimate relation to tribology. "Friction, lubrication and wear" as the three pillars of tribology play predominant roles in various functions of human and artificial joints. Therefore, the actual functions of human and artificial joints should be evaluated from the viewpoint of biotribology.

Synovial joints are usually driven by muscular forces under restriction with ligaments. As described in Chapter 2 on biomechanics, the force and deformation of cartilage/bone construct in human joints are adequately estimated by considering not only the geometry, structure, and material properties of cartilage/bone construct but also the mechanical situations of muscular forces, ligament forces, and inertia forces in related musculoskeletal system. Malalignment and edge loading in human and artificial joints can generate surface failure and severe wear, and therefore the biomechanical considerations are required to maintain the healthy situation.

Human synovial joints have excellent tribological performance such as extremely low friction coefficient, like 0.002 to 0.03 or so, even under high load at severalfold of body weight in hip, knee, and ankle joints in various daily activities with smooth movement for the entire lifetime. Human joints have common features to other vertebrate joints but may have some intrinsic evolved structure and property for bipedal locomotion. Furthermore, in upper limb, shoulder, elbow, wrist, and finger joints contribute to actualize synergistically complex motions and mobility. In the healthy natural synovial joints composed of bone ends covered with compliant biphasic articular cartilage layers lubricated with synovial fluid, their superior lubricating ability and excellent load carrying capacity with appropriate range of motion bring about the excellent function of the biological joints/bearing with the extremely low friction and minimal wear. Importance of tribology and various superior lubrication mechanisms in natural synovial joints are described in Chapter 3.

In some joints under severe loading/rubbing or diseased conditions, however, even native articular cartilage can be worn and/or deteriorated. For example, in the case of osteoarthritis with worsening of wear and deterioration, pathognomonic changes such

Biotribology of Natural and Artificial Joints. https://doi.org/10.1016/B978-0-12-823669-7.00006-5

as osteophyte, bone hardening, and scar in joint capsule take place (Fig. 1.2b). Advanced osteoarthritis lowers the ability of walking and produces severe pain. For these patients, the replacements by artificial joints (joint prostheses) have been applied, which have made them able to do their daily activities again and given them relief from pain. Lubrication modes and friction and wear behaviors in artificial joints are described in Chapter 4.

To improve the lubrication mode in joint prostheses, the application of compliant artificial cartilage materials has been proposed. Effectiveness of compliant and/or biphasic (hydrogel) artificial cartilage on the superior lubrication situation is described in Chapter 5.

As a medical treatment to repair focal chondral defects, the implantation of regenerated cartilage has been applied. Appropriate mechanical and tribological stimulations on chondrocytes are expected to be effective in improvement of functions of regenerated cartilage. Some examples of effectiveness are shown in Chapter 6.

In next section, the essential points are summarized for Chapters 1−6.

7.2 Summarization of biotribology of human and artificial joints

7.2.1 *Importance of biotribology*

Chapter 1 "Importance of Biotribology" is composed of four sections: 1.1 Introduction to biotribology, 1.2 Friction, 1.3 Roles of lubrication, and 1.4 Wear.

In introduction to biotribology, the origin of biotribology and various subjects/topics on biotribology are described. The relation between natural synovial joint and artificial joint is shortly mentioned. On friction, the friction laws and mechanisms under dry, unlubricated, and lubricated conditions are mentioned. The mechanism of rolling friction is described. Examples for low friction in natural synovial joints and high friction on finger skin are demonstrated. On roles of lubrication, the fundamental lubrication mechanisms and features for fluid film, elastohydrodynamic, boundary, mixed, and solid lubrications are explained. On wear, fundamental wear mechanisms for adhesive, abrasive, fatigue, erosive, and corrosive wear are introduced.

As described in this book, biotribology has important roles in biomechanical functions of natural and artificial joints, particularly on friction, lubrication, and wear mechanisms.

The field of biotribology is receiving great deal of attention now and the premier issues of *Biotribology* (Cann & Wimmer, 2015) and *Biosurface and Biotribology* (Zhou & Jin, 2015) were launched in 2015.

7.2.2 *Biomechanics of human and artificial joints*

Chapter 2 is composed of six sections: 2.1 Importance of biomechanics in human and artificial joints, 2.2 Anatomy and kinematics of human joints, 2.3 Loads in human joints, 2.4 Importance of kinematics in human joints, 2.5 Importance of alignment in human joints, and 2.6 Biomechanics in artificial joints.

The functions of both human and artificial joints are controlled by the biomechanical conditions. First, the anatomy and kinematics/motion of synovial joints are described from the viewpoint of biomechanics. Next, the estimation of the loads in human joints by static and dynamic analyses is described. Loading patterns of joint forces during walking in hip and knee joints are introduced. Peak loads as 2 to 4 times of body weight were confirmed in instrumented hip and knee implants during walking. Kinematic feature of rolling with sliding in tibiofemoral knee joint and the importance of alignment in lower limb are shown. The difference in contact conditions between hard-on-hard and hard-on-soft hip prostheses is discussed. For hard-on-soft hip prostheses with small radial clearance, the application of ball-in-socket model is recommended. To evaluate the contact conditions of total knee prostheses, in vivo measurements by single-plane fluoroscopy are described on (a) pattern matching, (b) image matching, and (c) 2D-3D registration methods. Finally the reproduction of physiological and active knee motion in novel knee simulator is discussed from the viewpoint of biomechanics.

7.2.3 Biotribology of natural joints

Chapter 3 is composed of five sections: 3.1 Importance of tribology in natural joints, 3.2 Articular cartilage, 3.3 Synovial fluid, 3.4 Lubrication mechanism in natural joints and 3.5 Wear and degradation mechanism in natural joints.

First the structure of articular cartilage composed of the extracellular matrix of proteoglycan reinforced by the collagen fibers (predominantly type II) and the chondrocytes is described in relation to their distribution in superficial, middle, and deep zones. Lubricious cartilage surface microstructure and AFM surface image are shown. The biphasic properties of the articular cartilage as load-bearing material are mentioned. The constituents, viscous properties, and roles of the synovial fluid as lubricant are discussed.

Next, the various lubrication mechanisms for the fluid film lubrication including soft-elastohydrodynamic lubrication (soft-EHL) and micro-EHL, the biphasic lubrication, the hydration lubrication, the boundary lubrication, and the gel film lubrication are described from the viewpoint of the adaptive multimode lubrication with low friction and minimal wear. Flattening of surface asperity in micro-EHL extended the ability of fluid film lubrication. It is shown in biphasic lubrication that the fluid load support maintains with low friction for migrating contact, but the fluid load support gradually decreases for stationary contact on cartilage. Cases with effectiveness of hydration lubrication are shown for several modes. If adequate water molecules are attracted by hydrophilic polymers fixed to the surface, the low friction would be sustained between hydrated layers at low velocity or even under static conditions. In boundary lubrication, adsorbed films of lubricin, phospholipids, and serum proteins appear to protect the rubbing surfaces with low friction. In pendulum tests for porcine shoulder joints, the addition of phospholipid and γ-globulin reduced friction in boundary or mixed lubrication regime. In reciprocating tests for articular cartilage against glass plate, the addition of single protein increases friction but the addition of simulated synovial fluid as HA solution containing 0.01 mass% DPPC, 1.4 mass% albumin, and 0.7 mass% γ-globulin reduced friction and wear.

In more severe reciprocating tests lubricated with saline solution, changes in transient friction after removal of adsorbed films are observed. The surface observation of AFM images indicates the protective roles by the proteoglycan gellike layer as gel film lubrication.

In summarization of the various lubrication mechanisms, the adaptive multimode lubrication mechanism is proposed with the synergistic combinations of various lubrication modes in wider operating conditions (Dowson, 1966–67; Murakami et al., 2017). Finally, wear and degradation mechanism in natural joints is discussed.

7.2.4 Biotribology of artificial joints

Chapter 4 is composed of five sections: 4.1 Importance of tribology in artificial joints, 4.2 Lubrication modes in artificial joints, 4.3 Boundary lubrication in artificial joints, 4.4 Wear in artificial joints, and 4.5 Improvement of tribological properties in artificial joints.

First, development of artificial joints and roles of tribology are described.

Second, lubrication modes in hip (hard-on-hard, hard-on-UHMWPE) and knee (hard-on-UHMWPE) joint replacements are discussed on the basis of experimental evaluation and numerical analysis on elastohydrodynamic lubrication. The lubrication mode is fundamentally discriminated as "fluid film lubrication," "mixed lubrication," or "boundary lubrication" according to the range of the film parameter (lambda ratio) as the ratio of theoretical minimum fluid film thickness to the composite surface roughness of both rubbing surfaces.

In experimental evaluation in hip joint simulator, frictional behaviors in Stribeck curve indicate the corresponding lubrication mode. In friction tests, it should be noticed that the proteinaceous constituents in lubricant can generate considerable frictional resistance as adsorbed films. For metal-on-metal total hip prostheses, the measurement of fluid film formation with an electric resistivity technique or an ultrasonic reflection method is performed. In mixed lubrication regime, the combination of protein-based boundary lubrication and super-imposed fluid film action with elastohydrodynamic lubrication appears to control the frictional behaviors.

In simplified estimation, minimum film thickness in total hip prostheses can be predicted using the formula by Hamrock and Dowson (1978) as an equivalent ball-on-plane model, but in detailed analysis, a ball-in-socket model is applied. In elastohydrodynamic lubrication analysis for 3D physiological loading and motion conditions, it is shown that the fluid film thickness during walking increased compared with the range under 1D loading and motion (Gao et al., 2009). Although thicker fluid film ($\Lambda > 3$) can be formed in CoC and MoM with larger femoral head radius and smaller radial clearance, predominant lubrication regimes for total hip prostheses are indicated as follows:

- Ceramic-on-ceramic (CoC): fluid film lubrication or mixed lubrication
- Metal-on-metal (MoM): mixed lubrication
- Hard (metal or ceramic)-on-UHMWPE: boundary or mixed lubrication

The fluid film formation in total knee prostheses in knee joint simulator was evaluated by the electric resistance method using metal-coated UHMWPE tibial component. Knee prostheses as fixed or mobile bearing have wide range of geometrical conformity. Considerable fluid film formation was observed depending on load and flexion conditions in conformal cylindrical knee prosthesis, but little fluid film was formed in anatomical knee prosthesis with low conformity during walking motion. Elastohydrodynamic analysis for anatomical knee prostheses with low conformity predicted minimum fluid film thickness $h_{min} < 0.1$ μm during walking. The predominant lubrication mode is suggested to be severe mixed or boundary lubrication regime.

Next, main features in boundary lubrication in artificial joints are mentioned. Synovia constituents form adsorbed film on rubbing surfaces in mixed or boundary lubrication regime and exert significant influences on friction and wear behaviors. Particularly, the elucidation of tribological behaviors of proteinaceous adsorbed films between rubbing surfaces is required with consideration of mutual interactions between lubricant constituents.

Minimization of wear in artificial joints is regarded as an important issue in prevention of not only surface damage but also generation of wear debris, because wear debris can induce osteolysis and the eventual loosening of joint prostheses. To reduce wear, it is required to identify the predominant wear mechanism from the various mechanisms such as adhesive, abrasive, fatigue, corrosive wear, and so on. The improvement of surface finish or surface texture and wear resistance of rubbing material or geometric design becomes effective. It is indicated for wear of UHMWPE that the considerations of the multidirectional motion, oxidation of UHMWPE, and influence of lubricants become important. Application of highly cross-linked polyethylene is considered to have remarkably improved wear resistance of UHMWPE. Advantage of ceramic-on-UHMWPE to metal-on-UHMWPE, and the occurrence of the stripe wear in ceramic-on-ceramic are described. In total knee prostheses, the delamination and pitting as fatigue wear are discussed on the basis of elastoplastic analyses. In knee prostheses with low conformity, excessive contact stress should be suppressed. Main mechanism on wear debris–induced osteolysis is mentioned to be caused by the biological response to wear debris, where the synthesized cytokines from macrophages stimulate the surrounding osteoclasts and activate bone resorption. As evaluation methods of wear properties, the essential terms in the screening tests based on ASTM Standards and the total joint simulator tests based on ISO Standards are described. Computer simulation for wear evaluation is expected to become a predominant tool to contribute to the optimization of joint prostheses.

Finally, the following cases are described as effective improvements of wear resistance in UHMWPE:

(a) cross-linking of UHMWPE,
(b) addition of vitamin E,
(c) wear reduction and prevention of osteolysis by phospholipid polymer grafted cross-linked polyethylene.

On the effect of phospholipid polymer grafting on cross-linked polyethylene, the superior results of long-term hip simulator testing were shown. The evaluation of the effects of dehydration and rehydration by loading-unloading procedure, and the evaluation of phospholipid polymer grafting with concentrated contact load are discussed. Superiority of biocompatible phospholipid polymer grafting in friction and wear is shown.

7.2.5 Biotribology of artificial joints with artificial cartilage

Chapter 5 is composed of six sections: 5.1 Effectiveness of compliant artificial cartilage, 5.2 Effectiveness of hydrogel artificial cartilage, 5.3 Role of biphasic lubrication in hydrogel artificial cartilage, 5.4 Roles of boundary lubrication, 5.5 Influence of hydrogel wear debris, and 5.6 Bionic design with biomimetic artificial cartilage.

First, the effectiveness of compliant artificial cartilage is mentioned on lubrication mechanism and low friction property. In knee joint simulator tests for walking motion, the significant fluid film formation was confirmed in the knee prosthesis composed of metallic femoral component and compliant tibial insert. In friction tests in hip joint simulator, soft-layered artificial hip joint showed remarkable low friction property in comparison with traditional hip joint with UHMWPE cup. In these cushion form bearings as joint prostheses with compliant articular layer, the fluid film lubrication appears to be attained by effective soft-EHL and micro-EHL mechanism during normal walking condition. However, under adverse lubrication conditions at slower motion, at shorter stroke length, at little movement, or starting and stopping, fluid film may not be sufficiently formed. Optimization with alternate lubrication mechanism and longer durability is required.

Second, the effectiveness of hydrogel artificial cartilage is discussed. Hydrogels are three-dimensional cross-linked hydrophilic polymer networks swollen by water. Articular cartilage is one of the natural hydrogels. Therefore, the application of compliant artificial cartilage of hydrogel is expected to mimic the natural articular cartilage and to improve lubrication by combined effect of soft-EHL, micro-EHL, biphasic lubrication, hydration lubrication, and boundary lubrication. For example, in knee joint simulator test, the knee prosthesis with artificial cartilage of biocompatible and hydrophilic poly(vinyl alcohol) (PVA) hydrogels of high water content exhibited significantly low friction in HA solution containing serum protein, while the knee prosthesis with polyurethane layer showed more than tenfold higher friction. For PVA surfaces, the adsorbed films of protein appear to play a protective role with reduction in friction.

For hydrogels, effectiveness of biphasic lubrication is clarified. The interstitial fluid pressure in reciprocating tests for the cylindrical PVA hydrogel layer moving specimen on the acrylic flat specimen was measured using microelectromechanical system (MEMS) pressure sensor in wide reciprocating speeds from 1.56 mm/s to 400 mm/s. At lower speed region than 10 mm/s, the characteristic pressure profile for interstitial fluid pressurization was observed, and friction coefficient could be estimated from the proportion of fluid load support.

For PVA hydrogels different in manufacturing methods as repeated freezing-thawing (FT), cast-drying (CD) and hybrid (a layered structure of FT and CD) gels, the reciprocating tests and biphasic finite element (FE) analysis were performed. FT gel is opaque, and has porous, fibrous, and rough surface and high permeability. CD gel of uniform structure is transparent and has smooth surface and low permeability. Reciprocating tests of these PVA hydrogels on glass plate in saline solution showed highly different frictional behaviors as hybrid (CD on FT) $<$ CD $<$ FT $<$ hybrid (FT on CD). In simulated synovial fluid, all friction levels were reduced but these orders in friction were same. These time-dependent frictional behaviors could be appropriately estimated from the fluid load support ratio and surface lubricity in biphasic FE analysis. Furthermore, hybrid (CD on FT) gel exhibited minimal wear.

The effectiveness of fiber reinforcement in PVA FT hydrogel on interstitial fluid pressurization and friction reduction is shown by sliding biphasic FE simulation and reciprocating ball-on-plate test. It is indicated that fiber reinforcement in parallel to the surface is effective to enhance the stiffness to compressive load and interstitial fluid pressurization. The enhancement of interstitial fluid pressurization and reduction in friction by fiber reinforcement are dependent on the depth location. Therefore, optimum arrangement of fiber reinforcement is required.

On the compliant artificial cartilage surfaces rubbed in mixed or boundary lubrication regime, significant direct contacts can occur. Therefore, to reduce friction and protect the compliant surfaces, adsorbed films or some lubricating layers should be formed. In reciprocating test of PVA hydrogel of high water content against itself lubricated with HA solution containing proteins under thin film conditions, the lubricant constituents and protein concentration influenced wear property of PVA hydrogel. It is noticed that wear of PVA increased with an increase in protein concentration in HA solutions containing single protein, but at certain combination of albumin and γ-globulin in HA solutions the wear was remarkably reduced. At appropriate total protein concentration of 2.1 mass% as 1/2 or 2/1 ratio of albumin and γ-globulin (A/G ratio), the wear of PVA was remarkably reduced as minimum values. The observation of the fluorescent images of adsorbed films indicated that the optimum adsorbed film as layered structure had been formed. In contrast, for adsorbed films of heterogeneous structure, severe wear occurred. In reciprocating test of PVA hydrogel on glass plate, HA solution containing 0.01 mass% DPPC, 1.4 mass% albumin, and 0.7 mass% γ-globulin (simulated synovial fluid) showed extremely low friction and minimum wear, which is similar to the case for articular cartilage.

Next, the influence of PVA hydrogel wear particles on the immune response of THP-1-derived macrophages by assaying two cytokines IL-1β and TNF-α was examined. Here, cytokine syntheses of IL-1β and TNF-α in both wear particle-stimulated groups for PVA-FT and PVA-CD gels were significantly lower than that of the lipopolysaccharide(LPS)-stimulated positive control. Therefore, it is suggested that PVA hydrogel wear particles do not affect the macrophage immune response.

Finally, bionic design with biomimetic artificial cartilage is discussed. Various biocompatible compliant artificial hydrogels as complex hydrophilic polymer network have been developed to mimic natural articular cartilage as one of the hydrogels. To improve the durability, the realization of enhancement of wear resistance and mechanical properties including gamma-irradiated cross-linking, superior lubricity, suppression of elastic friction with polymer brush, fiber-reinforcement, adjustment of water content, and other related technologies are expected with self-repairing ability. In addition, the suppression of immune response of hydrogel debris should be confirmed for new hydrogels.

7.2.6 Biotribology of regenerated cartilage

Chapter 6 is composed of three sections: 6.1 Medical treatments to repair focal chondral defects, 6.2 The influences of mechanical stimulations on regenerated cartilage tissue, and 6.3 The influences of tribological stimulations on regenerated cartilage tissue.

First, main medical treatments to repair focal chondral defects are described under the osteochondral autograft transplant (mosaicplasty), the autologous chondrocyte implantation (ACI), and the implantation of regenerated cartilage tissues derived from native chondrocytes, mesenchymal stem cells, ES cells or iPS cells.

Next, the influences of mechanical stimulations on regenerated cartilage tissue are mentioned, like effects of dynamic compression loading and effects of shearing. Dynamic compression stimuli for chondrocyte-agarose construct increased the stiffness and enhanced the extracellular matrix (ECM) network of inner tissue. On intermittent cyclic shearing stimulation, lower amplitudes increased the matrix synthesis.

The influences of tribological stimulations on regenerated cartilage tissue during sliding motion and frictional behaviors of regenerated cartilage in different cultivation conditions are discussed. The addition of sliding motion for the chondrocyte-polyurethane construct cultured under dynamic compression loading enhanced stiffness and reduced friction and exhibited strong immunoreactivity for lubricin at the top surface. Furthermore, the upregulation by multidirectional articular motion was indicated.

Finally, the frictional behaviors of the regenerated cartilage as chondrocyte-fibroin sponge construct against a stainless steel plate were examined in physiological saline solution in reciprocating apparatus. It is noticed that at early stage during sliding test, friction coefficient decreased remarkably after 7 days' cultivation, where surface layer containing proteoglycan had been newly formed. It is indicated that the newly formed regenerated surface layer was partially removed with rubbing and in consequence friction gradually increased.

7.3 Future directions of biotribology of human and artificial joints

Various daily activities of human body with smooth movements have been attained with the aids of superior functions of natural synovial joints or alternative artificial joints in musculoskeletal system. In this section, future directions of biotribology of human and artificial joints are discussed for a new era after a half century since the proposal of biotribology.

7.3.1 Biomechanics

Mechanical severity of human joints and artificial joints such as at edge loading and mal-alignment are controlled by biomechanical conditions in musculoskeletal systems. To sustain healthy situation in human and artificial joints, the biomechanical analysis and evaluation can contribute to elucidation of many cases in clinical joint problems.

Actual evaluation with in vivo measurement of kinematics of total knee prostheses by single plane fluoroscopy is useful for identification of position/orientation and contact situations. Appropriate selection of pattern matching, image matching, and 2D-3D registration methods is required. Edge loading and lifting should be modified on the basis of these in vivo data with the aid of biomechanical simulation.

The proposal of novel knee simulator to reproduce physiological and active knee motion is expected to be applicable in new knee simulator (Hirokawa et al., 2015).

In evaluation of functions of human and artificial joints in computational modeling such as muscloskeletal multibody dynamics models, the integrated consideration for both the biomechanics and biotribology including contact mechanics with geometry and deformation properties for joints is expected to give realistic simulation (Chen et al., 2014).

7.3.2 Human joints

On the ingenious lubrication mechanism with extremely low friction and minimal wear in human joints, various lubrication mechanisms have been proposed and significantly clarified (Dowson, 2012). In various daily activities, the operating conditions widely change. During normal walking motion, the fluid film lubrication modes as soft-EHL and micro-EHL are considered to be effective. However, under thin film conditions with slow movement, high load and low viscous synovial fluid, various lubrication modes such as biphasic, mixed, weeping, hydration, boundary, and gel film lubrications are likely to be effective depending on the severity of operating conditions. This mechanism is called adaptive multimode lubrication (Murakami et al., 2017). Some unsolved issues on lubrication mechanism are the importance of sustaining cartilage motion at stroke end for cartilage/cartilage contacts, the mutual relationship between biphasic fluid pressurization and hydrodynamic pressurization (Horibata et al., 2018; Fujie et al., 2019), the real constituents/structures of boundary film, the actual behavior of surface gel layer, the effectiveness of collagen fiber orientation in the surface zone, and so on (Medley et al., 2021).

It is expected in future that biotribological knowledge on lubrication mechanisms with low friction and minimal wear should be effectively utilized for repairing and strengthening in suppression of joint diseases such as osteoarthritis and traumatic diseases.

7.3.3 Artificial joints

The successful replacement for worn, damaged, or diseased human joints with artificial joints became the greatest advance in orthopedic surgery during the past century. However, tribological problems such as wear/surface failure and wear debris-induced osteolysis and eventual loosening have not been completely solved, even in 21st century. The predominant lubrication modes of total hip and knee prostheses are considered to be mixed or boundary lubrication except fluid film lubrication mode for ceramic-on-ceramic hip joint during normal walking. Therefore, to solve these problems, various improvements in materials, designs, and operation systems have been performed . Application of robotic surgery for patient-specific arthroplasty is expected to contribute to appropriate replacement with improved component positioning and alignment (Jacofsky & Allen, 2016). On total hip and knee prostheses as hard (metal or ceramic)-on-UHMWPE, the applications of highly cross-linked polyethylene with addition of vitamin E and/or biocompatible phospholipid polymer grafting appear to have significantly improved the durability. In phospholipid polymer grafted cross-linked polyethylene, the hydration lubrication mechanism is likely to be effective, and its surface property and structure are attractive from the viewpoint of biotribology. Properties of metallic and ceramic materials have been improved but metallic components have the possibility of production of very fine wear debris and metallic ions. Appropriate surface modification for metallic material may be effective. In contrast, application of all-polymer or all-composite total joint prosthesis is expected. It is expected that the enhancement of fluid film formation can contribute to reduction in rubbing severity for various material combinations. Consequently, the optimum design with lubricating ability considering material combination and geometric design is required.

In the evaluation of wear property of new prosthesis design with preclinical simulation, various hip and knee joint simulators have been applied and additional functions are required to evaluate various human activities. On the other hand, not only the progress of wear but changes in pressure distribution and fluid film thickness can be evaluated in computational wear modeling (Gao et al., 2018). Hereafter, it is expected to evaluate appropriately the unknown wear behaviors of new joint prosthesis in computational wear modeling as the valuable tool for various daily activities (Mattei et al., 2011). However, clinical investigations for the observation of actual failure, wear debris and the interaction with lubricant constituents are required.

7.3.4 Artificial joints with artificial cartilage

The application of compliant artificial cartilage enhanced the fluid film formation by soft-EHL and micro-EHL mechanism. Furthermore, in artificial joints with artificial hydrogel cartilage, the biphasic lubrication mechanism contributes to improvement in lubrication. The interstitial fluid pressurization changes depending on the hydrogel properties like permeability, elastic properties, and surface lubricity. Therefore, fluid load support and friction properties can be predicted on the basis of biphasic lubrication theory including fiber-reinforcement effect. On PVA hydrogels, the laminated hybrid (CD on FT) gel and composite hybrid gel exhibited superior low friction and low wear properties on glass plate. In addition, PVA-hybrid (laminated) gel showed extremely low friction and minimum wear when rubbed against artificial joint materials such as Co–Cr–Mo alloy and alumina ceramic in biological environment (Yarimitsu et al., 2016). Therefore, biocompatible PVA-hybrid gel is a potential candidate material for artificial cartilage for joint prosthesis.

Various hydrogels with high strength have been developed, but the evaluations of the actual tribological behaviors including boundary film formation and durability in simulated operating conditions are required for new hydrogels. Furthermore, the changes in material properties of hydrogels with aging should be considered

(Murakami et al., 2021). In aged PVA FT gels, frictional behaviors were improved by lowering of permeability and increasing of elastic modulus with aging, which were caused by the formation of hydrogen bonding in amorphous region. Function improvement with aging and/or self-repairing ability is expected.

7.3.5 Regenerated cartilage

Clinical implantation of regenerated cartilage tissue has been conducted in qualified hospitals and various technologies for clinical applications different in cell sources such as native chondrocytes, mesenchymal stem cells and iPS cells, scaffolds, and culture conditions (environment) including signaling factors have been developed. As optimal scaffolds, the future clinical application of the decellularized extracellular matrices are expected (Sun et al., 2018). The appropriate biomechanical and biotribological stimulations on chondrocytes in culture process are expected to be effective as described in Chapter 6. Systematic elucidation of reactive mechanism of chondrocytes to the different stimuli from the viewpoints of biotribology is required.

References

Cann, P., & Wimmer, M. (2015). Welcome to the first issue of biotribology. *Biotribology, 1.*

Chen, Z., Zhang, X., Ardestani, M. M., Wang, L., Liu, Y., Lian, Q., He, J., Li, D., & Jin, Z. (2014). Prediction of in vivo joint mechanics of an artificial knee implant using rigid multi-body dynamics with elastic contacts. *Proceedings of the Institution of Mechanical Engineers, Part H: Journal of Engineering in Medicine, 228*(6), 564–575. https://doi.org/10.1177/0954411914537476

Davim, J. P. (Ed.). (2010). *Biotribology*. ISTE Ltd and John Wiley & Sons, Inc.

Dowson, D. (1966–67). Modes of lubrication in human joints. *Proceedings of the Institution of Mechanical Engineers, 181*(Pt3J), 45–54.

Dowson, D. (2012). Bio-tribology. *Faraday Discussions, 156*(1).

Dowson, D., & Wright, V. (1973). Bio-tribology. In T. C. Davenport (Ed.), *The rheology of lubricants* (pp. 81–88). Applied Science Publishers Ltd.

Dumbleton, J. H. (1981). Tribology of natural and artificial joints. In *Tribology series (issue 3).* Elsevier Sci. Publishing Co.

Fujie, H., Horibata, S., & Yarimitsu, S. (2019). Synovial fluid pressurization enhances biphasic lubrication in articular cartilage. *Proceedings of Biotribology Sendai 2019*, 56–59.

Gao, L., Hua, Z., Hewson, R., Andersen, M. S., & Jin, Z. (2018). Elastodynamic lubrication and wear modelling of the knee joint replacements with surface topography. *Biosurface and Biotribology, 4*(1), 18–23. https://doi.org/10.1049/bsbt.2017.0003

Gao, L., Wang, F., Yang, P., & Jin, Z. (2009). Effect of 3D physiological loading and motion on electrohydrodynamic lubrication of metal-on-metal total hip replacements. *Medical Engineering and Physics, 31*(6), 720–729. https://doi.org/10.1016/j.medengphy.2009.02.002

Hamrock, B. J., & Dowson, D. (1978). Elastodynamic lubrication of elliptical contacts for materials of low elastic modulus. Part 1: Fully flooded conjunction. *Transactions of the ASME Journal of lubrication technology, 100*(1), 236–245.

Hirokawa, S., Murakami, T., Kiguchi, K., & Fukunaga, M. (2015). Development of the novel simulator that reproduces physiological and active knee motion. *Transactions of the JSME (In Japanese), 81*(824), 14–00488. https://doi.org/10.1299/transjsme.14-00488

Horibata, S., Yarimitsu, S., & Fujie, H. (2018). Influence of synovial fluid pressure on biphasic lubrication property in articular cartilage. *Tribology Online, 13*(3), 172–177. https://doi.org/10.2474/trol.13.172

Jacofsky, D. J., & Allen, M. (2016). Robotics in arthroplasty: a comprehensive review. *The Journal of Arthroplasty, 31*, 2353–2363. https://doi.org/10.1016/j.art.2016.05.026

Mattei, L., Di Puccio, F., Piccigallo, B., & Ciulli, E. (2011). Lubrication and wear modelling of artificial hip joints: A review. *Tribology International, 44*(5), 532–549. https://doi.org/10.1016/j.triboint.2010.06.010

Medley, J. B., Murakami, T., & Jin, Z. (2021). My personal stories and anecdotes with the late Professor Duncan Dowson. *Biosurface and Biotribology, 7*(2), 43–52. https://doi.org/10.1049/bsb2.12013

Murakami, T., Sakai, N., Yarimitsu, S., Nakashima, K., Yamaguchi, T., Sawae, Y., & Suzuki, A. (2021). Evaluation of influence of changes in permeability with aging on friction and biphasic behaviors of artificial hydrogel cartilage. *Biotribology, 26*, 100178. https://doi.org/10.1016/j.biotri.2021.100178

Murakami, T., Yarimitsu, S., Sakai, N., Nakashima, K., Yamaguchi, T., & Sawae, Y. (2017). Importance of adaptive multimode lubrication mechanism in natural synovial joints. *Tribology International, 113*, 306–315. https://doi.org/10.1016/j.triboint.2016.12.052

Sun, Y., Yan, L., Chen, S., & Pei, M. (2018). Functionality of decellularized matrix in cartilage regeneration: a comparison of tissue versus cell sources. *Acta Biomaterialia, 74*, 56–73. https://doi.org/10.1016/j.actbio.2018.04.048

Yarimitsu, S., Sasaki, S., Murakami, T., & Suzuki, A. (2016). Evaluation of lubrication properties of hydrogel artificial cartilage materials for joint prosthesis. *Biosurface and Biotribology, 2*(1), 40–47. https://doi.org/10.1016/j.bsbt.2016.02.005

Zhou, Z. R., & Jin, Z. M. (2015). Biotribology: Recent progresses and future perspectives. *Biosurface and Biotribology, 1*(1), 3–24. https://doi.org/10.1016/j.bsbt.2015.03.001

Index

Note: 'Page numbers followed by "*f*" indicate figures and "*t*" indicate tables'.

I

K

Y

FSC
www.fsc.org
MIX
Papier | Fördert gute Waldnutzung
FSC® C083411

Kolibri 360 GmbH
Ferdinand-Jühlke-Straße 7
99095 Erfurt, Deutschland
produktsicherheit@kolibri360.de